THE ECOLOGY OF SOIL DECOMPOSITION

This book is dedicated to
James D. Berger
Tom Cavalier-Smith
David C. Coleman
Max (F.J.R.) Taylor

The Ecology of Soil Decomposition

Sina M. Adl
Department of Biology
Dalhousie University
Halifax, Nova Scotia
Canada

CABI Publishing

CABI Publishing is a division of CAB International

CABI Publishing
CAB International
Wallingford
Oxon OX10 8DE
UK

Tel: +44 (0)1491 832111
Fax: +44 (0)1491 833508
E-mail: cabi@cabi.org
Website: www.cabi-publishing.org

CABI Publishing
44 Brattle Street
4th Floor
Cambridge, MA 02138
USA

Tel: +1 617 395 4056
Fax: +1 617 354 6875
E-mail: cabi-nao@cabi.org

A catalogue record for this book is available from the British Library, London, UK.

Library of Congress Cataloging-in-Publication Data
Adl, Sina M., 1964–
The ecology of soil decomposition / Sina M. Adl.
p. cm.
Includes bibliographical references and index.
ISBN 0-85199-661-2 (alk. paper)
1. Soil microbiology. 2. Soil ecology. I. Title.
QR111.A34 2003
577.5′7--dc21

2003001601

ISBN 0 85199 661 2

Typeset in 10pt Baskerville by Columns Design Ltd, Reading.
Printed and bound in the UK by Cromwell Press, Trowbridge.

Contents

Preface

This book is about the trophic interactions among species that live in the soil. These interactions are responsible for the decomposition of previously living cells and tissues, for pedogenesis (from a biological perspective, it is the accumulation of organic matter into the mineral soil) and for biomineralization (making the decomposing matter available as nutrients in the soil solution). The biological interactions between species that carry out the decomposition of organic matter are our primary focus.

In this endeavour, we must consider the ecology of species that live in the soil habitat. The ecology of these species must take into account both the biological interactions among edaphic species and the abiotic environmental parameters which affect species composition and activity patterns. Over time, changes in the soil profile due to transformations of the organic matter affect species composition. In turn, changes in species composition affect the chemistry of new litter and decomposing organic matter. These changes are reflected in the spatial heterogeneity of the soil and in succession patterns over longer time scales.

The organisms that are mostly responsible for decomposition include saprotrophic bacteria and fungi, soil protozoan species, symbiotic protozoa in saprotrophic invertebrates, fungi in symbiosis with plant roots, Oomycetes and many specialized invertebrates (such as nematodes, rotifers, mites, Collembola, earthworms and enchytraeids). These species are mostly microscopic, with the notable exception of earthworms and several less abundant invertebrate groups. In this respect, the study of decomposition in the soil habitat is also the study of soil microbial ecology in its true sense, i.e. the soil ecology of all those organisms too small to see without magnification, the microbes.

Here, I recognize that phylogenetically related species are biochemically more similar to each other. This is an important consideration when comparing metabolism, regulation of behaviour and general biology of the organisms. However, from an ecological perspective, the functions of species in the habitat are more related in terms of what they feed on, and their spatial and abiotic preferences. One must distinguish between categories that group species according to evolutionary relatedness (systematics) and categories that group species according to ecologically relevant functional groups. It is clear that taxonomic groups implicated in decomposition each contain species that belong to different functional categories. This distinction must be carried through into schematic representations of food webs and ecosystem function. Moreover, it is the amount of organic matter flux between these functional categories that is more significant than the actual amount of biomass in each category. Many aspects of the interactions between species within functional groups, and of nutrient flux rates between functional groups, are poorly quantified. In the chapters that follow, I try to provide a description of the biology of taxonomic groups, and a structure for organizing these species into functional groups. From here, we must try to complete our investigation of decomposition by quantifying neglected interactions and including neglected taxa. There lies before us an exciting period for exploration and discovery, of the dark and cryptic reticulate matrix where decomposition completes the recycling of nutrients.

The recycling of nutrients through decomposition is one side of the coin in ecology, with autotrophic primary production being the other. Photosynthesis-driven primary production drives much of the ocean surface and above-ground ecology. From this primary production-driven ecology, litter and partially decomposed matter accumulate as soil organic matter. The soil organisms feed on and excrete this organic matter, as well as feeding on each other by predation. In terrestrial systems, these trophic interactions release a steady trickle of soluble nutrients in the soil solution, which are plant root available. They also release carbon dioxide through metabolic respiration, which provides the carbon source for photosynthesis. Therefore, terrestrial ecosystems can be subdivided into two subsystems, each with a distinct set of species that are spatially and functionally segregated. This was recognized by Peterson (1918) who proposed the idea of a photosynthesis- and consumer-based subsystem (which drives primary production through photosynthetic carbon fixation), and a second decomposition-based subsystem which decays the organic matter. The interface between these two subsystems is tight. The links are maintained through mycorrhizae, fine-root grazing, surface litter species, species with life cycles that bridge both subsystems, and parasites. As Elton (1966) noted, the task of ecologists is to integrate our understanding of both subsystems.

In this age of global climate change, with excessive pollution, exploitation and erosion of our soil resources, we cannot afford to model biogeochemistry and nutrient cycles based on the dynamics of one subsystem only. It is pertinent to understand how the soil harbours decomposition, in enough detail to apply successful management and remediation protocols for the future. I hope this book stimulates discussion and research to fill the gaps in our understanding and to correct the errors in my conclusions. I leave this book in your hands, with a quote from François Forel (1904) who conducted the first recorded survey of an ecological system, at Lake Léman, to conclude 'Ce que je vois de plus admirable quand je contemple la nature, c'est sa simplicité. Au premier abord tout paraît compliqué; à l'étude, tout s'ordonne et s'unifie.'

Sina Adl
August 2002

Elton, C.S. (1966) *The Pattern of Animal Communities*. John Wiley & Sons, New York.

Forel, F.A. (1904) Le Léman. *Monographie Limnologique* (Lausanne).

Petersen, C.G.J. (1918) The sea-bottom and its production of fish-food. A survey of the work done in connection with valuation of the Danish waters from 1883–1917. *Reports of the Danish Biological Station* 25, 1–62.

Acknowledgements

I must express my deepest gratitude to Tim Hardwick and the staff at CAB International for accepting this project; to my friends and colleagues at the Institute of Ecology (University of Georgia) for hosting me while this book was being researched; and to NSERC-Canada for their financial support. This book would not have been possible without the support and generosity of David Coleman during this time. I am very grateful to David Coleman, Dac Crossley and Paul Hendrix for countless conversations and a great deal of information. I also thank several friends and colleagues who have read and commented on various parts of this book: J.D. Berger, D.C. Coleman, D.A. Crossley, B.S. Griffiths, B.L. Haines, P.F. Hendrix, G. Kernaghan, D.A. Neher, D. Porter and G.W. Yeates, and Angela Burke for copy editing the manuscript. Finally, I thank Colleen for her unswerving understanding and help.

The Saprotrophs 1

'The extent to which progress in ecology depends upon accurate identification, and upon the existence of sound systematic groundwork for all groups of animals, cannot be too much impressed upon the beginner in ecology. This is the essential basis of the whole thing; without it the ecologist is helpless, and the whole of his work can be rendered useless.' (From Elton, 1966, *Animal Ecology*, first published 1927.)

In order to understand the ecology of an ecosystem, it is fundamental to understand the biology of the organisms implicated. Not knowing the biology of species is like trying to do chemistry without prior knowledge of the properties of the elements in the Periodic Table. The level of the science remains rudimentary and often unpredictable. Similarly, without adequate background in the biology of species involved in an ecosystem, the science remains primitive. In order to proceed beyond natural history, whereby one simply observes and describes, towards analytical and experimental science, it is necessary to have accumulated the available knowledge on the species and their interactions with each other, in the ecosystem. That is necessary in order to manipulate the biological communities and to measure correctly the parameters under observation. Understanding the biology is essential to interpreting the data and postulating mechanisms for the results.

Most students will be more familiar with species associated with primary production-driven above-ground ecology, rather than those associated with decomposition. It is therefore necessary to provide an overview of the biology of taxa implicated in decomposition. However, an ecology book cannot be a substitute for learning the biology of the relevant taxa. It can only be used to guide towards the relevant taxa and to provide an idea of their biology. The dominant soil organisms involved in decomposition are found in the protists (protozoa, chromista, fungi), invertebrates (microarthropods, nematodes, earthworms) and bacteria.

Eukaryotic Cells

Eukaryotic cells are distinguished from viruses and from prokaryotes (bacteria and archea) by their intracellular organization and genetic complexity. Eukaryotic cells exist as protists or multicellular organisms. The principal characteristics of a eukaryotic cell can be summarized as follows (Fig. 1.1).

The cell membrane. The existence of cells relies on the separation between the external environment and an internal cytoplasm which carries out the chemistry of life. This boundary is maintained by a membrane which provides a barrier to the solution outside the cell, and holds the cytoplasm molecules and organelles inside. The cell membrane provides a hydrophobic boundary, which consists of phospholipids and sterols, embedded with proteins. The details of the composition of cell membrane lipids and sterols vary with nutritional state and temperature, and the specific molecules vary between taxa. For example, a characteristic sterol in plant cell membranes is stigmasterol, it is cholesterol in animals, tetrahymenol in ciliates and ergosterol in fungi. The cell membrane allows free passage of gases such as oxygen and carbon dioxide, and of small polar but uncharged molecules such as water through the proteins. It is relatively

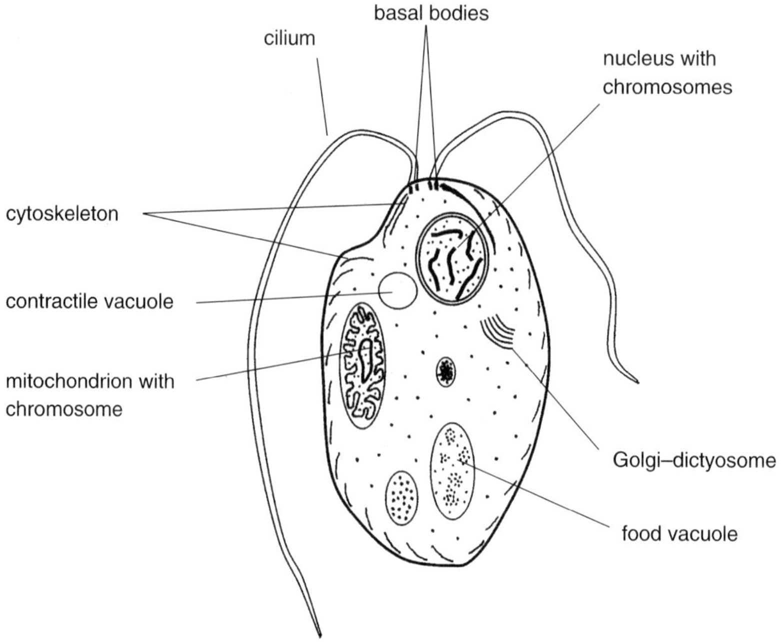

Fig. 1.1. Generalized eukaryotic cell organelles.

impermeable to all ions and most nutrients. The cell membrane proteins regulate the interaction between the external and internal environment. These proteins are responsible for binding molecules outside the cell, modifying the physiological state of the cytoplasm and responding to the external molecules. Some membrane proteins are actively regulating ion transport into and out of the cell. This exchange creates a potential difference across the cell membrane, with the cytoplasm often being at about –30 mV relative to the outside solution in a resting cell. Several responses can be stimulated through changes in the cell membrane. This can be a directional locomotion response (**chemotaxis**) towards or away from the stimulus. The response can be ingestion of the molecules by initiating **phagocytosis** of particles which are digested as food. **Pinocytosis** involves the ingestion of tiny vesicles of external solution. Specific large molecules can be internalized by binding receptor proteins in the membrane by **endocytosis**. Some membrane proteins act as hydrophilic channels that selectively permit entry of molecules, especially certain nutrients and ions. Other membrane proteins are enzymes that digest complex molecules in the external solution or inactivate toxins. The optimal functioning of the cell membrane proteins depends on the external solution composition, including pH, ion concentrations and osmotic pressure, because they affect the potential difference across the membrane. There are also a variety of external peripheral proteins attached to membrane proteins, and outward-facing oligosaccharides attached to the proteins and lipids. These oligosaccharides vary between individuals, between species and between higher taxonomic hierarchies. They can provide a signature for each mating type, population allele and species.

One molecule that crosses the cell membrane relatively freely is water. The passage of water through the membrane proteins but the selective exclusion of solutes is called **osmosis**. This is important for understanding periods of organism activity and inactivity. When the soil solution contains few solutes relative to the cytoplasm of cells, water will flow into the cell. This tends to happen after rain, when there is more water and the soil solution is diluted. In contrast, during soil desiccation, the soil solutes become concentrated in increasingly less water. At some point, the soil solution will be more concentrated with solutes than the cytoplasm. This will draw water out of the cytoplasm. Different species have different tolerances of and preferences for external solution composition. For the cytoplasm to function optimally, it requires a constant concentration. Therefore, species have developed various means of **osmoregulation** that permit adjustments and buffering of cell water content, against changes in the external solution. Terrestrial protozoa and chromista have a **contractile vacuole** which removes excess water from the cytoplasm and excretes it together with soluble nitrogenous wastes and other solutes (Patterson, 1980). The role of contractile vacuoles in osmoregulation is similar to that of nephridia and renal cells in multicellular species. However, the cell's ability to buffer these changes is limited. If the external solution becomes too concentrated or too dilute, the cell becomes inactive and shuts itself off from the environment. If the soil is dried or diluted too rapidly, or too much, cell lysis or death is caused as the cell membrane ruptures. Fungi do not have contractile vacuoles and desiccate when soil is too dry. They are protected from lysis by the cell wall. The filaments become active again when the cytoplasm absorbs soil water. Invertebrates regulate their water content through renal cells and nephridia, which couple osmoregulation with nitrogenous waste excretion.

Food that will be used in the cell must enter through the cell membrane. Food is obtained by **phagocytosis** and **osmotrophy** (pinocytosis, receptor-mediated endocytosis and active transport through the membrane). Food particles that enter by phagocytosis need to be digested in food vacuoles with digestive enzymes from lysosomes. The enzymes available vary between species, so that different proportions of the food vacuole organic matter will be digested. The remaining undigested portion must be excreted from the cell by **exocytosis**, i.e. the transport of the vesicle to the membrane and evacuation of its contents out of the cell. The digestion of food material releases usable nutrient molecules and indigestible molecules. Nutrient molecules that have entered the cytoplasm become the source of energy for metabolism, cell growth and replication of the cell. For these processes to occur, a sufficient supply of balanced nutrients is required.

Cellular metabolism. For a cell to grow and divide, the nutrients it acquires must be used to provide both a source of chemical energy for reactions and substrate molecules for these reactions. Catabolism refers to the chemical breakdown of nutrients to provide a source of chemical energy for cellular reactions. Anabolism refers to the cellular reactions that use the chemical energy and the nutrients as substrate molecules to build the more complex biological molecules and organelles. The details of metabolism of nutrients vary between species. The amino acids, carbohydrates, lipids, nucleic acids and mineral ions, which are obtained from digestion and from the external solution, are the building blocks for the rest of the cell. The energy for carrying out these synthetic (or anabolic) reactions is obtained from the respiration pathway. The respiration pathway varies between species depending on the presence of mitochondria, hydrogenosomes, intracellular symbiotic bacteria or protists, and the panoply of enzymes available to the species. Therefore, species that have lost a mitochondrion during evolution are **anaerobic**, but those with mitochondria are **aerobic** or **facultative anaerobes**. Anaerobic species are common symbionts in the digestive tract of multicellular species, but free-living species also occur. The mitochondrion-independent pathway (glycolysis and fermentation) uses glucose molecules which are oxidized to two molecules of pyruvate, with a net yield of two adenosine triphosphate (ATP) molecules. Pyruvate can be oxidized further to smaller molecules (such as acetate, lactate or ethanol) to yield further energy. Glycolysis does not require oxygen and does not yield CO_2. The glucose and other substrates for this set of reactions is obtained from digestion of food, from *de novo* synthesis from other molecules, or transport into the cell in species where that is possible. Normally, it is assumed that most free-living protozoa do not transport glucose through the membrane, but acetate and other soluble 2–3 carbon molecules often are. In species with mitochondria, a further sequence of reactions is possible which produce >30 ATP molecules from each glucose equivalent. The mitochondrion provides the cell with enzymes for the Krebs cycle, the electron transport chain and oxidative phosphorylation. These respiration reactions require oxygen as a final electron acceptor, pyruvate and acetyl-CoA as substrates from the cytoplasm, and release CO_2 and H_2O as end-products. The rate of these reactions varies with the supply of oxygen. For this reason, eukaryotic cells have a variety of oxygen-scavenging molecules such as haemoglobin and other haem-containing molecules. For example, plant roots produce leg-haemoglobin when the soil is water saturated and short of oxygen, and many protozoa have haemoglobin-like proteins. The efficiency of these oxygen scavengers at low partial pressure of oxygen in the solution (pO_2) determines tolerance to near-anaerobic conditions (anaero-tolerance). Intermediate molecules produced during glycolysis and mitochondrial respiration reactions are used as substrates for the *de novo* synthesis of amino acids, nucleic acids, lipids, carbohydrates and complex macromolecules, which together form the cytoplasm and organelles. The release of chemical energy from the γ-phosphate bond in ATP, to adenosine diphosphate (ADP) or adenosine monophosphate (AMP), is used to drive the metabolic pathways to synthesize, modify or degrade molecules in the cell. ATP is the main energy currency of the cell. The total amount of cellular ATP can be an assay of metabolic activity because it is short lived and unstable outside cells, in the soil solution.

Secretion and exocytosis. One important difference from prokaryotes (later in this chapter) is that eukaryotes are able to target large amounts of diverse enzymes and other molecules for **secretion**. The endomembrane network links the ribosomes, endoplasmic reticulum, Golgi and dictyosome to vesicles where the molecules and enzymes targeted for secretion accumulate. The vesicles are varied and named according to their function. Some contain digestive enzymes (lysosomes) that will fuse with ingested food vacuoles (or phagosomes). Others are destined to be secreted out of the cell (secretory vesicles) for external digestion of food resources. These secretory vesicles often accumulate near the cell membrane until a stimulus triggers their release *en masse*. For example, in fungal cells, when a suitable substrate is detected, large numbers of secretory vesicles with digestive enzymes fuse with the cell membrane and release the enzymes externally (exocytosis). Secretion of individual proteins out of the cell also occurs, but it is not as significant as exocytosis in releasing large amounts of material from the cell. (In prokaryotes (this chapter), secretion of small single proteins is the only means of secretion. A single eukaryotic secretory vesicle is many times larger than a whole prokaryotic cell.) **Excretion** of undigested remains of the food vacuoles occurs by targeting the food vacuole for fusion with the cell membrane and release of the contents outside. Often the remains of several food vacuoles will fuse before being excreted. Exocytosis is not restricted to vesicles with enzymes or old food vacuoles. Many species of protists and cells of multicellular species can secrete mucopolysaccharides, or release defensive organelles or chemicals, and pheromones (mating hormones). The release of mucus (from mucocysts) and defensive organelles (trichocysts) is triggered in defence against predation, and chemical or mechanical stimuli. Gland cells in invertebrate epithelium secrete mucus for sliding, absorbing water or providing a protective barrier.

Cytoskeleton. The shape of eukaryotic cells is determined by their external cell wall or internal cytoskeleton. In animal cells and protists which lack a cell wall, the shape is determined by **stable microtubule networks** and associated fine filaments which support the cell membrane and hold the nucleus and basal bodies in position. The basal bodies have an important role in orienting and organizing the cytoskeleton. The exact arrangement of these stable cytoskeletal networks varies between phyla and classes. The stable microtubular networks in the cell provide a frame with distinct anterior–posterior and left–right axes. The remaining cytoskeletal elements are organized from this. The **cilium** (or 'eukaryotic flagellum') is an extension of the cell membrane, supported by microtubular and associated elements (the axoneme) that extend from basal bodies. The cilium is responsible for pulling or pushing the cell through liquid films and free water. Hair-like protein extensions from the cilium are called **mastigonemes** and occur in certain taxa, especially in many chromista. The cilium is anchored at its base by a basal body which is composed of stable cytoskeletal proteins. The **basal body** (or kinetosome) consists of a core of circular microtubules and several anchoring elements extending into the cytoplasm. The basal body and its associated anchoring cytoskeleton are called a **kinetid**. Basal bodies occur in pairs in most taxa, and not all basal bodies are ciliated. In some taxa (such as Ciliophora, Hypermastigea and Opalinea), the entire cell is covered with cilia. The details of the kinetid structure are key to the correct identification of families and genera of protists. The **transient**

cytoskeleton is more dynamic, continually changing in distribution. It is composed of transient microtubules, actin filaments and associated fine filaments. It is implicated in the amoeboid locomotion and shape changes of cells, as well as in directional transport of vesicles and organelles in the cytoplasm. Most protists without an external cell wall are capable of at least some amoeboid locomotion. Some taxa also have **supporting elements** under the cell membrane. For example, the euglenid species have a proteinaceous thickening (pellicle strips) that reinforces the membrane. Some classes of Ciliophora have additional membranes (the alveolae) and cytoskeleton which support the cell membrane. Lastly, many soil species also have a **cell wall** outside the cell membrane. In Testacealobosea, the cell wall consist of chitinous and proteinaceous secretions. In the Difflugida (Testacealobosea), the test is reinforced further with mineral soil particles cemented or embedded in the wall material. In several families (such as the Paramaoebida, Paraquadrulidae and Nebellidae) and in the Euglyphid (Filosea), the wall is composed of mineralized secreted plates of various shapes. In the Oomycetes and Hyphochytrea, the cell wall consists of tight fibrils of β-glucans and cellulose deposited outside the cell membrane. In the fungi, it is composed of chitin, glucans and a variety of polysaccharides that vary between taxa. Species with an external cell wall are less able to change shape and squeeze through spaces. However, they are more resistant to abrupt changes in osmotic pressure and desiccation, which are more frequent in surface soil and litter.

Differentiation. In response to an environmental stimulus, protists can differentiate to an alternative morphology or functional role. The stimulus usually involves dilution or decrease of food resources, desiccation of soil, and changes in pH, temperature or salt composition. The **differentiation** response involves a dramatic replacement of the mRNA by a new population of mRNA molecules from activated genes, over a few minutes to hours. The result is a new set of proteins and enzymes that are synthesized to carry out a new set of reactions. The cell switches its function from growing and dividing into a non-feeding sexual pathway or into a dormant cyst stage. In some taxa, there is no known sexual pathway. For example, it is uncommon or at least unreported in the Amoebaea but observed in many Testacealobosea and Ciliophora. Some taxa respond to poor conditions by differentiating to a dispersal stage. For example, the Percolozoa are normally amoeboid cells that become ciliated dispersal cells when resources are reduced. However, probably all soil species are capable of **encystment** to resist desiccation or an unsuitable environment. Encystment involves a partial or complete resorption of the stable cytoskeletal network and loss of cell shape. The cell secretes a protective cell wall that is usually proteinaceous and/or chitinous. Encystment in soil species usually involves a gradual excretion of water through the contractile vacuoles that is important to confer chemical resistance by dehydration of the cell. Desiccation of the cyst in certain species can proceed to a near crystalline state which makes molecular manipulation and extractions difficult. The excystment stimulus varies between species, and involves the synthesis of new mRNA populations required for morphogenesis, metabolic and cell growth regulation proteins.

In multicellular species, differentiation is a continuous process that begins with egg fertilization or parthenogenesis, and through successive cell divisions during development into the mature adult. It results in differentiation of tissues and of cells within tissues, each with a distinct function. In these species, cells do not separate after cell division and remain attached, forming a multicellular organism with several tissues. Each tissue has cells with specific functions, such as osmoregulation in renal cells, contraction in muscles, nutrient endocytosis in the gut, enzyme secretion in digestive glands, and so on.

Protists

The term **protist** is not a taxonomic designation, but groups together species that are unicellular eukaryotes. Unicellular organisms exist in several morphological forms. Many are single independent cells, which may or may not be motile. However, some form filaments as in many algae and fungi, others are held together in colonies by a secreted sheath or a common secreted stalk, and in a few classes cells remain attached in single or multilayered sheets or filaments. In general, all cells remain identical, even in species that form colonies, filaments or otherwise attached cells. In cases where cellular differentiation occurs, it is limited to forming resistant inactive stages (such as cysts) and sexual reproduction, so that reproductive cells can have a different morphology as well as a different function.

As a unicellular organism, each protist cell is capable of – and as an individual, responsible for – its own function, protection and propagation. As individuals, each is as physiologically complete and independent as are multicellular organisms. Each protist cell interacts directly with the external environment, where it obtains its food and returns its waste products. Cells must also be able to respond to the hazards of changing conditions in their environment, such as desiccation, temperature extremes and fluctuations in pH and ion concentrations. The physiological functions necessary for these adaptations are motility, chemotaxis, food acquisition, digestion–excretion, osmoregulation, encystment–excystment, cell growth and division, and in some species cell–cell recognition and species recognition. Species from three kingdoms of protists are implicated in decomposition: the protozoa, the chromista and the fungi.

Protozoa

The kingdom **protozoa** comprises eukaryotic cells that are ancestrally phagotrophic, without cell differentiation between vegetative cells. General reference sources for the following information and for further descriptions of taxa and biology can be found in Grassé's Encyclopaedia (Grassé, 1949–1995), *The Illustrated Guide to the Protozoa* (Lee *et al.*, 2001), Haussman and Hullsmann (1996), Page (1976), Patterson (1996) and as noted in the text. The systematic names and ranking used for the descriptions in this chapter are according to Cavalier-Smith (1998). An overview of phyletic relationships can also be obtained at http://phylogeny.arizona.edu/tree/eukaryotes/eukaryotes.html/ More web resources can be accessed through http://megasun.bch.umontreal.ca/protists/protists.html

Retortomonadea and Trepomonadea (phylum Metamonada)

These species lack mitochondria, are variably aero-tolerant and inhabit environments with low oxygen where they feed on bacteria (Fig. 1.2). They are found in low-oxygen environments as commensals in the intestinal tract of animals or free living. Retortomonads consist mostly of small organisms 5–20 μm long, with two pairs of basal bodies with usually four emerging cilia (e.g. *Chilomastix*). The basal body pairs are perpendicular to each other and connected by a striated fibre. The recurrent trailing cilium runs through the cytoplasm and emerges posteriorly through

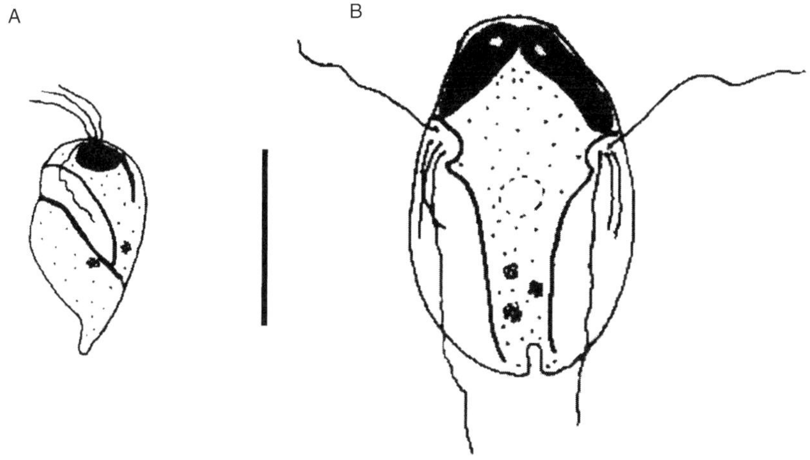

Fig. 1.2. Representative Retortmonadea and Trepomonadea. (A) *Chilomastix,* (B) *Trepomonas.* Scale bar 10 μm.

a groove which functions as the cytostome. The recurrent cilium and its associated ribbons of cytoskeleton are more closely associated with the nucleus, and form an elaborate organelle that contributes to motility and cell rigidity. The cytostome is a long funnel into the cell and is used for phagocytosis of bacteria. The cell membrane and cytostome are lined by supporting microtubules. There is no Golgi, and the endomembrane network is poorly developed. The nucleus is held tightly by the cytoskeleton close to the anterior of the cell. Mitosis in Retortomonadea is closed, with an intranuclear spindle. The resistant cyst stage is transferred between host individuals, and sexual stages are not known.

The Trepomonadea resemble the Retortomonadea on most points, with the following distinctions. There are additional supporting cytoskeletal fibres extending from the basal bodies to the nucleus (*Enteromonas*, *Trimitus* and *Caviomonas*). In the Trepomonadea, mitosis is semi-open, with a mostly intranuclear spindle. Some genera (*Trepomonas*, *Hexamita*, *Spironucleus*, *Octomitus* and *Giardia*) are double cells that are a mirror image of each other. Mitosis is a semi-open type with an intranuclear spindle. There are free-living aero-tolerant species (in *Hexamita* and *Trepomonas*), but most are intestinal parasites or commensals. The latter have lost a functioning cytostome and rely on osmotrophy or pinocytosis.

Oxymonadea (phylum Metamonada)

The oxymonads also lack both the Golgi and mitochondrion. These species are larger (mostly ~50 μm), strictly symbiotic anaerobes in the digestive tract of wood-eating insects (mostly known from termites and cockroaches). Typical genera include *Monocercomoides*, *Oxymonas*, *Pyrsonympha* and *Saccinobacculus* (Fig. 1.3). They lack a cytostome and ingest fragmented wood chips and microdetritus from the host gut, by phagocytosis from the posterior region of the cell. Pinocytosis occurs over the whole cell surface. There are two pairs of anterior basal bodies connected by a thick fibre. The cilia are trailing posteriorly and spiral-wrap around the cell body. From one basal body, a sheet of tightly cross-linked microtubules (the **pelta**) overlays the nucleus and supports the anterior region. A ribbon of cross-linked microtubules and other supporting cytoskeletal elements combine to form the **axostyle**. It is a characteristic organelle which runs the length of the cell and participates in locomotion. It is a strong organelle that bends the cell and contributes to its wriggling sinusoidal locomotion. There is no microtubule network supporting the cell membrane, which is therefore flexible. Newly divided cells can be seen swimming in the gut, but eventually attach at the pelta to the host digestive tract lining. Mitosis is closed, with an intranuclear spindle. Sexual stages with meiosis occur. The surface of the Oxymonadea is covered by other symbiotic bacteria, which also participate in the digestion process inside the gut.

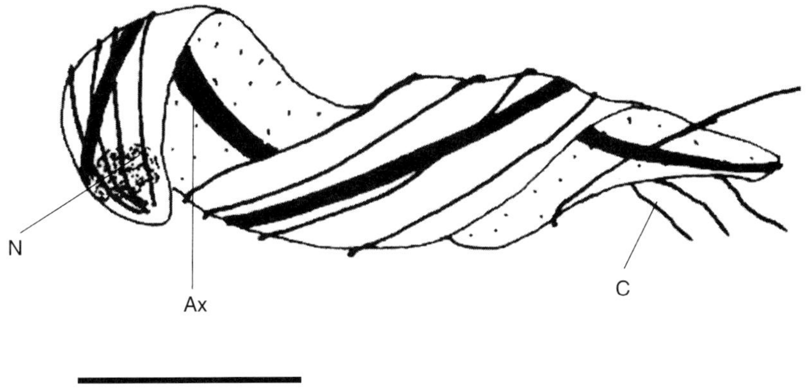

Fig. 1.3. A termite gut oxymonad, *Pyrsonympha,* with four cilia (C) spiralling around the body, an internal axostyle (Ax) that extends the length of the cell and an anterior nucleus (N). Scale bar 10 μm.

Trichomonadea and Hypermastigea (phylum Trichozoa)

The trichomonads and hypermastigotes belong to the parabasalian protozoa, which are primarily endozoic symbionts or parasites. About half of the genera occur in wood-eating insects, such as the roach *Crypotcercus*, but two free-living species are known from water solutions rich in dissolved organic matter. Their internal structure has an elaborate fibrous cytoskeleton supporting the basal bodies, the nucleus and the anterior of the cell. They are amitochondriate anaerobes that feed by amoeboid phagocytosis on microdetritus in the digestive tract of host organisms. Their metabolism is enhanced by hydrogenosomes which carry out respiration of pyruvate to acetyl-Co A, CO_2 and H_2, producing ATP. The storage material is glycogen. Mitosis is closed, with the chromosomes attached to the nuclear membrane, and an extranuclear spindle. The Parabasala derive their name from an elaborate system of fibrous cytoskeleton, that emanates from basal bodies to support the nucleus and the anterior of the cell, and hold in place long Golgi bodies adjacent to the basal bodies. These species are therefore capable of extensive protein synthesis and secretion through the endomembrane network. Many parabasalian species, particularly the Hypermastigea, are important to decomposition for their role in the digestion of wood microdetritus in the digestive tract of wood-eating cockroaches and the lower termites. Different species of Trichozoa are found in different populations and species of wood-eating insects.

The basic structure of a Trichozoa can be described from a typical Trichomonadea such as *Trichomonas* (Fig. 1.4). There are four basal bodies, with one trailing and three anterior cilia. The recurrent cilium is intracellular and held by an extension of the cell membrane (it appears

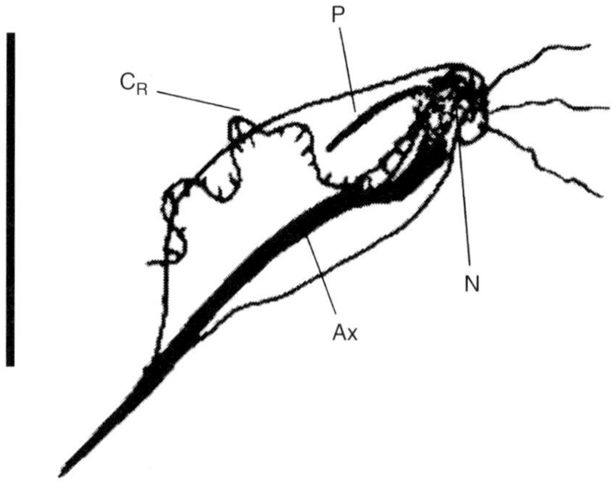

Fig. 1.4. The Trichozoa *Trichmonas*, showing a recurrent cilium (C_R) held by a membranelle, an axostyle (Ax), pelta (P), anterior nucleus (N) and three anterior cilia. Scale bar 10 μm.

as an undulating membrane). The **costa** is a contractile rod supporting the trailing cilium. It extends from the basal body and contributes to motility. The anterior of the cell is reinforced by a pelta, and an axostyle wraps at least partially around the nucleus and extends the length of the cell. It appears as a central cone that extends out of the cell in some species. A fibrous rootlet (**rhizoplast**) from a basal body extends to the nucleus. From the anterior-most basal body, cytoskeletal elements (the parabasal fibres) extend to the nucleus and support two elongated Golgi bodies. There is no cytostome and no supporting microtubules under the cell membrane. The cell is therefore very flexible, and phagocytosis occurs along the cell membrane, ingesting bacteria and larger microdetritus. Most species of trichomonads are 5–25 μm intestinal parasites in animals. Some such as *Dientamoeba* and *Histomonas* do not have emerging cilia and are amoeboid. Others are symbionts in termites, such as *Mixotricha paradoxa*.

The Hypermastigea are larger cells, of approximately 100 μm, that resemble *Trichomonas* or *Monocercomonas* but with the cytoskeletal structures replicated many times across the cell, i.e. there are multiple pairs of basal bodies at the anterior with cilia spiralling along the body towards the posterior, or emerging as tufts at the anterior. The axostyle and other cytoskeletal elements from the basal bodies merge into a single large unit with complex modifications. Typical genera are *Barbulanympha*, *Joenia*, *Lophomonas* and *Spirotrichonympha* (Fig. 1.5). The Polymonadea are similar to other hypermastigote Parabasala, but with multiple nuclei, axostyle and parabasal bodies. They resemble hypermastigotes without the merged organelles (e.g. *Calonympha*, *Coronympha* and *Snyderella*). Species of hypermastigotes are found exclusively in

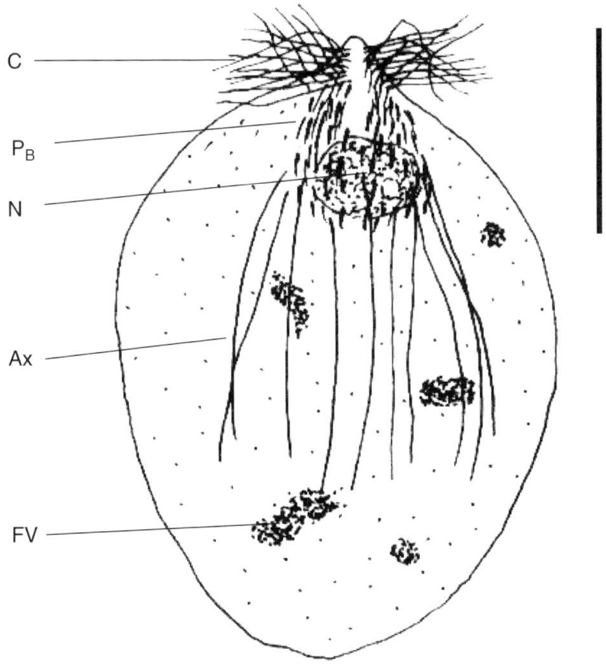

Fig. 1.5. The hypermastigote *Barbulanympha* with many axostyles, extensive anterior cytoskeleton and numerous anterior cilia. Cilia (C), parabasal bodies (P_B), nucleus (N), axostyle (Ax), food vacuoles (FV). Scale bar 25 μm.

wood-eating cockroaches and lower termites, where sometimes they are accommodated in specialized termite gut chambers. As with other termite symbionts, they are associated with extracellular surface spirochetes (bacteria) which contribute to motility, as well as intracellular symbiotic bacteria. These species cooperate together in the digestion of woody microdetritus ingested by the insect.

Percolozoa (phylum Percolozoa)

These are transiently ciliated species (usually in the dispersal phase in an unfavourable or no food environment), with normally amoeboid cells. The amoebae move with eruptive pseudopodia, and feeding by phagocytosis occurs in both amoeboid and ciliated stages. Species are assumed to be bacterivorous, but osmotrophy occurs at least in some species because they can be cultivated axenically, and cytotrophy is possible in others. The number of cilia on the swimming dispersal cells depends on the species, e.g. *Tetramitus* (tetraciliated), *Naegleria* (biciliated) and *Percolomonas* (always ciliated amoeba). Accurate identification of families and species requires descriptions of the amoeba, swimmer, cyst morphology and mitosis, and confirmation of mitochondria with discoid cristae. The Percolozoa include 14 genera, of which several are very common in soil, particularly in the Vahlkampfiidae (order Schizopyrenida), such as the *Adelphamoeba*, *Didascalus*, *Naegleria*, *Percolomonas*, *Tetramitus* and *Vahlkampfia* (Fig. 1.6). Also encountered in the soil are *Stachyamoeba*, which is occasionally limax, and *Rosculus* in forest litter. Species may also be encountered in anaerobic or fermentative habitats, as they seem to be efficient at scavenging oxygen. Several species are opportunist pathogens with infectious strains that will penetrate nervous tissues and the brain through nasal passages, causing death in a few days. Infection occurs from soil on the hands and face, so caution is required in handling soil.

Euglenids (phylum Euglenozoa: Euglenoidea)

Euglenoidea consist of about 1000 species known mostly from aquatic habitats, but they are common in moist soil habitats. Modern treatment of this group can be found in Triemer and Farmer (1991) and in Leander and Farmer (2001). Cells are typically 10–50 μm long, elongate or spindle shaped, with a deep (but sometimes inconspicuous) anterior invagination (Fig. 1.7). The invagination often functions as a cytostome and always contains two cilia. One cilium emerges

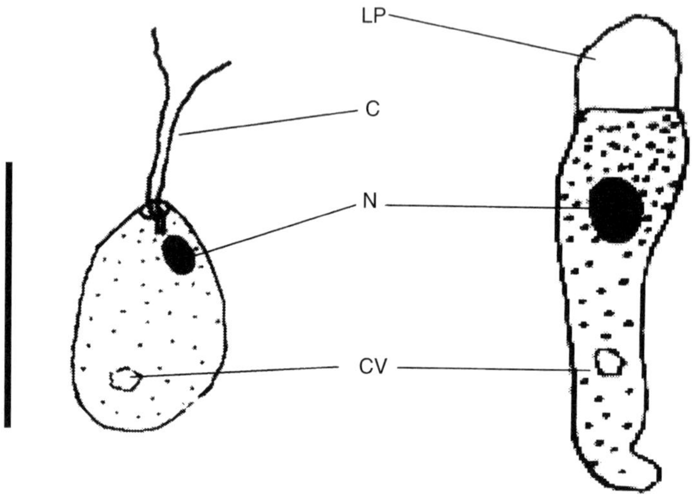

Fig. 1.6. Representative species of a Percolozoa, an *Adelphamoeba* showing the ciliated phase and the amoeba phase with eruptive clear lobopodia. Cilia (C), contractile vacuole (CV), lobopodium (LP), nucleus (N). Scale bar 25 μm.

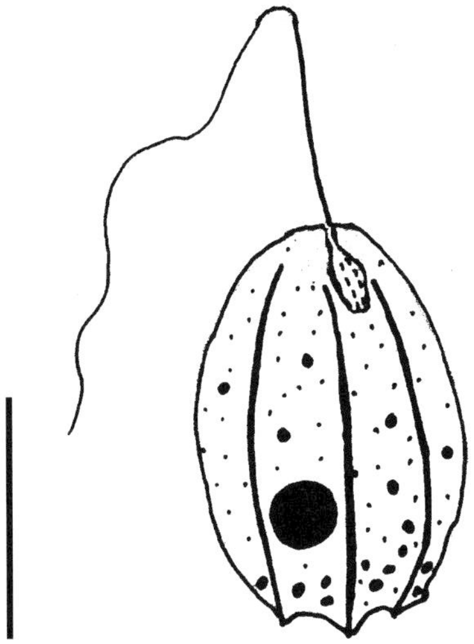

Fig. 1.7. A large euglenoid, *Petalomonas*, with an anterior pocket holding two cilia, only one being emergent and visible by light microscopy. Scale bar 25 μm.

from the cytostome, and the other may be very short and not emerging from the cytostome. The longer cilium often has a paracrystalline proteinaceous rod alongside the axoneme. An eye-spot of lipid vesicles occurs adjacent to a swelling at the base of the emergent cilium and the cytostome. The cilia bear non-tubular mastigonemes. In a few predatory species (cytotrophy on other protists), the cytostome can be a very elaborate organelle with mobile tooth-like elements. The storage material, paramylon, is an unusual molecule composed of β(1,3)-glucan polymer. The mitochondria have discoid cristae with constricted bases. The cell membrane is reinforced by strong interlinked strips (the **pellicle**) that wrap the cell. The pellicle provides stiffness and strength to the cell shape. Pellicle strips can move and bend in some species, allowing a form of amoeboid motion called metaboly. Permanently condensed chromosomes can be seen by light microscopy. Mitosis is closed, with an internal spindle, persistent nucleolus and slow asynchronous separation of chromosomes at anaphase. Conjugation with meiosis occurs in some species. Soil species are usually bacterivorous, osmotrophic or cytotrophic. About one-third of described species are photosynthetic, and some can be isolated from moist surface soil with adequate sunlight, where they appear in green patches. A group with few species, related to the euglenids, the Hemimastigophora (e.g. *Hemimastix* and *Spironema*) are about 10–60 μm in length, vermiform, and have two spiral rows of cilia in a groove. Food vacuoles form at the anterior by phagocytosis, and the posterior can attach to the substrate.

Bodonids (phylum Euglenozoa: Kinetoplastea)

The class includes many human parasitic species that cause Chagas disease, leishmaniasis, and African sleeping sickness, others that are animal blood trypanosomes, fish tissue parasites and the *Phytomonas* which are parasites in plant latex vessels. The kinetoplastid name comes from a large clump of mitochondrial DNA, the kinetoplast, located anteriorly near the basal bodies. The kinetoplast DNA consists mostly of DNA minicircles (0.2–0.8 μm long) or longer circles more typical of mitochondrial DNA (8–11 μm). There are about 600 species of kinetoplastids. Most are parasitic such as the *Trypanosoma* and *Leishmania*, or commensal and parasitic bodonids, but many are free-living Bodonea found in soil and freshwater habitats. These bodonids are small species (3–15 μm) with two heterodynamic anterior cilia emerging from an invagination (e.g. *Bodo*, *Dimastigella* and *Rhynchomonas*) (Fig. 1.8). Both cilia have a supporting proteinaceous rod that extends to the basal bodies. Supporting cytoskeletal elements extend from the basal bodies as ventral and dorsal fibres, along the length of the cell. The cell membrane is supported by a layer of longitudinal microtubules. A cytostome region, usually a well-developed invagination, is reinforced on one side with microtubules from the basal bodies. Certain genera, such as *Bodo*, have non-tubular mastigonemes in tufts on the anterior cilium. Cells have a contractile

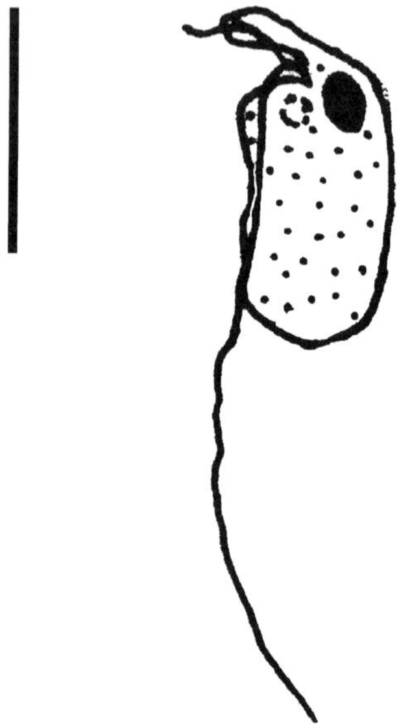

Fig. 1.8. *Rhynchomonas*, a free-living bodonid. Scale bar 5 μm.

vacuole which empties into the invagination that holds the cilia, a Golgi–dictyosome and a single mitochondrion with flat cristae constricted at the base. Mitosis is closed, with an internal spindle, and sexual stages are not known for free-living bodonids. Soil species are bacterivorous by phagocytosis at the cytostome, but osmotrophy should not be excluded.

Amoebae (phylum Amoebozoa)

This is a diverse phylum with groups of pseudopodial species which move mostly by ameoboid locomotion with lobose or conose pseudopodia and feed by phagocytosis with the help of pseudopodia. The pseudopodia may have short filose subpseudopodial extensions. Classes in this phylum possess mostly tubular mitochondrial cristae, with a well-developed Golgi–dictyosome. If ciliated species occur in a class, there usually is only one cilium per kinetid, and rarely more than one kinetid. The Amoebaea and Testacealobosea are abundant in the soil, and both classes are speciose.

Amoebaea (phylum Amoebozoa: Lobosa)

The lobose Amoebaea consist of families in the Gymnamoebae, which represent the true naked amoebae (Fig. 1.9). Microtubules do not appear to be involved in pseudopodial stiffening or motility (Hausmann

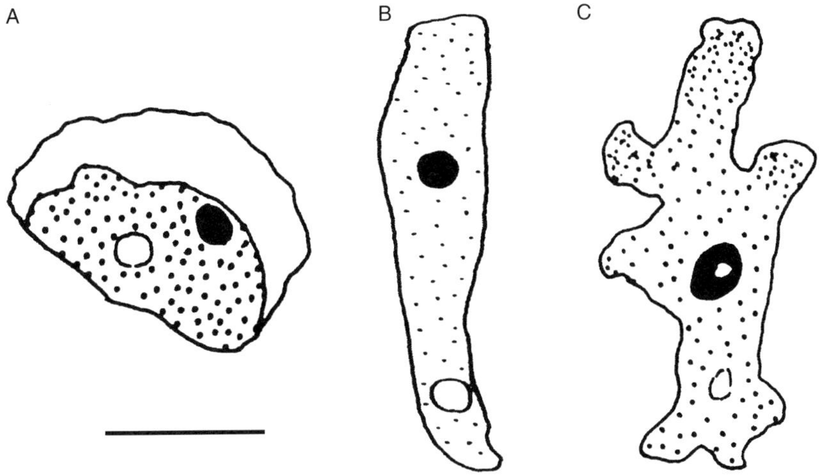

Fig. 1.9. Typical lobose amoebae (Gymnamoebae, Amoebozoa). (A) *Platyamoeba*, (B) *Hartmanella*, (C) *Deueteramoeba*. Scale bar 10 μm.

and Hulsmann, 1996). Cilia and centrioles are missing, and microtubules are limited to the mitotic spindle. Mitochondria possess branching tubular cristae. The cell membrane is naked, but species tend to have a thick glycoprotein layer (**glycocalyx**) and some genera have microscales (*Dactylamoeba* and *Platyamoeba*), glycostyles (*Vanella*) or a thickened cuticle (*Mayorella*), which are seen by transmission electron microscopy only. The free-living soil Gymnamoebae are essentially bacterivorous, supplemented by pinocytosis. Certain larger species may be cytotrophic on protozoa or other amoebae. Many genera are facultative fungivores (Old and Chakraborty, 1986) but some are obligate fungivores, such as *Deueteramoeba*. Cells can be only 5 μm in diameter but more usually 8–20 μm. Exceptionally, some aquatic species of amoeba can be up to several millimetres. Gymnamoebae are usually recognized as belonging to a particular family by their overall shape during locomotion. The principal families encountered in soil samples are presented in Table 1.1, and Lee *et al.* (2001) provide a useful six step identification key to families by light microscopy. Further identification and species designation require ultrastructure and DNA sequencing, as failing to do so will probaby lead to argument rather than agreement between systematists. However, many genera are tentatively identifiable with light microscopy. There clearly are species that are early colonizers and r-selected (such as *Acanthamoeba*), whereas other species appear later in the succession of species in decomposing litter (such as *Mayorella*), and yet other species are encountered only in more stable environments. Some species are particularly adapted to dry conditions, such as the tiny *Platyamoeba*, when other species are no longer active. Most field sites have a small number of abundant species that vary in their activity period with the seasons and daily abiotic conditions, and a large number of rarer species.

Table 1.1. Selected families of class Amoebaea with edaphic genera.

Families	Selected genera
Acanthamoebidae	*Acanthamoeba, Protacanthamoeba*
Amoebidae	*Deueteramoeba*, mostly aquatic genera
Cochliopodiidae	*Cochliopodium*
Echinamoebidae	*Echinamoeba*
Gephyramoebida	*Gephyramoeba*
Hartmannellidae	*Glaeseria, Hartmannella, Saccameba*
Hyalodiscidae	*Flamella, Hyalodiscus*
Paramoebidae	*Dactylamoeba, Mayorella*
Thecamoebidae	*Thecamoeba, Thecochaos, Sappinia, Sermamoeba*
Vannellidae	*Discamoeba, Pessonella, Platyamoeba, Vanella*
Leptomyxidae	*Leptomyxa, Rhizamoeba* (these are eruptive monopodial species)

Testacealobosea (phylum Amoebozoa: Lobosa)

The testate amoebae in this class are separated from the testate amoebae in the Filosea (Cercozoa) on the basis of molecular phylogenies, ultrastructure and morphology of pseudopodia (Table 1.2). The Testacealobosea consist of an amoeboid cell with lobose pseudopodia, inside a test with a single aperture (opening, or pseudo-cytostome). The cytoplasm contains one nucleus (rarely more), a zone rich in endoplasmic reticulum and Golgi–dictyosomes. Mitochondria have branched tubular cristae and tend to accumulate at the posterior of the cell. A contractile vacuole is present, and food vacuoles often predominate at the anterior. The test morphology consists of a secreted chitinous or proteinaceous material, which may have embedded mineral particles (Fig. 1.10). The embedded particles are cemented by an organic compound which could be similar to the secreted test material. Calcareous scales or a siliceous test occur in certain genera. Otherwise, the embedded particles in the Difflugina consist of mineral particles of more or less uniform size (see Fig. 1.11A), diatom tests or scales of digested prey. The morphology and details of the test architecture are key to initial identification, at least to the family level. However, there are serious limitations to test-based descriptions and species identification, as outlined below with the Filosea (Cercozoa) testate amoebae.

Table 1.2. Families of class Testacealobosea with selected edaphic genera.

Suborder	Families	Selected genera
Arcellina	Arcellidae	*Arcella*
	Microchlamyiidae	*Microchlamys*
	Microcoryciidae	*Diplochlamys*
Difflugina	Centropyxidae	*Centropyxis, Proplagiopyxis*
	Difflugiidae	*Difflugia, Schwabia*
	Distomatopyxidae	*Distomatopyxis*
	Heleoperidae	*Heleopera, Awerintzewia*
	Hyalospheniidae	*Hyalosphenia*
	Lamptopyxidae	*Lamptopyxis*
	Lesquereusiidae	*Quadrulella*
	Nebelidae	*Nebela, Apodera, Porosia, Argynnia, Schoenbornia*
	Paraquadrulidae	*Paraquadrula, Lamtoquadrula*
	Plagiopyxidae	*Protoplagiopyxis, Plagiopyxis, Geoplagiopyxis, Paracentropyxis, Bullinularia, Hogenraadia, Planhoogenraadia*
	Trigonopyxidae	*Trigonopyxis, Cyclopyxis, Cornuapyxis, Geopyxella, Ellipsopyxis*
Phryganellina	Cryptodifflugiidae	*Cryptodifflugia, Wailesella*
	Phryganellidae	*Phryganella*
Incertae sedis		*Geamphorella, Pseudawerintzewia*

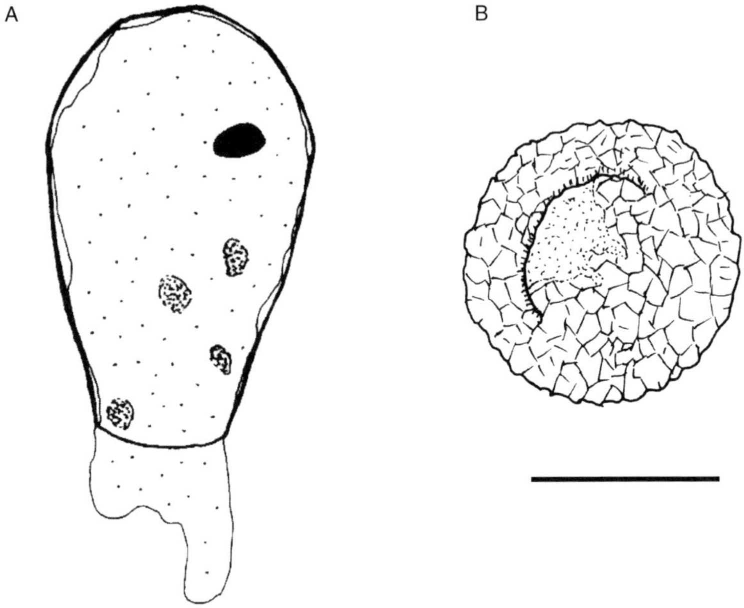

Fig. 1.10. Typical testate amoebae (Testacealobosea). (A) *Hyalosphaenia* with clear test. (B) *Plagiopyxis* with embedded soil mineral particles encasing a narrow slit-like opening. Scale bar 50 μm.

The distribution of Testacealobosea is primarily in the organic horizon. Many species are described from mosses and ground vegetation. However, as Foissner (1987) points out, most of these species are edaphic species that were sampled and transferred from the accompanying soil. None the less, there are species specific to the wet moss habitat that feed on the vegetation or prey in that environment. The feeding habit of Testacealobosea is varied, with smaller species being bacterivorous or saprotrophic, while some larger ones are saprotrophic, cytotrophic, fungivorous and predacious on microinvertebrates, microdetritivorous or omnivorous.

Archamoebae (phylum Amoebozoa: Conosa, Pelobiontida)

These common amoebae, such as *Mastigella*, *Mastigina*, *Mastigamoeba* and *Pelomyxa*, are found active in soils that are transiently anaerobic and rich in dissolved organic matter, and in peat or bogs (Fig. 1.12). They are secondarily amitochondriate, and possess an atypical cilium that appears rudimentary (Walker *et al.*, 2001). The axoneme is reduced, always with a single and unpaired kinetosome. The amoeba in some species becomes rigid and loses its amoeboid motion when there is lack of food

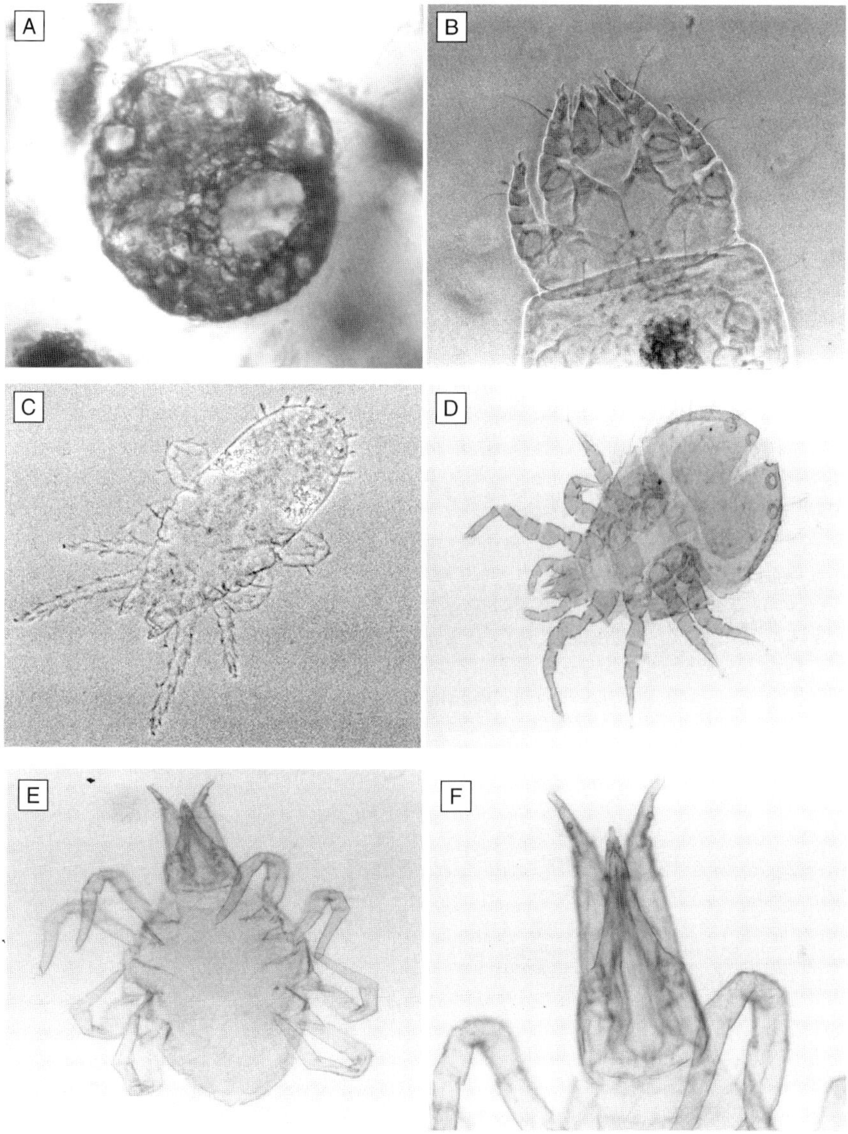

Fig. 1.11. (A) A common testate amoeba, *Difflugia* (110 μm diameter). (B) An Astigmatid mite showing mouth parts (180 μm wide). (C) A *Pachygnatidae* (Prostigmata) mite (280 μm long). (D) A *Zerconid* sp. (Prostigmata) mite (650 μm long). (E) A *Cunaxa* sp. (Prostigmata) mite (600 μm long). (F) Mouth parts of the same *Cunaxa* specimen as in (E). (G) A Mesostigmata *Rhodacarellus* sp. (390 μm long). (H) A Uropodidae (Mesostigmata) (body length 480 μm). (I) *Isotoma* (Collembola) (420 μm long). (J) Entomobryidae (Collembola) (1.3 mm long). (K) *Onychiurus* (Collembola) (470 μm long) – *continued overleaf.*

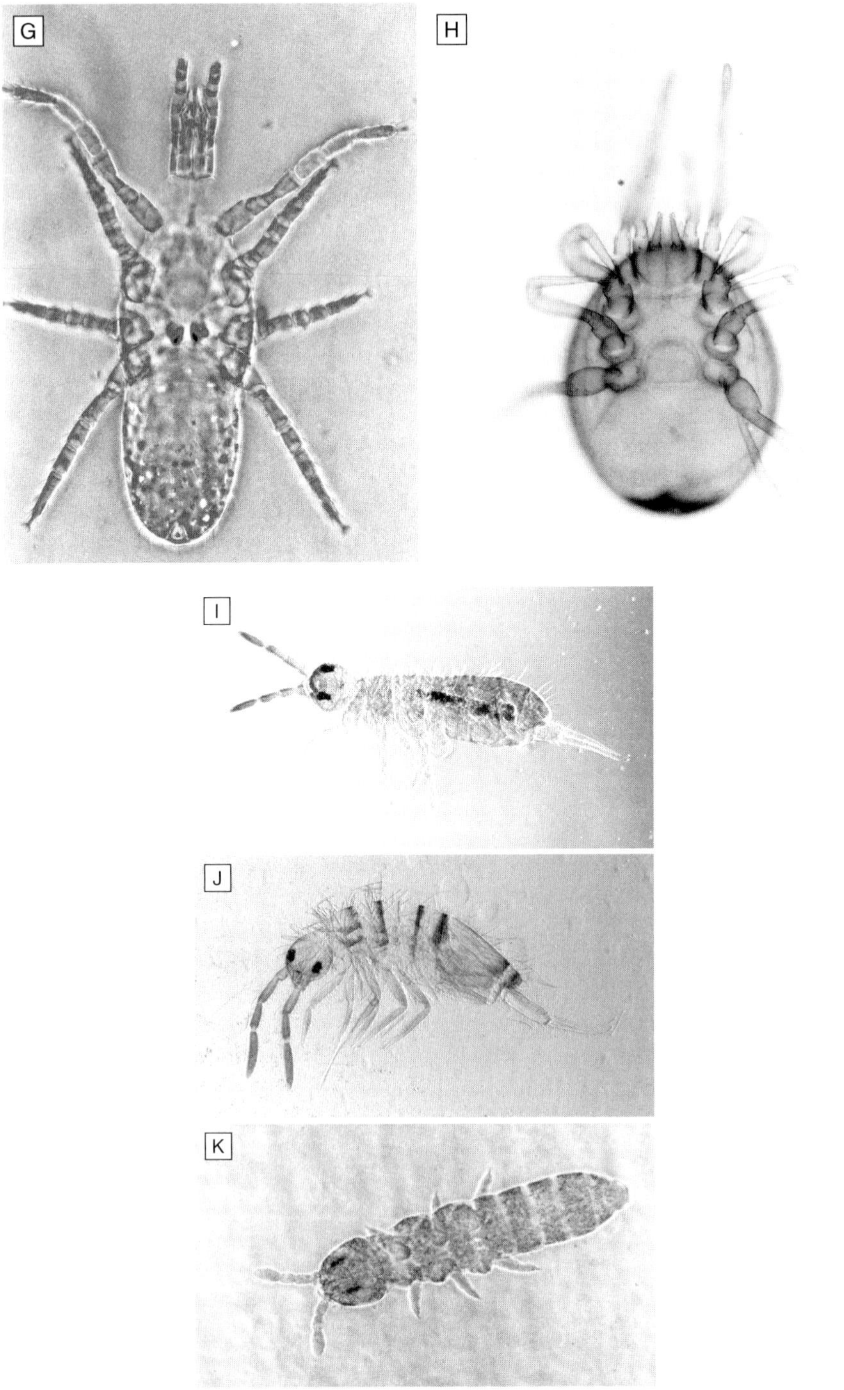

Fig. 1.11. *Continued.*

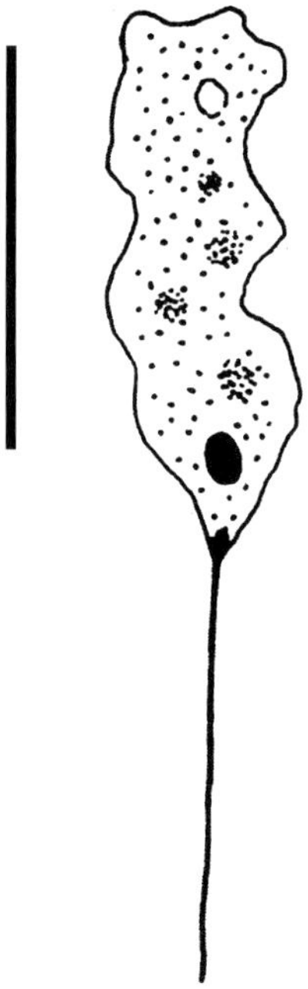

Fig. 1.12. A *Mastigamoeba* (Pelobiontida, Archamoebae) with single anterior cilium and flexible body. Scale bar 25 μm.

or conditions deteriorate. The nucleus is partially wrapped in a basket of microtubules emanating from the kinetosome. The storage material is glycogen, and mitosis involves an intranuclear spindle with persistent nucleoli. Species are osmotrophic on rich soil solution (and some can be grown axenically) supplemented with bacterivory.

Eumyxa (phylum Amoebozoa: Conosa, Mycetozoa)

The Eumyxa (or Myxogastria) are found in soil organic horizons, in surface litter including woody debris, in animal cellulosic dung and in tree bark. Specimens can spread across several centimetres and be 1–2 mm thick (e.g. *Physarum*). Species are brightly pigmented or clear. The cell is a multinucleate cytoplasm which continues to grow without cellular divisions and is referred to as a **plasmodium** (Fig. 1.13). Successive nuclear divisions are synchronous. The cytoplasm forms channels to increase diffusion, for moving cellular components, particularly food vacuoles and secretory vesicles. These channels appear like veins that branch across the plasmodium, and the movement of granules and cytoplasm is easily observed under the dissecting microscope. If desiccation occurs, the plasmodium secretes a siliceous material that hardens the dried cell and serves to protect the plasmodium, like a flat cyst. When food is scarce,

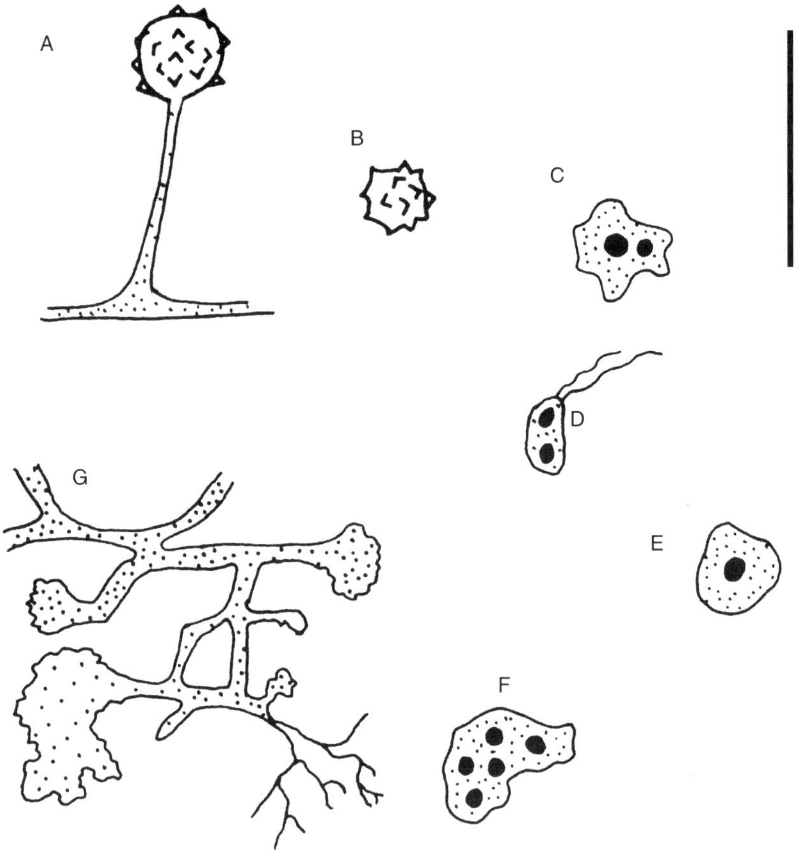

Fig. 1.13. Life cycle stages of a generalized Eumyxa (or Myxogastria). The sporangium (A) releases spores (B), from which an amoeba emerges (C). The amoeboid and ciliated morphologies (D) are feeding and dispersal haploid phases which lead to a diploid amoeboid cell (E) by fusion of complementary mating types. Nuclear divisions and growth of the diploid cells occur without cytokinesis, forming a large multinucleate cell (F and G) called a plasmodium. The plasmodium sporulates in numerous sporangia (A) where meiosis occurs. Scale bar (A) and (G) 1 mm, (B–F) 25 μm.

vertical stalks of cytoplasm extend upward. These become the site of meiotic divisions that lead to haploid walled spores accumulating at the apex inside sporangia, with an external protective wall. The spores are 5–20 μm and are released under suitable conditions. Light plays a role in initiating spore formation in certain species, so that some photoregulation exists. Once released, spores emerge as biciliated haploid cells for dispersal. The dispersal cells must fuse with another complementary mating type. The fusion forms a diploid cell which can grow into a new plasmodium. In Protostelids (e.g. *Protostelium*), the feeding phase consists of amoeboid cells with somewhat filose pseudopodia. These can fuse to form a plasmodium. Sporangia can form from single amoebae or fragments of plasmodium, but consist of one or a few spores only.

The activity of species is seasonal, with preference for wet or rainy periods, and species are restricted to preferred habitats. In temperate forests, different species are active during the spring, summer or autumn, but some may be active from ground thaw to frost. Substrate specificity is recognized in many species that prefer cellulosic dung or the bark of certain living plant species. There are five orders with nearly 600 species keyed from life history stages, habitat and sporangium morphology. However, accurate identification probably requires verification with gene sequences.

Dictyostelia (phylum Amoebozoa: Conosa, Mycetozoa)

The dictyostelid groups families which may be unrelated but that are the result of convergent evolution, towards aggregation of amoeboid cells into a **sorocarp**. There are two orders, Dictyosteliida (~70 species) and the poorly studied Guttulinida. The better known Dictyosteliida include *Acytostelium*, *Coenonia*, *Dictyostelium* and *Polysphondylium* (Fig. 1.14). A review of the most studied genus *Dictyostelium* can be consulted for its general biology (Kessin, 2001). The amoeboid feeding cells are bacterivorous with filose or lobose pseudopods, and are common in soils. When bacteria are reduced in numbers in Dictysteliida, secretion of cAMP or other short signal molecules initiates an aggregation response. Cells move by chemotaxis towards the source of the signal, and begin to release the signal themselves. The amplified response causes the aggregation of hundreds or thousands of cells. The aggregated cells can be seen with the naked eye on agar plates. Within the aggregate, cells differentiate into base cells, stalk cells which climb on

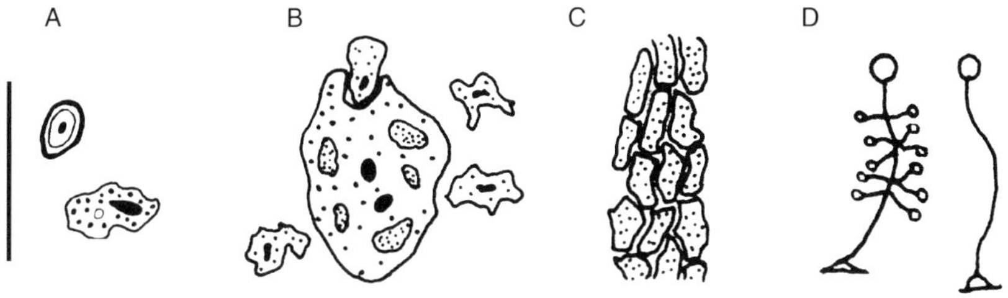

Fig. 1.14. Life cycle stages of a generalized dictyostelid (Amoebozoa, Mycetozoa). (A) Cyst and emerging amoeba. (B) Amoebae attracted to a large diploid cell which engulfs them and becomes a large macrocyte. (C) A stream of starving amoebae aggregating. (D) Typical sporangium morphology of aggregated amoebae. Scale bar (A–C) 25 μm, (D) 1 mm.

top of each other, and cells at the apex which produce cysts. The amoebae develop a cellulosic cell wall for support. The apex cells continue to divide while the stalk and base cells die. Release of the cysts under suitable conditions leads to amoeboid cells emerging from the cysts. Complementary mating types will pair under reduced food conditions. Under certain conditions, this leads to a macrocyst. The macrocyst secretes cAMP to attract other responsive amoebae, and feeds on them. The diploid cells can grow very large in some species. Encysted macrocysts undergo meiosis and multiple mitotic divisions to release haploid feeding amoebae.

In the order Guttulinida (e.g. *Acrasis*, *Copromyxa*, *Copromyxella*, *Fonticula*, *Guttilinopsis* and *Pocheina*), the amoeboid cells produce eruptive lobopodia reminiscent of the schizopyrenid amoebae (Percolozoa, Vahlkampfamoebae), and without ciliated dispersal cells (except *Pocheina flagellata*). There are species with flat cristae, and others with tubular cristae. These genera are poorly sampled and probably not so rare.

Cercozoa and Neomonada

These two phyla include many taxa of small free-living heterotrophic species <20 μm, many under 10 μm, with zero, one, two, three or four cilia. Most species previously were designated as 'nanoflagellates' or 'zoo-flagellates' and 'amoebo-flagellates'. The Neomonada are known mostly from marine samples and are not considered here.

The **Cercozoa** have tubular cristae and many have peroxisomes. Ciliated cells are variably amoeboid. There is no cytostome region, but filopodia are common. Some are gliding species in water films with a trailing cilium. There are anaero-tolerant genera which appear active in anoxic soils. Most species probably form cysts with a cell wall. Two classes of Cercozoa are important to soil ecology, the Sarcomonadea which includes the heterotrophic amoeboid-ciliated genera (e.g. *Cercomonas* and *Heteromita*), and the Filosea which includes the filose testate amoebae (such as euglyphid testamoebae) and several amoeboid reticulate genera (Fig. 1.15). The Filosea have a surface test with secreted siliceous plates. Although the correct phylogeny and systematics of the Cercozoa are uncertain, soil nanoflagellates are very abundant and unavoidably important in nutrient cycling. There is clearly more work required on understanding the biology and diversity of edaphic Cercozoa. The systematics of the Cercozoa are still evolving, and many genera and families will be reshuffled. A useful key to most sarcomonad genera was provided in Lee's illustrated guide (2001, pp. 1303–1307). The Sarcomonadea in the soil include many of the tiny ciliated species (<20 μm, but mostly under 10 μm) that are so abundant in fertile soil. They are primarily bacterivorous, with vari-

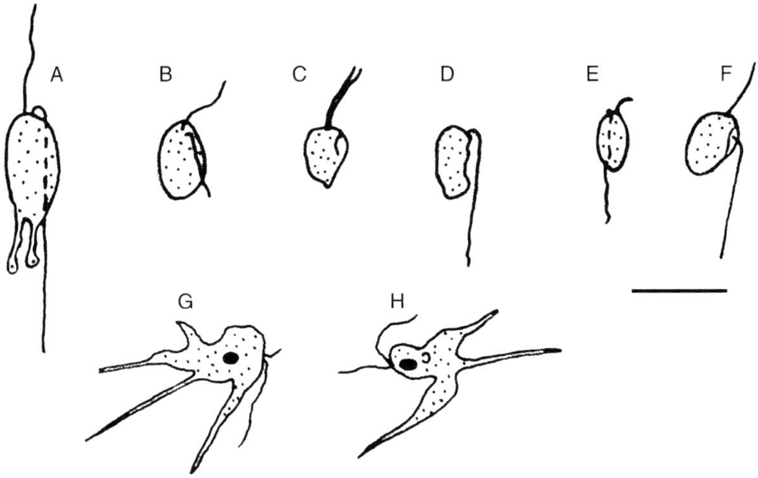

Fig. 1.15. Representative Cercozoa. (A) *Cercomonas,* (B) *Heteromita,* (C) *Adriamonas,* (D) *Allantion,* (E) *Allas,* (F) *Ancyromonas.* (G and H) Elongating filopodia of a cercozoan cell spread thin and flat against the substrate. (Certain genera tentatively assigned to Cercozoa.) Scale bar 10 μm.

able ability for osmotrophy. Selection of prey bacteria is probably common and depends on the bacteria's dimensions, shape, secondary metabolites, capsule size and composition. Filaments of bacteria are too long to ingest, and long or wide bacteria may be too difficult to phagocytose. Filose pseudopodia stretch through the soil to explore pores and surfaces, and participate in directing chemotaxis. Multiple branched or unbranched filopodia (<2 μm diameter) can explore pores that are inaccessible to the cell, by stretching out long hair-like strands of pseudopodia. Edaphic species are mostly gliding or amoeboid on surfaces, but in sufficient water they round up into a swimming cell rapidly (seconds) to change microhabitat. Gliding species usually possess an anterior cilium with a trailing posterior cilium. The sarcomonads offer little morphology for identification with light microscopy, and they are poorly described. Identification requires electron microscopy and gene sequence analysis, which is lacking for many soil isolates. Species may be confused with bodonids, Colpodellidae or Neomonada. Genera that may be Cercozoa and are encountered in soils include *Adriamonas* (Pseudodendromonadida), *Allantion*, *Allas*, *Amastigomonas*, *Apusomonas*, *Artodiscus*, *Cercomonas*, *Heteromita*, *Multicilia*, *Parabodo*, *Proleptomonas* (with chytrid-like behaviour), *Sainouron* and *Tetracilia*.

The dimorphids are encountered infrequently in soils but have 2–4 cilia and axopodia that are retracted if disturbed in swimming. They possess extrusomes and are bacterivorous, cytotrophic or ingest microdetritus.

The **Filosea** include testate amoebae with filose pseudopodia such as the euglyphids which secrete siliceous scales for their test (Fig. 1.16), and others such as *Pseudodifflugia* which bind soil mineral particles with a secreted cement to form the test. Both are common in soil litter. They were traditionally within the Testacealobosea but separated based on ultrastructure and molecular phylogenies (Table 1.3). Their biology and feeding habits are similar to those of other testate amoebae described above (see Amoebae).

Species descriptions for testate amoebae traditionally are based on details of test morphology. The terrestrial species are smaller than similar aquatic species, and some variation in test size occurs naturally in the soil. This variation may reflect an adaptation to drier periods as well as resource limitations. For example, *Trinema lineare* (Filosea) makes a smaller test under low soil moisture conditions (Laminger, 1978). Many species form a smaller test opening in drier periods, probably to minimize desiccation. Schonborn (1983) observed that the presence of spines and horns on the test can vary with culture conditions. Foissner (1987) states that identical test morphology can be found between samples, ranging in size from small to very large, raising the question of whether these could be different species or simply biogeographical variations. There are many reports of observed changes in test morphology occurring in the laboratory under culture conditions or naturally in the soil in response to changing environmental conditions and food availability

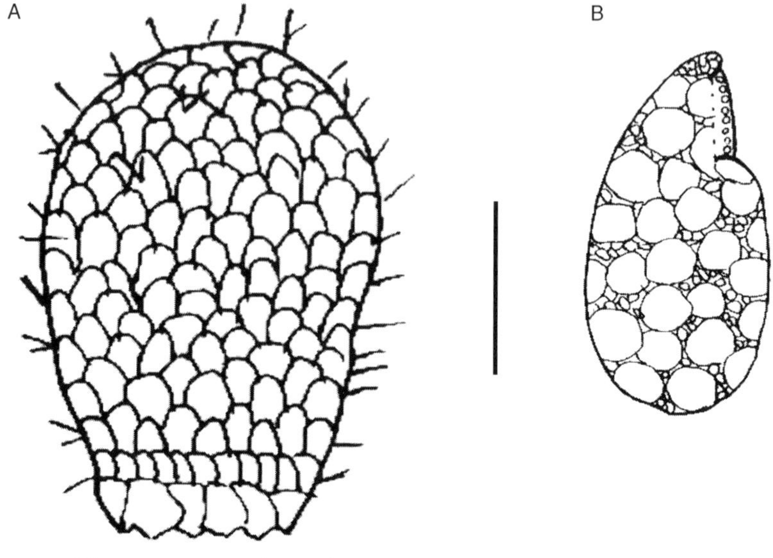

Fig. 1.16. Filose testate amoebae (Filosea, Cercozoa). (A) *Euglypha* with overlapping siliceous scales. (B) *Trinema* with large and small scales embedded in the test. Scale bar 25 μm.

Table 1.3. Families and selected genera of testate edaphic Filosea.

Orders	Families	Selected genera
Euglyphida	Euglyphidae	*Euglypha, Tracheleuglypha, Placocista, Assulina, Trachelocorythion*
	Trinematidae	*Trinema, Corythion, Playfairina, Pileolus, Deharvengia*
	Cyphoderiidae	*Messemvriella*
Incertae sedis		*Euglyphidion*
Other	Chlamydophryidae	*Rhogostoma*
	Pseudodifflugiidae	*Pseudodifflugia*
	Psamnobiotidae	*Nadinella, Psamonobiotus, Edaphonobiotus*
	Amphitremidae	*Amphitrema*
Incertae sedis		*Frenzelina*

(Andersen, 1989; Schonborn, 1992; Bobrov *et al.*, 1995; Foissner and Korganova, 1995; Wanner 1995, 1999). Verification of the validity of this test-centred taxonomy and whether it is useful at the genus or species level awaits phylogenic comparison of DNA sequences (Wanner *et al.*, 1997). Additional characteristics such as mating types and conjugation, life history details, food preferences and optimal abiotic conditions for activity would be useful. One common culture limitation may be the absence of suitable test synthesis material or sensitivity to changes in the soil solution composition (or culture medium). In the Difflugina (Testacealobosea) with the test composed of cemented soil mineral particles, there could be variation in shape or size reflecting the availability of different mineral particles between soil types.

The reticulate genera are not usually sampled, and it is difficult to estimate their abundance in soils. Their feeding habits are diverse and they tend to be larger species up to 500 μm in spread. The most common edaphic species are the Vampyrellidae, with numerous filopodia not usually anastomosing, and mitochondrion cristae that are best described as vesiculate. Cell divisions occur at the end of a feeding period in the cyst or at excystment. The best known in soils are the *Arachnulla* which can ingest bacteria and cysts, but are best known for puncturing fungal cell walls in hyphae and spores (Fig. 1.17). The Nucleariidae resemble vampyrellid amoebae with several ultrastructural differences. They have discoid–flat cristae with filose branching pseudopods. The pseudopods extend mostly from the sides of the cell and do not possess microtubular elements. Species feed on microdetritus or bacteria, or penetrate algal protists. Other reticulate genera include *Biomyxa*, *Leptophrys* (which feeds on algae and nematodes) and *Theratromyxa* (a multinucleate plasmodial genus which consumes nematode larvae and hyphae).

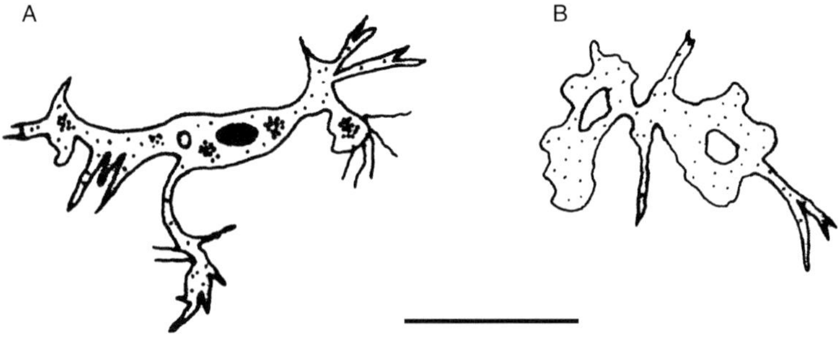

Fig. 1.17. Reticulate amoebae. (A) A vampyrellid amoeba (Vampyrellidae, Cercozoa). (B) A nuclearid amoeba, *Theratromyxa* (Nucleariidae, Cercozoa). Scale bar 50 μm.

The Colpodellidae (probably should be in phylum Sporozoa) are tiny cytotrophic predacious species that resemble free-living Sporozoa (also called Apicomplexa by some authors). The rostrum ultrastructure resembles the tissue-invading organelle of the parasitic species. The cells attack eukaryotic cells by bumping into the cell membrane until the invasive organelle is triggered. They proceed to ingest the prey cytoplasm. They may also aggregate around dying or wounded microinvertebrates along with other coprozoic species, such as the ciliate *Coleps*.

Ciliates (phylum Ciliophora)

The ciliates consist of a diverse phylum with >250 families of free-living species, in ten classes. Although common in the soil habitats, only one class is abundant in soil samples, while species in other families are active in small numbers only (Foissner, 1987) (Fig. 1.18). In general, ciliates are characterized by two functionally distinct nuclei and the presence of alveolar membranes under the cell membrane, long tubular cristae, extrusomes that are defensive or for predation, pluri-ciliated trophic or dispersal stages, cilia basal bodies organized into a coordinated system and a coated pit (parasomal sac) adjacent to cilia. The two nuclei are distinguished as the **micronucleus** which is inactive but participates in mitosis and meiosis, and the **macronucleus** which is polygenomic (with amplified copy number of the active genes). The macronucleus in most classes loses some or all transcriptionally inactive genes in the vegetative phase. The alveolar membranes were demonstrated to store calcium in some species. The **extrusomes** may secrete defensive mucus (mucocyst) which sometimes has degradative enzymes or toxins. Alternatively, they may be trichocysts, which discharge a lance into prey or predator. The somatic and cytostome ciliature is key to identification and requires silver staining or electron microscopy, espe-

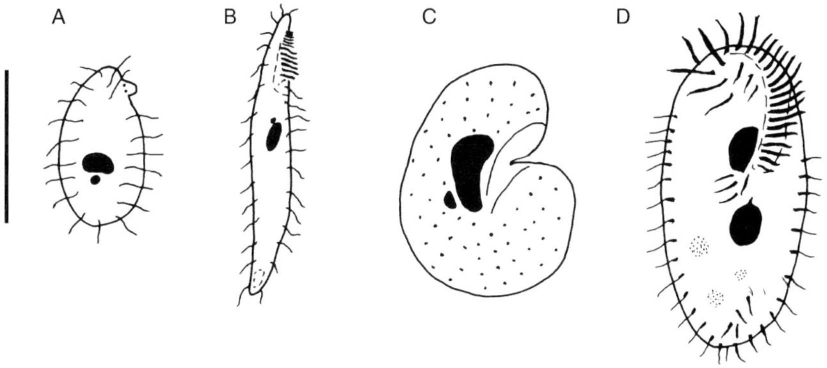

Fig. 1.18. Representative soil ciliates. (A) The fungivorous *Pseudoplatyophrya* (Grossglockneriididae, Colpodea). (B) A bacterivorous *Spathidium* (Spathidiidae, Litostomatea). (C) The cytotrophic *Bresslaua* (Colpodidae, Colpodea). (D) The omnivorous *Sterkiella* (Oxytrichidae, Stichotrichia). Scale bar (A and B) 25 μm, (C and D) 50 μm.

cially for the smaller species (Foissner, 1991; Lynn, 1991; Lynn and Small, 2002). Details of the ultrastructure of the kinetids are essential in family and genus identification. The kinetosome are arranged as single or paired somatic basal bodies. Not all carry a cilium. They are linked by cytoskeletal elements in longitudinal rows and beat in chronological sequence. The kinetosomes may be grouped in clusters of synchronously beating cilia. Most soil Ciliophora have a **cytostome**, specialized for the accumulation or apprehension of food particles by phagotrophy. The cytostome has specialized cilia and a shape which varies between families and genera. Some endosymbionts have lost the cytostome and feed by osmotrophy and pinocytosis. Others, such as several sand ciliates (*Kentrophoros*, *Trachelonema* and *Tracheloraphis*), do not have a cytostome, but phagocytosis occurs over a part of the somatic cell membrane. Many species are known to have endosymbiotic bacteria or protists. Ciliates are often sexual species, with two or many complementary mating types able to conjugate.

Contrary to the claim of early protozoologists, species found in marine sediments, soil and fresh water are distinct and habitat restricted. However, many edaphic species can be sampled from fresh water, especially as cysts, through surface runoff and leaching of water from the soil to streams. Particular adaptations of soil species are cysts resistant to desiccation (sometimes for decades), thigmotactic or surface-associated feeding, and a flexible and somewhat labile shape in thin water films. In general, about 50% of soil species are Colpodea and 37% are surface-associated Stichotrichia (Foissner, 1987).

The class Karyorelictea occurs in intertidal and marine sands, where most species feed on bacteria, although some are cytotrophic. Other ciliate taxa also occur in this habitat. Some species are associated with ectosymbiont bacteria, such as the sulphur bacteria on *Kentrophoros*. An extensive treatment of these intertidal species is provided in Dragesco (1960). The hypotrich subclass Stichotrichia contains many soil species, particularly in the family Oxytrichidae. Most are bacterivorous or cytotrophic on small protists, and many have a range of prey choice. Several species are known to change cell length and dimensions dramatically with changing resources, as well as changing preferred prey. Certain families across ciliates may appear in anaerobic water-saturated soil, such as species in the orders Armophorida and Odontostomatida. The class Litostomatea contains cytotrophic and predacious species which may be encountered occasionally. In surface litter during wet periods, Vorticellidae (Peritrichia) may occur in abundance. They are stalked ciliates that feed on bacteria by filter feeding. Gut endosymbionts of invertebrate saprotrophs (centipedes, millipedes, insects and oligochaetes) occur in the order Clevelandellida. The subclass Astomatia contains families that are endosymbionts (or parasitic) of annelid guts, particularly in oligochaetes and other invertebrate taxa. The class Colpodea are primarily soil species and are the more usually encountered ciliates in the soil. There are six orders with 12 families (Foissner, 1993; Lynn and Small, 2002). Cell division occurs inside cysts in the Bryophryida and Colpodida. Most species of Colpodea are bacterivorous, but some with larger cytostome are known to be cytotrophic, such as *Bresslaua* (Colpodidae). The Grossglockneriididae (Colpodida) are small species of 10–15 μm which puncture hyphae and yeast cell walls with an extensible cytostome to ingest the cytoplasm. The single genus of the Sorogenida is unusual, in forming cell aggregates which form a sorocarp reminiscent of the Dictyosteliida (Ameobozoa), and for feeding on other colpodids.

Chromista

The chromists include the brown algae, the haptophytes and the cryptophytes, as well as several heterotrophic groups in the **zoosporic fungi** (pseudo-fungi), in the Heterokonta and in the opalinids. The particularity of chromista is the location of the plastid inside the lumen of the endoplasmic reticulum. The plastid itself is surrounded by an additional periplastid membrane which in some species still retains three chromosomes in a second nucleus (the nucleomorph), as the remnant of the ancestral photosynthetic eukaryotic symbiont. The heterotrophic species have secondarily lost the plastid, and most species retain a remnant of this ancestral endosymbiosis. Most ciliated species exhibit one longer cilium bearing tubular mastigonemes and a shorter one without, but there are many variations.

Bicoecea, Labyrinthulea (phylum Sagenista)

The Labyrinthulea (Thraustochytrids and Labyrinthulids) are osmotrophic marine saprotrophs, associated with vegetation in early stages of decomposition. They are found exclusively in estuarine and coastal wetland environments and will not be discussed here. The Bicoecea are sampled primarily from marine environments.

Oomycetes and Hyphochytrea (phylum Bigyra: pseudo-fungi)

The Oomycetes are a commercially important class of fungi-like species, because many species are plant parasites that cause extensive damage to crops. There are species adapted to marine or brackish water, but most species are found in terrestrial soil and freshwater habitats (Fig. 1.19). Species grow as parasites on a variety of protist, animal or plant structures, or as saprotrophs on litter. Most species have a heterokont biciliated dispersal stage, although some terrestrial species (in the Saprolegniales and Lagenidiales) do not. The anterior cilium has bilateral tubular mastigonemes, but the posterior trailing cilium is smooth. The cilia kinetosome is anchored by a characteristic rootlet. There is an external cell wall of β-glucans (70%) and cellulose (10%), with additional minor components such as mannose. Chitin is not normally found, except in several Leptomitales and *Achlya*. There are mitochondria with tubular cristae. There is a closed mitosis with an internal mitotic spindle, but centrioles remain outside the nuclear membrane. The growth phase

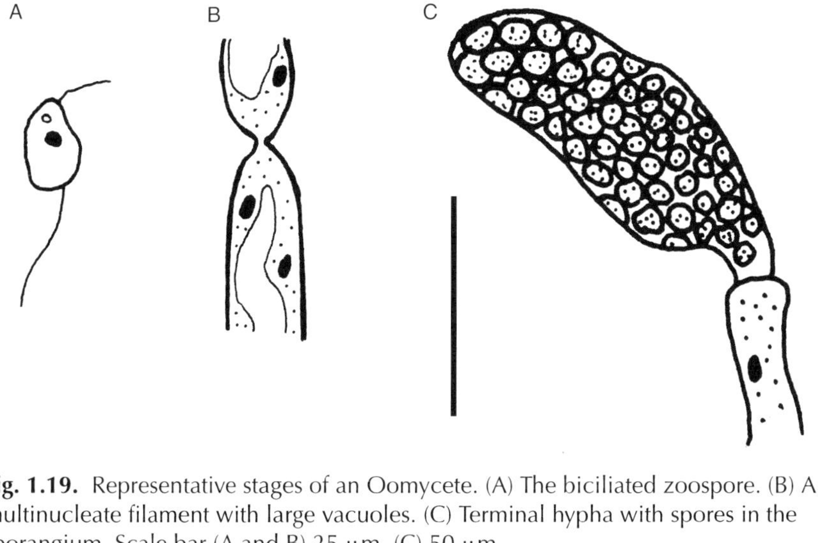

Fig. 1.19. Representative stages of an Oomycete. (A) The biciliated zoospore. (B) A multinucleate filament with large vacuoles. (C) Terminal hypha with spores in the sporangium. Scale bar (A and B) 25 μm, (C) 50 μm.

is diploid, with filamentous branching hyphae into the substrate. Feeding is by secretion of extracellular digestive enzymes in response to substrate, and osmotrophy of dissolved nutrients. Three types of vesicles with glycoproteins are synthesized by the Golgi bodies. One type contains storage material, while the other two are regulated secretory vesicles (Dearnaley and Hardham, 1994). The primary storage material are mycolaminaran and mycolaminaran phosphates (β-glucans). Tip growth requires actin filaments, elongated branched vacuoles, inward Ca^{2+} fluxes and directional polar transport of cell membrane and wall components. Turgor pressure has been discredited as a mechanism of substrate or host penetration (Money, 1997). The growing hyphae are vulnerable to damage but have a wound response, initiated within seconds. Callose deposition with calcium was demonstrated to form a plug in *Saprolegnia ferax* (Levina *et al.*, 2000). In response to poor resources or unsuitable habitat, hyphal tips form diploid sporangia that release motile biciliated dispersal cells. Release of the motile cells requires water with low solute concentration and it is inhibited by high external osmotic pressure. The ciliated cells encyst and are stimulated to excyst by nutrients or an appropriate stimulus. The contractile vacuole appears in the ciliated cells and disappears during encystment as the cell dehydrates into a cyst (Mitchell and Hardham, 1999). There are sexual stages with separate male and female gametangia. The haploid female in *Achlya* secretes the steroid pheromone antheridiol to initiate male antheridia (Carlile, 1996). Haploid nuclei migrate from the male polynucleate antheridia to the female gametes through a fertilization tube. The resulting zygote is a resting cyst which can be dispersed and stimulated to excyst. The role of these species is better known from aquatic habitats, where they are found on decomposing leaves, woody debris, cadavers or other organic matter such as excreta. The order Saprolegniales consists mostly of saprotrophic species with large diameter hyphae (>10 μm). The morphology of the sporangium and zoospore release are used in determining genera, and the sexual stages are used for species identification. The order Peronosporales also includes terrestrial saprotrophic species, as well as the parasitic Peronosporaceae (downy mildews) and Albuginaceae (white rusts). Other orders are primarily parasitic.

The Hyphochytrea, or hyphochytrids, are a small group with about 25 species, some of which are found in soil (Sparrow, 1973; Karling, 1977).

Fungi

Soil fungi carry out three main functions in the ecosystem: (i) as saprotrophs that digest and dissolve litter and detritus; (ii) as predators and parasites of soil organisms; and (iii) as symbionts of plants (as mycorrhizae or lichens) and insects. The organelles of fungal cells vary with phyla and classes as outlined below. In general, most groups have a cell wall that consists of chitin and other polysaccharides. The chitin polymer is synthesized from uridine-diphospho-*N*-acetylglucosamine precursors, which are added to the existing chain of chitin monomers. In the Archemycota taxa, the cell wall contains chitin, chitosan and polyglucuronic acid. In the Ascomycetes and Basidiomycetes, the cell wall contains chitin, glucans, gluco- or galacto-mannans and mannoproteins. These are all cross-linked together by hydrogen bonds into a protective matrix that forms a barrier about 0.2 μm thick. The cell wall is composed of 80–90% carbohydrates. However, there are numerous important proteins that need to be exposed to the environment. These include recognition proteins for mating types, strains and species, as well as structural proteins and enzymes for substrate digestion. The cell membrane must include substrate receptors and carrier proteins. The role of the cell wall in supporting the cell is important during periods of water stress, as in plant cells, but it also serves as a defensive barrier to predation. The main reserve materials are glycogen and lipids. Mitochondria with plate-like cristae and peroxisomes are present in the Ascomycetes and Basidiomycetes, but they are absent in some Archemycota, such as the gut endosymbiont Enteromycetes species which contain hydrogenosomes instead. Mitosis is closed, with a persistent nucleolus which divides. Chromosome do not align along the spindle in a typical metaphase, as there are often too many chromosomes. The endomembrane network is prominent in saprotrophic species, especially in active hyphae where secretory enzymes are synthesized and accumulated. The Golgi–dictyosome are small with 2–5 cisternae. Some vacuoles also participate in osmoregulation and storage of amino acids, ions or other soluble nutrients. Many vesicles containing cell wall material are directed to the growing tip. In saprotrophic species, secretory vesicles containing digestive enzymes fuse with the cell membrane to release enzymes. Enzymes of up to 20 kDa may pass through the cell wall into the substrate. Common fungal exoenzymes are diverse and include proteases, amylases, cellulase (cellobiohydrolase), xylanases, pectin-degrading enzymes (lyase, esterase, pectate lyase and polygalacturonase), ligninases (lignin peroxidases and manganese peroxidases) and other secreted enzymes. In all species, feeding is by osmotrophy from dissolved nutrients outside the cell wall. Haploid spores form as a result of sexual reproduction between two complementary karyotypes. The spore wall contains additional layers and it is thicker than in the vegetative cells. It holds an inactive cytoplasm with storage vesicles. The conidiospores are also resistant dispersal cells from successive mitotic divisions.

Three modes of growth are recognized in fungi (Fig. 1.20). One mode is called **yeast growth** and consists of cell division and separation

Fig. 1.20. Yeast form of growth in a fungus. Scale bar 10 μm.

of cells into independent, but sometimes loosely attached cells. This mode of growth occurs in many Basidiomycetes and Ascomycetes, but also occurs in some Archemycota. A second form of growth occurs in Chytridiomycetes, Enteromycetes and Allomycetes, where one individual cell extends cytoplasmic branches into the substrate and forms a mononuclear **thallus**. The third mode of growth is called **hyphal growth**. This mode of growth produces a long extending filament without cytokinesis (cell separation) between successive nuclear divisions. Cell walls form completely, partially, rarely or not at all, depending on the taxonomic group. The repeated branching of growing hyphae forms a three-dimensional mass called the **mycelium**. In many orders, especially in the Ascomycetes and Basidiomycetes, hyphae of the same species, strain or individual are able to fuse. This is called **anastomosis** and requires a growing tip fusing with a complementary hypha. Some species of fungi may grow as hyphae under some conditions and as yeast under different conditions. The mechanism of growth is by tip elongation. In yeasts, a short hypha-like extension (the bud) forms which separates at cytokinesis into an independent cell with one nucleus and other organelles. Behind the growing tip of hyphae, the cell wall consolidates into a rigid structure. However, at branch initiation behind the growing tip, the cell wall must be loosened to allow for an extension of the cell membrane and a new elongating tip to form.

Yeast forms are primarily osmotrophic, with little or no secretion of digestive enzymes. Very few species can hydrolyse hemicelluloses, cellulose and pectins (unlike most filamentous saprotrophic fungal species). They are therefore very limited saprotrophs and depend on released nutrients in the soil or litter solution. Species have nutrient preferences, and there are soil/leaf specificity and niches. They are generally found in assemblage of species, not large colonies of one species.

In filamentous fungi, the hyphal cytoplasm is often separated by a cross-wall called a **septum** (Fig. 1.21). The details of septa can be a class characteristic. For example, in Ascomycetes, there are frequent septa that are perforated by one or more pores which allow passage of cytoplasm and organelles including nuclei. The septa separate the cytoplasm along the hyphae into compartments with one or more nuclei, but permit translocation of nutrients to other parts of the mycelium. In some species, a proteinaceous Woronin body will plug the septa if the adjacent cell is damaged. Basidiomycetes have septa that separate the hyphae after each nuclear division into compartments (cells) with one nucleus (homokaryon) or two nuclei from different mating types (heterokaryon). These septa are perforated by a single large pore. In some classes of Basidiomycetes, the pore is associated with modified endoplasmic reticulum closely appressed on both sides, and called **dolipore septa**. The dolipore ultrastructure is complex and regulates passage of material between the cells; notably, nuclei do not pass through. The role of septa is crucial in containing damage after a hypha is broken or invaded by a predator or parasite. Septal pores are blocked by cell wall deposition, and a new branch initiates a growing tip. The Archemycota generally have fewer septa. When they are present, the septa tend to be very perforated.

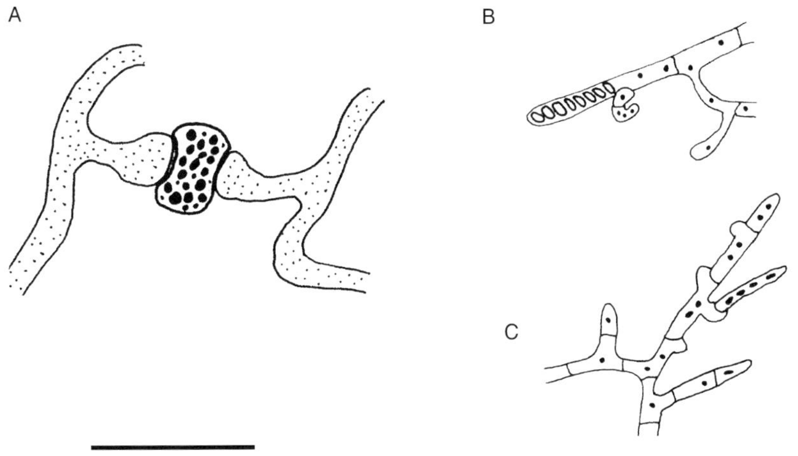

Fig. 1.21. Filamentous growth in fungi. (A) The H-junction between hyphae of complementary mating types in Zygomycetes (Archemycota), with a diploid spore at the junction. (B) Ascomycetes septate hyphae with terminal ascospores and a hook junction forming. (C) Basidiomycetes septate hyphae, with monokaryotic and dikaryotic cells. Scale bar 50 μm.

Species identification in fungi involves description of the morphology of spores and reproductive structures. Species producing only asexual reproductive structures traditionally have been placed in the deuteromycetes, an informal group of mitosporic fungi, until placed in their corresponding class. The taxonomy of fungi is complicated further, because many species with both yeast and hyphal growth forms were described as separate species and therefore have two species designations. From field samples, the morphology of asexual hyphae (or sterile hyphae) does not provide sufficient characters for identification. Most cannot be assigned to a taxonomic group without sequencing of DNA regions for molecular phylogenic analysis. When reproductive or dispersal structures form in culture, identification becomes possible, otherwise one is forced to rely on analysis of DNA sequences.

For further details on the biology and morphology of fungi, the student is directed to the literature cited in the text below, and to general texts such as Carlile *et al.* (2001), Kurtman and Fell (1998) and Moore (1998), or to web sites such as http://mycology.cornell.edu/

Chytrids (phylum Archemycota: Chytridiomycetes, Enteromycetes, Allomycetes)

These species are composed of motile cells with one posterior cilium without mastigonemes (Fig. 1.22). The individuals are often noticed by

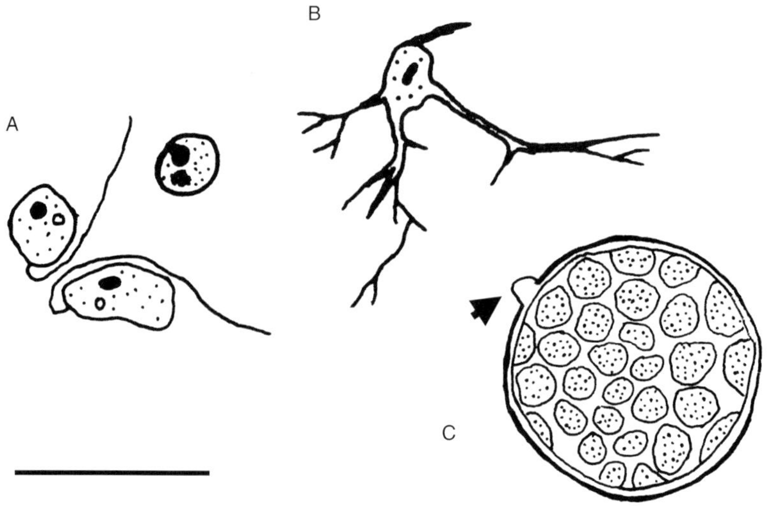

Fig. 1.22. Typical morphological stages of a chytrid. (A) Cyst and monociliated zoospore dispersal phase. (B) Extending branched thallus from a cell growing on a substrate. (C) Enlarged chytrid with numerous spores, each holding a zoospore, about to discharge. Scale bar 25 μm.

darting cells which change direction quickly. The monociliated cell can be partially amoeboid when on a surface or thin water film. Chytrids are ubiquitous as a group, found from the arctic to tropical climates. They have been most studied in aquatic habitats (Sparrow, 1960; Powell, 1993). They are very common in soil samples but are usually overlooked. Soil species are also found active in bogs, riparian and stream habitats. Many species occur on microinvertebrates and on aerial parts of plants. Some species are parasitic on a variety of animals, fungi and plants, including commercially important crops and domestic animals. Many soil species are predatory on microinvertebrates (nematodes, rotifers, tardigrades and eggs), on larger invertebrates and on protists, including fungal spores and other chytrids. Most soil species are saprotrophs on pollen, chitin, keratin, cellulose and substrates that are difficult to digest. Their role in facilitating the digestion of these compounds is largely ignored. It is believed that they facilitate the entry of other organisms, including bacteria and fungal hyphae, into the litter enclosed by these protective polymers by loosening and softening the protective walls. For a general introduction to the literature, students are referred to Fuller and Jaworski (1987), Powell (1993), www.botany.uga.edu/zoosporicfungi/ and www.botany.uga.edu/chytrid/

Structurally, the cell wall of chytrids consists of chitin and β-glucan polymers, similar to other fungi. The mitochondria have flattened cristae, and the storage products are glycogen and lipid vesicles. Several organelles are particular to this group. There is a nuclear cap of dense ribosomes with endoplasmic reticulum closely associated. There is a microbody–lipid globule complex at the cell posterior, which could be involved in lipid metabolism. Adjacent to this microbody, the rumposome consists of flattened membrane vesicles which have been implicated in sensory orientation and taxis in certain species. The rumposome often has bridge connections to the cell membrane and to the kinetosome microtubular rootlet. It stores Ca^{2+} ions which may mediate the direction of taxis. A cytoplasmic centriole exists in addition to the basal body of the kinetosome. Sex probably occurs in most species and involves male and female gametes produced in separate antheridia and oogonia on terminal hyphae. Motile gametes are released and their conjugation involves pheromones.

Chytrids find their prey or substrate by chemotaxis towards specific molecules (see list in Sparrow, 1960). The initial detection is from leakage of soluble nutrients from dead cysts, spores, litter cells and decomposing tissues. Once detected, changes in the frequency of turning and distance of forward movement (kinesis) direct the cell towards the source. The activity of motile cells varies with moisture, temperature, soil solution composition and amount of stored food reserves. Upon making contact with substrate, its recognition and attachment involves surface molecules and strong binding to prevent accidental detachment. The motile cell (zoospore) encysts and loses the cilium, and grows cytoplasmic extensions into the substrate or prey. The cytoplasmic extensions (**thallus**) resemble hyphae and can be extensively branched. The growing thallus is not the result of cell divisions, does not have a cell wall as fungal hyphae would, and is capable of phagocytosis at least in some species (Powell, 1984), as well as extensive secretion of extracellular digestive enzymes and osmotrophy. When the food resources are exhausted, dispersal spores (or resistant cysts) are formed by repeated mitotic divisions. Release of the spores occurs at an opening away from the substrate and thallus. It may require rehydration after desiccation and sometimes partial decomposition of the chitinous wall by bacteria. The emerging motile cells (zoospores) can encyst several times until a suitable substrate for attachment and thallus growth is found. The cysts of chytrids are very resistant to long periods of desiccation, chemical attack and anaerobiosis. This is particularly bothersome with parasitic species which can reappear after decades, from infected soil.

Many of the taxonomic characteristics for preliminary investigation are obtained by light microscopy with Nomarski optics. Based on the ultrastructure of the motile cells (zoospore), morphology, development and ecology, the four orders are separated into two lineages (Barr, 1983) (Table 1.4). The Chytridiales and Monoblepharidales contain mostly (but not exclusively) aquatic genera, whereas the Spizellomycetales and Blastocladiales are primarily soil species. For example, *Chytriomyces* (Chytridiales) can be isolated in soil and freshwater samples where they digest chitin. The Monopblepharidales, although aquatic, are found in decomposing insect parts, plant litter and woody debris, similar to the habitat of their soil counterparts. The Spizellomycetales are particularly adapted to soil (e.g. *Entophlyctis*, *Karlingia*, *Rhizophlictis*, *Rozella* and *Spizellomyces*). The motile stage lacks the ribosome cluster, and the rumposome is absent, with microtubules from the kinetosome oriented more randomly and associated with the nucleus.

Table 1.4. The organization of the Archemycota phylum in the fungi.

Phylum	Class	Order
Archemycota	Chytridiomycetes	Chytridiales, Monoblepharidales, Spizellomycetales, Neocallimastigales
	Enteromycetes	Eccrinales, Amoebidiales
	Allomycetes	Blastocladiales, Coelomomycetales
	Bolomycetes	Basidiobolales
	Glomomycetes	Glomales, Endogonales
	Zygomycetes	Mucorales, Mortierellales, Dimargaritales, Kickxellales, Piptocephalacea, Cuninghamellales
	Zoomycetes	Entomophthorales, Zoopagales, Harpellales, Asellariales, Laboulbeniales, Pyxidiophorales

The Blastocladiales (Allomycetes) are active in fresh water and moist soils (e.g. *Blastocladia* and *Blastocladiella*). The metabolism of many saprotrophic Blastocladiales species is fermentative and carboxyphilic, producing lactic acid as an end-product. Characteristically, motile cells aggregate the organelles in a complex in one part of the cytoplasm. Germination from the inactive spores is bipolar, with each emerging hypha repeatedly branching into finer hyphae to produce the mycelial thallus.

Although the literature refers to infections and parasitism when discussing chytrids, in many cases they are predacious and osmotrophic. The distinction between parasites and predatory species can be made by looking at the invading thallus. In cases of parasitism, the chytrid and host membranes remain intact and separate the two cytoplasms. In all other cases, the invading chytrid aims to destroy and digest the prey. Once a host cell is penetrated, lysis occurs within seconds. For example, *Sorochytrium milnesiophthora* is saprotrophic on dead rotifers, nematodes and tardigrades, but it is also predatory and capable of penetrating the cuticle of living individuals. The prey eventually dies and is digested, until release of motile zoospores propagate the individual to other prey or substrate in the microhabitat. Saprotrophic species are attracted to adequate substrates in the litter. The thallus grows into micro- and macrodetritus. The Chytridiomycetes provide good examples of a taxon with intermediate species between free-living species, opportunistic and obligate parasites.

The Enteromycetes are endosymbiotic chytrids in the intestinal tract of animals, found in ruminants. They are anaerobic species with a broad spectrum of digestive enzymes to digest plant cell walls and lignocellulosic debris. The chytrid thallus grows into the chewed macrodetritus from ingested plant debris.

Zygomycetes (phylum Archemycota: Zygomycetes and Zoomycetes)

The Zygomycetes and Zoomycetes consist of hyphal species that lack a ciliated stage, as well as centrioles and sporocarp. Hyphae contain mitochondria and peroxisomes, but the Golgi cisternae are unstacked. The Zygomycetes include mostly saprotrophs which produce zygospores and aerial asexual spores on stalked sporophores (Table 1.4). The better known species belong to the Mucorales (Zygomycetes) which are ubiquitous in soils and on herbivore excreta, fruit and decomposing mushrooms (aerial reproductive structures of higher fungi). The mucors are early colonizers of substrates that depend on the more soluble substrates such as sugars and amino acids. Species tend to lack digestive enzymes for substrates such as cellulose and chitin that are more difficult to break down. Some species are parasites of other Mucorales and unculturable without the host. Spores are stimulated to germinate by moisture, adequate solution and abiotic conditions. The spore absorbs water, and several hyphae may emerge over several hours. Hyphae of mucors are large and grow rapidly into the substrate, ahead of other fungi. Under anaerobic conditions, some species grow as yeast, but others remain hyphal, while others are strict aerobes. Dispersal occurs from aerial or vertical hyphae that differentiate at the end (sporangiophore) into about 10^5 sporangiospores, each with several nuclei. The wall of sporangiospores consists of sporopollenin, an oxidized and polymerized form of β-carotene. Some species also produce chlamydospores, which are thick-walled cysts inside terminal hyphae. Conjugation between complementary mating types occurs when sporangiophores fuse, after growing towards each other by autotropism or chemotaxis. In some cases, elaborate resistant zygospores may form from sporangiophores.

The Zoomycetes consist of species that are predatory or parasitic on protozoa and animals (mostly microinvertebrates). The Zoopagales and Entomophthorales are encountered in soils, where many are predacious on amoebae, ciliates, nematodes and other small invertebrates. These species do not produce sporangiospores for dispersal but release spores from propulsive conidia. The hyphae are syncytial, with septa at intervals. The other orders consist of species that are symbiotic with, or parasites of mandibulate arthropods, either in the gut or on the cuticle.

Glomales (phylum Archemycota: Glomomycetes and Bolomycetes)

There are about 150 species of Glomomycetes and Bolomycetes which consists of species that do not produce sporangiospores and zoospores (Table 1.4, Fig. 1.23). The cytoplasm of hyphae contains mitochondria, peroxisomes and unstacked Golgi cisternae. The Glomomycetes (two orders: Endogonales and Glomales) do not have centrioles, and hyphae are not septate and do not produce conidia, aerial spores or stalked sporophores. A sclerotium-like sporocarp contains chlamydospores in the Glomales, or frequent sporocarps form in the Endogonales with zygospores inside. The Bolomycetes have centrioles, septate hyphae and a single large propulsive conidium on unbranched conidiophores. The zygospores are not produced inside a sporocarp. Bolomycetes consist of one family of saprotrophic species.

With the exception of some saprotrophic species in the order Endogonales, the Glomomycetes form close symbiotic associations with plant roots, by extending hyphae between root cells and growing specialized hyphae, the haustoria, into root cells, where they branch profusely. It is this branched arrangement of hyphae that is referred to as the arbuscule inside root cells. Species are obligate symbionts that, strictly defined, depend on their host for organic nutrients. However, this statement has not been completely verified physiologically for all

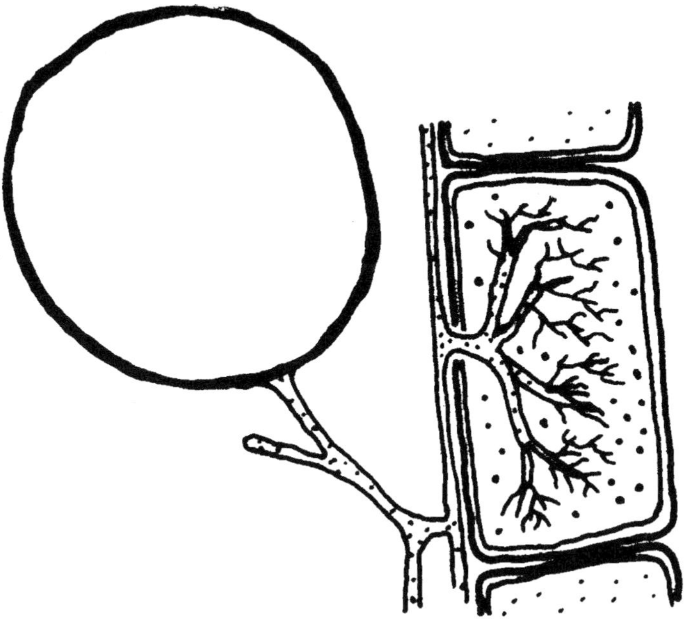

Fig. 1.23. Glomales (Glomomycetes, Archemycota) growing into a plant root cell (arbuscular mycorrhizal association), and a large chlamydospore in the soil. Not drawn to scale.

species. Most hyphae are intercellular between root cells of the primary cortex and epithelium, that do not penetrate the endodermis, vascular tissues or aerial plant organs. Some hyphae (5–10 μm diameter) do extend into the soil from the root. Anastomosis of hyphae in the root and substrate occurs. Species are probably asexual and produce large dispersal spores <800 μm in diameter. The order Endogonales (*Endogone* and *Sclerogone*) consists of saprotrophic species or those that form endomycorrhizae. The order Glomales consists of two suborders, Glomineae which form arbuscules and vesicles (*Glomus*, *Sclerocystis*, *Acaulospora* and *Entrophospora*), and Gigasporineae (*Gigaspora* and *Scutellospora*) which form only arbuscules. The latter seem not to have β(1,3)-glucan in the chitin cell wall, unlike other Glomomycetes. Vesicles and spores both store lipids. According to Smith and Read (1997), there is no clear evidence of specificity between the fungus and host root species, so that an association is possible between the fungus and any species capable of forming arbuscular mycorrhizae. However, this is not to say that the extent of root colonization and the effectiveness of the association are equivalent between species and strain combinations. Only a small number of families and genera of terrestrial plants do not form associations with Glomales.

Ascomycetes (phylum Ascomycota)

These species by definition form meiotic products that become endospores (called ascospores) separated by a cell membrane, or have secondarily lost a sexual phase. The phylum includes yeast, lichen and filamentous species. There are an estimated 14,000 lichens and >18,000 other species that clearly belong to Ascomycetes. The phylum is organized into seven classes (Discomycetes, Endomycetes, Geomycetes, Loculomycetes, Plectomycetes, Pyrenomycetes and Taphrynomycetes) and 46 orders. The sexual species produce ascospores by meiosis held inside a structure called the **ascus**. The ascus is a terminal hyphal tip that holds the meiotic products (**ascospores**) inside the hyphal tip cell wall. In yeast forms, the parent cell wall holding the meiotic products is the ascus. Similarly, terminal hyphae can produce dispersal spores by mitosis and are called conidiospores, held within the conidium. Modern taxonomy requires several growth characteristics to be considered as well as the morphology of the reproductive and dispersal structures, supported by molecular phylogeny. Many yeast forms produce ascospores in the mycelium phase. However, some yeasts have completely lost the ability for hyphal growth (several hundred species) and some are not sexual species. The Pezizales includes the morels, and Tuberales the truffles, which occur in forest soil organic horizon, woody debris and animal dung. The Sordariales includes species that are cellulolytic, such as *Chaetomium* in plant debris, and others which are good colonizers of macrodetritus or animal dung, such as *Neurospora*, *Podospora* and *Sordaria*. These species produce large numbers of wind-dispersed ascospores that become early litter colonizers. The Loculoascomycetidae (Loculomycetes) include species that colonize senescent leaves and fresh leaf litter, such as *Alternaria alternata* (or *Pleospora infectoria*) and *Cladosporium herbarum* (or *Mycosphaerella tassiana*), which are common allergens. The Eurotiales, such as *Aspergillus* and *Penicillium*, grow well in low moisture conditions and colonize fruit and rich organic matter. The Erysiphales include the parasitic powdery mildews.

Basidiomycetes (phylum Basidiomycota)

These species by definition form exospore meiotic products (basidiospores) by budding of the cell membrane and new cell wall deposition. There are about 22,000 known species distributed in four classes (Septomycetes, Ustomycetes, Gelimycetes and Homobasidiomycetes) and 41 orders. Most species are filamentous, forming hyphae with septa, and extensive mycelium in soil, litter or dead tissues. Notable exceptions are the Ustilaginales (Ustomycetes) and the Uredinales (Septomycetes) commonly known as the smuts and rusts which parasitize aerial tissues of plants. Growing hyphae of Basidiomycetes are often dikaryotic, with nuclei from two complementary mating types existing in the same hypha. After nuclear divisions and cell wall formation (septa), a clamp connection may form to redistribute the nuclei (Fig. 1.21). The clamp connection allows the passage of one nucleus to the new cell over the septa to maintain the dikaryotic state. It is noteworthy that there are species where clamp connections are unknown or where they occur only under certain growth conditions. Hyphae of filamentous Basidiomycetes (as those of Ascomycetes) generally vary in size from 3 to 10 μm. Typical Basidiomycetes include those commonly known as the agarics such as the common cultivated *Agaricus* (and related fungi such as *Boletus*, *Coprinus* and *Laccaria*), bird's nest fungi, bracket fungi, coral fungi, the jelly fungi, puff-balls and stink-horns. The Poriales (*Lentinus*, *Coriolus*, *Fomes* and *Pleurotus*) are wood decomposers and include predatory species that capture small invertebrates and protozoa, such as the oyster mushroom (*Pleurotus ostreatus*). Many Boletales (*Rhizopogon* and *Suillus*) and Agaricales (*Cortinarius* and *Russula*) are mycorrhizal symbionts, and many other Basidiomycetes decompose wood. In general, Basidiomycetes (and to a large extent Ascomycetes) constitute the bulk of the hyphal biomass in soil horizons and are responsible for most of the primary decomposition of organic matter, when they are present and active.

From an ecological perspective, the role of both Ascomycetes and Basidiomycetes in decomposition depends on their contribution to substrate digestion. The panoply of enzymes supplied by each species for external digestion of litter and soil organic matter would define their functional role. In addition, the periods of hyphal activity through the year and substrate specificity permit differentiation between species niches in the habitat through the seasons. In mycorrhizal species, niche distinctions also require consideration of plant host specificity and water and nutrient translocation patterns between the plant–fungus association, especially during periods when one or more nutrients or water are limiting.

Invertebrates

Several taxa of invertebrate organisms need to be considered. The role of these species in the soil interstitial space is inseparable from decomposition processes. They interact with protists and bacteria as grazers, prey or predators. They are presented in three sections as: (i) the nematodes, tardigrades, rotifers and gastrotrichs; (ii) earthworms and enchytraeids; and (iii) microarthropods, particularly the mites and Collembola. More detail on the structure and function of these taxa can be found in standard invertebrate zoology texts (see Grassé, 1949–1995; Dindal, 1990; Ruppert and Barnes, 1994).

The nematodes, tardigrades, rotifers and gastrotrichs share several characters that are ecologically and functionally similar. Each of the terrestrial species survives extremes of freezing and desiccation, by **anhydrobiosis**. Individuals can retain as little as 15% of their normal water content. After a dry period, individuals require minutes to hours to recover their functional hydrated state. The body is avascular, consists of very few cell layers, and is covered by a protective cuticle. There are considerable changes in body volume with osmoregulation and soil moisture content. The juveniles of these taxa emerge from the egg as a miniature version of the adult, and simply grow larger through several moults. The juveniles, therefore, have access to much narrower soil pore spaces than the adults. The number of cells in adults is constant for species of nematodes and rotifers. This developmental character is termed **eutely**, and also occurs in the tunicates and Myxozoa (both Animalia). In all the following taxa, the epidermis secretes a protective cuticle, there is a nervous tissue and longitudinal musculature. The internal organization consists of two tubes, the intestinal canal and the reproductive canal, inside a cavity that is filled with a serous solution.

Nematodes (phylum Nemathelminthes: Nematoda)

There are >20,000 morphotypes of free-living interstitial nematodes that are found in terrestrial habitats and along a continuous gradient, into the deep-sea sediments. Many more species remain to be described, particularly from marine sediments. To these we must add about 2000 plant parasites, that can devastate agricultural productions. Nematodes are ubiquitous in soils and are an integral component of decomposition ecology. For general reference on nematode structure and function, students are referred to Grassé's encyclopaedia (1965) and Perry and Wright (1998), and several general nematology texts are available as an adequate introduction.

The general body plan of nematodes is simple, consisting of a small number of circular cell layers (Figs 1.24 and 1.25). The cuticle is <0.5 μm in most soil species and covers the organisms like a protective skin. It also lines the anterior and posterior portion of the digestive tract (stoma and pharynx, rectum and anal pore). The cuticle is secreted by the epidermis cells (or hypodermis). It is elastic but strong, so as to allow the body to bend, but provides an exoskeleton against which the musculature can work. The cuticle permits gas exchange and osmoregulation (water balance and excretion of soluble by-products such as urea and ammonia), and provides a physical protection against the habitat. It is subdivided into four sublayers. The epicuticle is the outermost sublayer and consists of a glycocalyx 6–45 nm thick. It could be required for species and mate recognition but does not have a known function. The exocuticle (<200

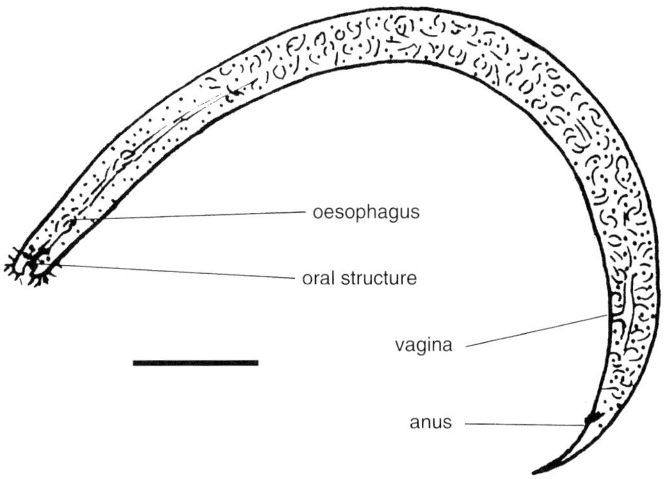

Fig. 1.24. A nematode with anterior mouth, posterior tail with anal pore and reproductive pore. Scale bar 50 μm.

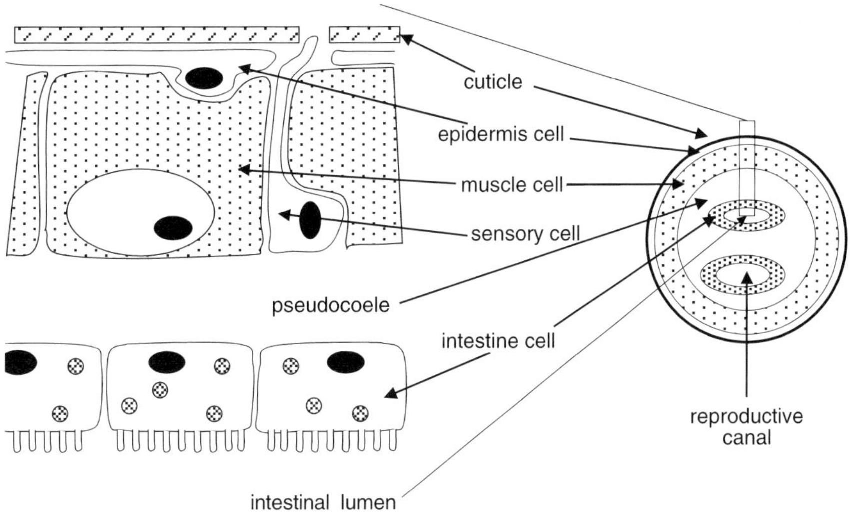

Fig. 1.25. Cross-section of a nematode showing the arrangement of the three cell layers and the cuticle.

nm) resembles keratin proteins in structure and is particularly thick in parasitic forms. The mesocuticle resembles collagen fibrous proteins in structure. It is the thickest layer of the cuticle and the most varied in structure between species. It is a compressable layer of fibrous proteins that becomes thickened in the anterior cephalic region. It is presumed to function as a mechanical buffer, or bumper, as the organism moves through soil or forces its way into tissues. The endocuticle is a thin sublayer of thin irregular protein fibres that resembles a basal membrane. It chemically links the epithelium to the cuticle. Nematodes grow from the juvenile form to the adult form by four consecutive moults. Each time the new cuticle is secreted by the epithelium, the old (outer) cuticle is shed.

The longitudinal muscle cells present in a single layer are striated, branched and have extensions to the nerve cells. Their contractions bend the body in sinusoidal waves that are characteristic of nematode locomotion. There is no circular muscle layer. The nerve cell extensions form a net throughout the body, with cell bodies concentrated into bundles (nerve cords). The larger nerve cord is ventral, with a smaller dorsal cord and several smaller ones across the periphery. The nerve cells are poorly integrated, so nematodes lack a true central nervous system. The sensory cells that extend between epithelium cells, through extensions of the cuticle and contact the outside environment are chemosensory and participate in chemotaxis. These sensory extensions are called **papillae**. Some do not break through the cuticle and probably detect touch and temperature. Several papillae occur about the anterior cephalic region and guide in food detection. Specialized sensory cells and secretory cells occur inside extensions of the cuticle around the reproductive organs, and sometime elsewhere on the body. They help in sensing and holding the partner during sex. Caudal secretory cells (the caudal gland cells), when present near the posterior ventral surface, allow the organism to attach its posterior end on to the substratum. This seems to permit more control over movement and direction. The caudal gland cells are secondarily lost in some species.

The variety of sensory cells along the nematode body permits chemotaxis, mate recognition, pheromone response and, at least in some species, CO_2 level sensing. This is not an exclusive list, but demonstrates the extent of directional locomotion possible. One important sensory response is to temperature (Dusenbery, 1989). Nematodes can respond to temperature gradients as low as 0.001°C cm (Pline *et al.*, 1988).

The intestinal tract is a central tube, extending from the pharynx to the rectum, composed of a single layer of cells, bearing microvilli on the absorptive surface. Digestion occurs mostly in the middle intestine, although secretory cells also occur around the upper intestine. Some cells participate in secretion of digestive enzymes and contain lysosomes, but all are absorptive. As older intestinal cells are lost, they are replaced by cell division with new cells. It seems that newly divided cells are only absorptive, and develop the ability to secrete digestive enzymes with age. Endocytosis has also been reported. These cells are capable of storage, and can accumulate protein granules, glycogen and lipids. Undigested material and whatever has not been absorbed by the intestinal cells reaches the rectum and is excreted through the anal pore back into the habitat.

One or more cells (renal cells) may participate in the accumulation and excretion of soluble metabolic wastes from the internal body cavity, through a narrow pore. The internal cavity of the nematode, the pseudocoel, is filled with a serous solution that contains absorbed nutrients from digestion and metabolic by-products from cells, as well as some free haemoglobin. The fluid participates in distributing nutrients and gas exchange, as contractions of the organism contribute to moving and mixing this solution. The older literature describes free-moving phagocytic cells sometime reported in the pseudoceolom. These were speculated to engulf invasive bacteria and to have a defensive role in protecting the organism. The modern literature admits there are 1–6 of these cells depending on the species, called **pseudocoelocytes**. However, these cells are claimed by some authors to be immotile and non-phagocytic, though branched with granular inclusions (Meglitsch and Schram, 1991). One can doubt the verity of these claims, based on comparisons with other related invertebrates, but their role remains uncertain in nematodes.

The pseudocoel and renal cells participate in osmoregulation and permit survival during soil desiccation. Soil water solutions of 15 mM NaCl equivalent (hypotonic) can increase to >300 mM NaCl equivalent (hypertonic). As water in the habitat becomes limiting, water is lost from the body and volume decreases. Water also becomes limiting inside the body, so that excretion of ammonia and diffusion of metabolites and gases become difficult. Urea and purine then become alternative modes of nitrogen excretion. The increase in internal osmotic pressure initiates dehydration and dormancy, called **anhydrobiosis** (see Perry and Wright, 1998).

The reproductive cells of adults can account for half of the body weight, and spermatogenesis in the males, or oogenesis in the females, diverts a large portion of the digested and absorbed nutrients. Male spermatozoa do not have a cilium but are amoeboid, and vary greatly in shape. Males deposit spermatozoa inside the female, and fertilization is internal. After fertilization, the egg cell wall acquires a protective chitinous mid-layer and the eggs are deposited outside. Development from a fertilized egg to completion of oogenesis in the new adult requires several days and varies with temperature. As an example, in well-fed conditions, a species may require 14 days at 14°C, 8 days at 20°C and 4 days at 28°C to complete the cycle from egg to egg. These values vary between species, as each has different durations of development and different optimal growth temperatures. In a small number of species, male spermatozoa are required to initiate fertilization, but the spermatozoa are not functional and do not fertilize the egg cell. The egg proceeds with development by parthenogenesis. In a few species, females are strictly parthenogenic, in that males are not required at all for fertilization, or development of the egg to embryo and adult.

Morphologically identical isolates are not necessarily the same species. It has been known for a long time that identical organisms from geographically distant locations do not necessarily mate. These may not recognize each other as compatible, or may conjugate and produce sterile offspring (Maupas, 1900, 1919). In some case, isolates of identical morphotype are each restricted to a different environment (Osche, 1952). These observations may demonstrate divergence between isolated populations of one species, or different species within one morphotype. It is clear that there are limits to morphological descriptions of nematode species.

Functionally, and for ecological purposes, nematodes can be separated into groups based on the structure of the stoma and pharynx (see Yeates *et al.*, 1993). The anterior region sensory extensions, as well as the details of the pharynx and stoma, are important morphological characters in species identification. The cuticle lining of the pharynx may be simple as in most species, but it can be reinforced or 'armoured' with thickened extensions. This armature may appear as rows of teeth (denticles, dentate pharynx), as larger tooth-like extensions ('mandibles') or a single spear-shaped stylet. The stylet is hollow in some species, like a syringe needle. The denticles, mandibles and stylet armature rub against ingested food particles with the contractions and suction of the pharynx. Thus, armoured species can ingest more diverse and tougher food particles than species with a simple pharynx. More than one form of pharynx armature is found in some species. In many armoured species, the stylet or mandibles are also extensible out of the stoma, and attached to muscle cells. These species can use more force to penetrate cells and tissues of prey. Functionally, the free-living nematodes are divided into four basic groups based on what they can ingest.

1. Those with a simple and narrow stoma feed by suction alone and remove small particles from their habitat. These include bacteria but also microdetritus. Some taxa are the Oxystomatidae, Halaphanolaimidae, Draconematina and Desmoscolecidae.
2. Species with prism, conic or cupuliform stoma, with or without a denticle, create a suction accompanied by more powerful peristalsis of the oesophagus muscles. These species can feed on larger particles that include diatoms, cysts, spores, invertebrate eggs and non-filamentous protists in general. The more common taxa include the Rhabditidae, Axonolaimidae, Terpyloididae, Monhysteridae, Desmodorina, Comesomatidae, Chromadoridae, Cyatholaimidae and Paracanthonchinae.
3. Those species with a denticle (or stylet) can succeed in penetrating the cellulosic walls of fine roots, plant tissues and algal filaments, or the chitinous wall of fungal hyphae and small invertebrates. Once the physical barrier is penetrated, the cytoplasm of cells is sucked out. Some common taxa are found in the Paracanthonchinae, Camacolaimidae, Tylenchidae and Dorylaimidae.

4. Species with an 'armoured' stoma and more powerful denticles also depend on oesophageal peristalsis for suction. Although some may feed on bacteria and protists in film water, it would not be sufficient to maintain growth and reproduction. These species can penetrate the cuticle of other nematodes and small invertebrates and can be effective predators, or parasitic in roots and invertebrates. Some taxa include the Enoplidae, Oncholaimidae, Choanolaimidae and Eurystominidae.

However, it is worth noting that free-living species may switch their food preference as they grow from juvenile forms to adult and, though some species tend to be less specific about their food preferences (omnivorous), others may be very specific. This method of classifying nematodes is faster and requires less knowledge than species identification. It is not as accurate or informative, and for more serious studies one always needs to attempt genus or species identification. This should be supplemented with fixed specimens, photographs and DNA extract.

Rotifers (phylum Acanthognatha: Rotifera)

Commonly found in forest litter and surface soil of riparian areas, only a few families are important to terrestrial soil. They are better known in streams and some marine sediments. However, when abundant on land, their numbers exceed those normally encountered in water. Description of some terrestrial species and their ecology in litter can be found in Donner (1949, 1950), Schulte (1954), Pourriot (1965) and Nogrady (1993). About 1800 species are known, and adults range in length between 100 and 2000 μm (Fig. 1.26). For our purposes, their internal body plan is elaborated from the Nematoda in the following ways. Several tissues can be syncitial, as opposed to the nematodes where only the epidermis of certain species is. Rotifers lack a cuticle, but the epithelial cell membrane is supported by actin filaments, and reinforced externally with an extensive glycocalyx. The pseudocoel space contains a loose network of amoeboid cells. The renal osmoregulation system is present and drains the whole body. The orientation of muscle cells is more diverse and more elaborate than in nematodes. Both striated and smooth muscle cells are present. The nervous system and sensory organs are more elaborate, and a brain can be identified. Some species have pigmented eye cells in the head region. The intestinal cells can be ciliated and store oil vesicles. The pharynx, called the mastax, is a muscular suction chamber with jaws, the trophi, that are used in species identification. The movable jaws of different species are used to grind food, pierce prey cells or as forceps for gripping prey. Four families of the order Monogomontes are worth mentioning. These are the Dicranophoridae, Asplanchnidae, Notommatidae and Atrochidae. They are never permanently fixed to a substratum. They crawl with 'toes' and head, in a characteristic looping fashion, in search of prey. Prey items depend on the size of the organism, but will include protists, nematodes and small invertebrate larvae. The Atrochidae become predators as adults, but the juveniles have oral ciliature and obtain food particles by filter feeding. The order Bdelloides contains many families, with terrestrial species that are active in the surface soil water films, within the litter. They depend on water currents generated from their oral ciliature, or corona, to obtain prey items. These include bacteria, protists and microdetritus of sizes that can be filtered out and ingested by the pharynx. There seems to be an indiscriminate ingestion of particles in the correct size range. Each organism attaches itself to a substratum with the posterior caudal gland mucus secretion, and vacillates in the water film, drawing a water current towards its pharynx. The corona is also used in swimming when not attached to a substrate. The Bdelloides are mostly parthenogenic species. For rotifers in general, development from fertilized egg to reproductive adult lasts 1–5 days, so that brief population explosions are possible. Lastly, with natural fluctuations in the environment, rotifers display morphological variations, referred to as **polymorphism**. These involve changes in body dimensions, surface ornamentation and pigmentation, and are caused by changes in temperature or diet, or are predator inducible.

Gastrotrichs (phylum Acanthognatha: Monokonta: Gastrotricha)

Being in the same phylum as rotifers, the body plan is very similar (Fig. 1.26). Unlike the nematodes which have secondarily lost cilia, both rotifers and gastrotrichs make use of their ciliature in motility, food acquisition and fertilization. The cuticle is covered with plates or spines. The ventral surface epidermis cells are monociliated, as well as the intestinal cells. Locomotion is mostly by swimming as they are seen darting forward. They are also capable of using their 'toes' and head to loop forward like rotifers. The mastax and trophi are used to obtain detritus, protists and bacteria. The order Chetonoides occur in fresh water and terrestrial litter. There are only females, which reproduce by parthenogenesis. They can be found in anaerobic muds or in stagnant pools, where some species do well, even in the presence of H_2S released from anaerobic bacteria. Gastrotrichs tend to occur as isolated individuals, so they are never numerous in samples. The order Macrodasyoqdes includes species that can be found in shallow marine sands. Reference to the ecology of gastrotrichs in terrestrial litter has been made in Varga (1959). Eggs of terrestrial species survive desiccation and frost, and at least some adults form a cyst stage. About 450 species are known, which range in adult length from 75 to 500 μm.

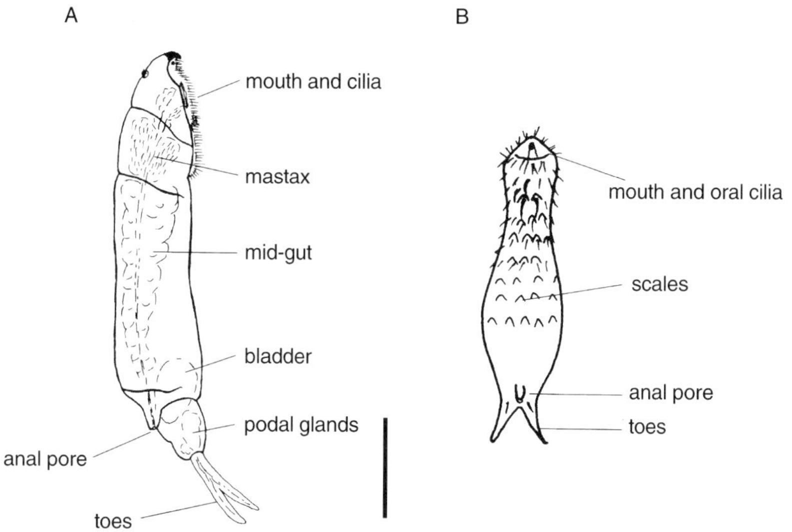

Fig. 1.26. Rotifers and gastrotrichs. (A) Rotifer (Monogomontes) with posterior toes, anterior mouth and muscular mastax. (B) Gastrotrich (Gastrotricha) showing posterior toes, anal pore, surface scales and cilia, and anterior mouth. Scale bar 100 μm.

Tardigrades (phylum Lobopoda: Onychophora and Tardigrada)

The **Tardigrada** inhabit terrestrial surface soil and tree bark, marine sands and sediments, and they have been reported in deep sea sediments at 5000 m (Cuénot, 1949b). They can be abundant in riparian areas, especially if the sediment is rich in primary producer protists, such as diatoms (Chromista) and chlorophyte algae (Plantae). There are about 600 species, ranging in adult size from 50 to 1000 μm (Fig. 1.27). The general body plan differs from that of the previous groups in the following ways. The cuticle of tardigrades is reinforced in some species with plates, which can be ornamented. There are four pairs of extensions used in locomotion, the podia. Claws at the end of the podia are useful in species identification. The musculature consists of smooth muscles oriented in longitudinal, dorsal, ventral and diagonal bundles. Attachments of the end of the bundles to the cuticle permit more complicated movement. There are sensory papillae covering the head, dorsal surface and podia. The nervous system of the segments is coordinated centrally and through a brain. Pigmented cells are located on the head and form two simple eyes. The pseudocoel contains a serous solution with free cells containing storage granules and vacuoles. It participates in gas exchange, distribution of absorbed nutrients, nitrogen waste elimination and osmoregulation. Terrestrial tardigrades are sensitive to oxygen deprivation and, despite unusual resistance to desiccation and cold temperature, will die without sufficient oxygen. The

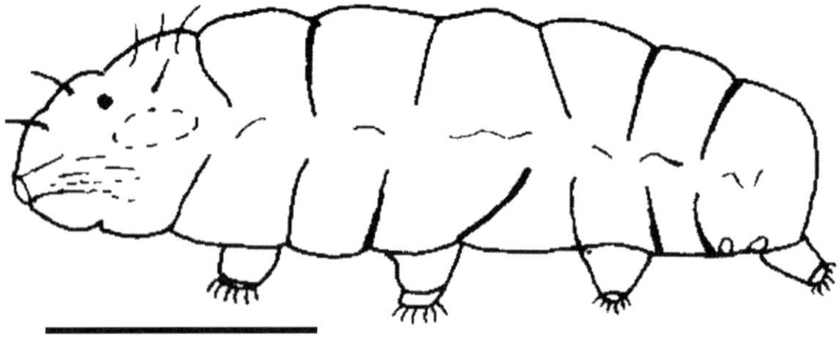

Fig. 1.27. Side view of a tardigrade showing one of each pair of legs, anterior mouth and head structures, and posterior pores. Scale bar 100 μm.

stoma has lips which facilitate suction created by the pharynx musculature. The cuticle of the pharynx is reinforced with six denticles that participate in 'chewing'. They are called macroplacoids and are made of chitin and calcium. Individuals search for organic matter and cells that can be sucked in. A stylet on each side of the oral cavity participates in ingestion. It is extensible in some species and used to pierce living cells of plants, mosses, lichens, nematodes, rotifers, other tardigrades and similar sized organisms in the habitat. Some such as *Milsenium* species are primarily predators. Copulation is unusual in that one or more males deposit spermatozoa through openings of the old moulting cuticle, or through the vaginal pore. After a long delay, 1–30 eggs are deposited in packets, often inside shed cuticles or debris. Eggs of some species bear protective scales. Juveniles emerge as miniature adults. As in previous taxa, some species seem to lack males and reproduce by parthenogenesis, whereas in others there is sexual dimorphism, with smaller males.

The **onychophorans** are similar to tardigrades (Fig. 1.28), but have more segments and a more complicated behaviour (Cuénot, 1949a). They also exhibit sexual dimorphism, with the larger females having more segments. Females lay up to one egg daily, but development of the juveniles to sexually mature adults lasts several months. Some species are known to care for the juveniles during this period. Only about 75 species are known. Their geographic distribution seems to be limited to the south of the Tropic of Cancer. They are found at high elevations as well as in lowlands throughout that region. Species vary in length from a few millimetres to 15 cm, and consist of from 14 to >40 segments, each with a pair of podia. However, individuals can become very thin and squeeze through narrow gaps and explore the top soil. Unlike the previous taxa, onychophorans require dark and humid habitats, and anhydrobiosis is not known. Their food preferences vary, with species that follow invertebrate tunnels to feed on the excreta or the invertebrate, and others which ingest rotting wood. The predators on small

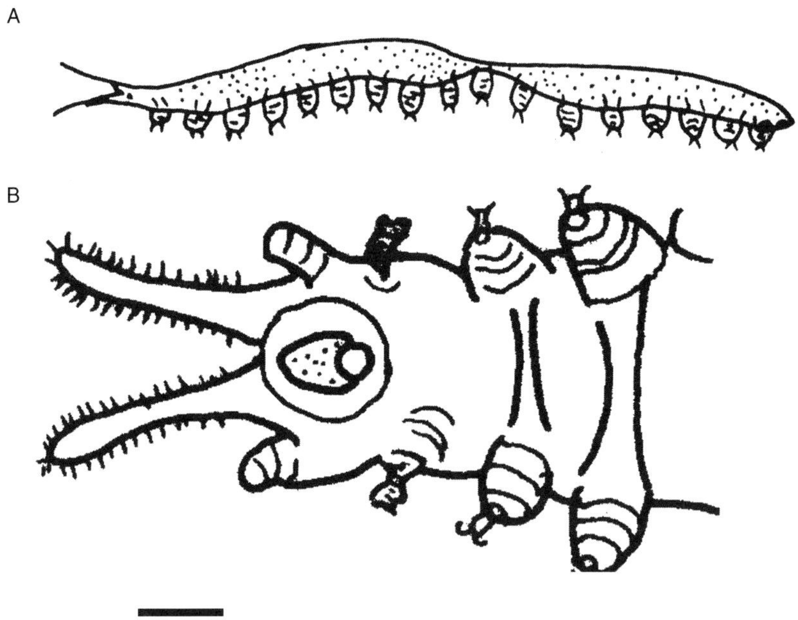

Fig. 1.28. (A) Side view of an onychophoran showing multiple pairs of legs, anterior appendages and posterior pores. (B) View of the anterior region from below, showing the location of mouth parts. Scale bar 1 mm.

invertebrates in the meso- and macrofauna are effective and more common. The prey are captured in a glue that is ejected from oral tentacles. The semi-liquid glue in larger species is ejected 30–50 cm and solidifies instantly on to the prey. They seem to be relatively indiscriminate, as laboratory cultures have been known to be cannibalistic when hungry. The stoma is bordered by a pair of claws and several labia. The oral cavity has a 'tongue' organ, with denticles and mandibles on each side of it for chewing. The pharynx and stoma are very muscular and provide the suction to ingest chewed tissue into the intestinal canal. The nervous tissue forms a brain, with a central nervous system that coordinates the segments. There are sensory papillae over the dorsal surface, as well as exterior of the podia, and a pair of antennae on the cephalic segment. A pair of eyes are used primarily to avoid light. The musculature consists of smooth muscles only, organized in several bundles. The role of the pseudocoel, or central cavity, in diffusion of gases and nutrients, nitrogen excretion and osmoregulation is facilitated by the following tissues. There is a urinary system that consists of nephridia, used in osmoregulation and excretion of uric acid. A dorsal tube functions as a blood vessel and heart to circulate the blood. Open at both ends, each contraction expels blood through both ends into the pseudocoel cavity. During relaxation, ostiole valves along the heart open and let blood flow back into the tube. The blood consists of a serous solution with nutrients, which facilitates gas exchange and has a cellular component. The cells are small corpuscles and active larger phagocytes. An elaborate tracheal system with openings in each segment supplements gas exchange and osmoregulation through the cuticle.

Earthworms (phylum Annelida: Clitellata: Oligochaeta)

There are >7000 species known from aquatic and terrestrial habitats, and a small number of species are known to inhabit marine sediments. Most species belong to about 20 families adapted for a terrestrial interstitial habitat. Their distribution is limited by absence of soil organic matter and litter on which they feed. They are absent or rare in arid or cold regions, for climatic as well as nutritional reasons. Their present geographical distribution seems to carry the mark of previous glaciation events, so that they probably disappeared from regions covered by Pleistocene ice sheets (Michaelsen, 1903). Their northward recolonization had not reached many regions of Canada when European colonization of North America began. Human travel and migrations in recent times have expanded the range of many species. These exotic species are called **peregrine**, because their dispersal was caused by human activity. Species that were already locally present are called **endemic** species. Some peregrine species have been effective at displacing endemic species, particularly in disturbed or agricultural regions. We will first consider the suborder Lumbricina and then briefly the suborder Enchytraeina.

For general reference and more details on aspects of morphology and ecology, the reader is directed to Edwards (1998), Edwards and Bohlen (1996), Lee (1985) and Grasse's encyclopaedia. The physiology of earthworms is treated further in Laverack (1963). Compared with the onychophorans, for our purpose and in general, the main differences are the elaboration of the circulation system with a capillary network, of nephridial excretion and of the central nervous system. These permit a larger body and more complex behaviour. The body consists of repeated segments, the somites, which are partially independent. There are more organs, and tissue differentiation is more extensive.

Lumbricina, or true earthworms

The cuticle of earthworms is thin and flexible, so that juveniles grow into the adult size without moults. As in previous taxa, it is secreted by the epidermis; however, it is not chitinous, but collagenous and only 1–4 μm thick (Jamieson, 1981). The epidermis, as in other tissues of annelids, is elaborated, with more diverse differentiated cells and often additional layers of cells. The epidermis and cuticle are traversed on the dorsal side by numerous canals from the coelom to the exterior. These normally are closed but permit passage of water into and out of the body for osmoregulation. The coelomic space of each somite is lined by a sheet of cells that form a septum between somites. The septum isolates the coelomic space between somites. The coelum is filled with a serous solution derived from the blood and the digestive tract, as well as by various types of free and motile coelomocytes, which are phagocytic and play a role in immunity. Their role is important because of the potential for bacterial invasion through the coelomic pores and wounds.

The coelom and nephridia function together in osmoregulation. Earthworms can lose up to 70% of their water content through the coelomic canals and cuticle, with soil desiccation. Under such soil hypertonic conditions, cells and tissues also lose water, and internal fluid electrolyte balance is affected. This may be important in initiating anhydrobiosis and periods of dormancy. In more dilute solutions or wet soils, the electrolyte balance is maintained. Rehydration occurs through the cuticle and restores cellular functions. There is one pair of nephridia in each somite, which clears coelomic fluid of soluble by-products and nitrogenous waste. The nephridia help to remove excess electrolytes, soluble organic molecules and generally maintain the physiological solution of the coelom. The concentration of the excreted urine varies with the hydration state of tissue fluids. As water becomes limiting, the proportion of excreted ammonia decreases and that of urea increases. Urea also predominates in starved individuals of *Lumbricus* and, in *Eisenia*, urea is always the main excreted nitrogenous waste. There is therefore some variation between species adaptations. In some species adapted to arid soils, such as *Pheratina posthuma*, urine from the nephridia is excreted into the intestinal tract. This conserves water for the tissues by re-absorption through the intestine. The intestine efficiently removes most of the water and releases a dry excrement through the anus.

There is a true closed blood circulation system, which enhances gas exchange and distribution of nutrients from intestinal cells to body tissues. A dorsal blood vessel pumps blood anteriorly and through a capillary network which supplies the tissues. The blood flows posteriorly through the ventral vessel. There are blood pigments which bind oxygen reversibly and remove CO_2. The pigments are varied in annelids but tend to be haemoglobin in Oligochaeta. This permits a larger body size, higher activity and aerobic respiration under low oxygen tensions to 3% O_2. Some species that live in limnetic mud are anaero-tolerant and have a metabolism adapted to tolerate anaerobic respiration, for up to 2 days. They have increased vascularization and modifications of the glycolysis to lactic acid pathway. The musculature consists of an outer circular layer with several inner circular and longitudinal muscle layers. They also articulate small chitinous setae, or podial extensions. These extensions are retractable, and used for grasping soil during locomotion and for burrowing into soil. Locomotion is facilitated by secretion of mucus from epidermal cells, through cuticle pores. The mucus absorbs water and maintains a wet slippery surface between the body and the soil. As it solidifies in the soil, it provides a rigid support to tunnel walls. The role of this secretion is also important in soil structure because its stickiness binds particles together and contributes to soil aggregate stability. This secretion is responsible for half of the total nitrogen released from the body.

Locomotion and behaviour of earthworms are coordinated by an elaborate nervous system. It consists of a central nervous system coordinated by cerebral ganglia in the brain, and paired ganglia in somites. The somite ganglia are connected by the ventral nerve cord. There are no specialized extensions on the head segment, as an adaptation for burrowing. The eyes are limited to photosensitive cells scattered on the head. The cuticle is highly innervated with touch-sensitive cells. Annelid burrowing forms often have gravitaxis detection, although they are not known in Oligochaeta. However, they are still able to right themselves if they fall on their back. Oligochaeta are capable of escape behaviour and memory. Therefore, there is a limited capacity for learning. The anterior head and mouth are highly innervated with chemosensory cells, which participate in directional search for food, avoidance of unpalatable particles and avoidance of toxicity. Therefore, food selection and migrations away from polluted sites occur. Food selectivity is affected by the concentration of phenolic, alkaloid and other cell wall molecules in plant litter or in the soil. For instance, leaf tissue can be made more palatable by first washing out these components (Mangold, 1953; Edwards and Heath, 1963; Satchell, 1967). Chemical sensitivity is well known from agricultural fields, where earthworms are kept out by many pesticides, herbicides and other chemical applications. A list of sensitivity to chemicals is provided by Edwards (1998). Sensitivity to chemicals is exacerbated by increased soil moisture, as chemosensory irritation is enhanced.

Food is ingested by suction created by the musculature of the mouth and pharynx. In some species, part of the pharynx is extensible for grasping. The ingested particles are then coated with a viscous mucus secretion. Food reaches the intestine by peristalsis of the intestinal wall musculature. There are, in sequence, one or more pouches for food storage, the crop and a gizzard. However, the crop is absent in some species. The gizzard has areas of reinforced cuticle that help macerate the food. This process helps to mix together the cellular tissues with mineral soil particles ingested at the same time. Digestion begins in the crop and gizzard with secretion of enzymes such as amylases, proteases and a lipase. The pH varies along the length of the intestine, from a slightly acidic anterior to a slightly basic (pH 7–8) middle and posterior. The intestine of some species contains chitinase and cellulase activity, but it is unclear to what extent these are contributed by intestinal bacteria and ciliate symbionts which can be present. Digestion of cellulose and chitin is normally attributed to **endosymbiont** activity. There is a gland in the upper intestine, the organ of Marren with chloragocytes, which contributes to the removal of ingested calcium and other abundant cations. These are precipitated as calcite and other crystals by a locally increased acidity. The crystals are excreted with the undigested remains of digestion. It is an important function, because ingestion of clays and plant matter can increase mineral concentrations to pathogenic levels, if they are all absorbed. The gland also has functions similar to the vertebrate liver, in that it supports blood detoxification and stores glycogen. The chloragocytes also contribute to removal of heavy metals (Jamieson, 1981).

Reproduction is by reciprocal cross-fertilization of mating pairs, because individuals have both male and female organs – they are hermaphrodites. Many species can also reproduce by parthenogenesis. However, the offspring of parthenogenesis are often less virile and less fertile. Fertilized eggs are wrapped for protection in a cocoon secreted by the clitellum. The cocoon is remarkably resistant to decomposition. Depending on species, 1–30 eggs are laid, but this number varies with the age and with how fed the individuals are. The number of cocoons produced annually also varies between species, from five to 1000. Most species produce a few hundred, depending on temperature and food availability. In general, deep-burrowing species produce fewer cocoons, and near-surface species more of them. In some species, cocoon production is seasonal, as an adaptation to the climate. Development of the egg to adult requires several weeks or months, depending on abiotic variables. Since genetic polyploidy is common and varied (as in some plants), the consequences on speciation and in population genetic studies need to be considered.

Oligochaeta have an exceptional ability to regenerate from mechanical damage. This is demonstrated by two processes. The first is the ability to regenerate broken sections at the anterior or posterior. Cut off somites are regenerated by cell divisions of tissues from damaged somites. The coelomocytes are particularly active in this process, and many lost somites can be regenerated. The wound healing process requires hours to days, but recovery of lost somites may take weeks. During this time, mobility is limited and, especially if the anterior is lost, feeding is also impeded. The second is the ability to fragment into two or more sections, when the individuals reach a particular size. Lumbriculus can fragment spontaneously into as many as eight sections, which then regenerate at both ends to form complete, but smaller, individuals. This is a mechanism of asexual reproduction, reminiscent of clonal expansion in unicellular organisms. There are also two adaptations to unfavourable conditions. Some species have seasonal periods of dormancy, or diapause, during which they are inactive, and do not feed. This may last through cold or dry periods of the years. The individual remains inside a cyst which consists of hardened mucus, where they are still capable of fragmentation. Most species are also able to become temporarily quiescent, by anhydrobiosis, through unfavourable conditions. This reversible process involves both partial tissue dehydration and physiological quiescence. However, individuals in anhydrobiosis are not resistant to the same extremes of temperature and desiccation as Tardigrada or even nematodes.

Enchytraeidae, or potworms

Enchytraeids are morphologically similar to the Lumbricina, but several differences need to be highlighted. This single family of the suborder contains about 600 species, which have a broader geographic distribution than the Lumbricina (Didden, 1993). Individuals are much smaller, with most species being a few millimetres or less, though some can reach 5 cm. They do not form permanent burrows as do some earthworms, but travel through the soil organic layers. They can be found in subarctic regions as well as under snow and glacier ice. They are more abundant in forest soils and soils with a rich organic layer, and less abundant in pastures and agricultural fields. Their food preferences have been studied (Dosza-Farkes, 1982; Kasprzak, 1982; Toutain *et al.*, 1982). They prefer macrodetritus or particulate organic matter that has been predigested by fungal and other saprotroph activity. Protists and mineral soil are ingested along with this litter. Interestingly, some species seem to have difficulty digesting cellulose, so that humus, hyphal and protist digestion may be more significant than plant tissue biomass ingested. Development of eggs in cocoons requires 2–4 months, varying with temperature and species. Parthenogenesis, self-fertilization and fragmentation are also common. Aspects of their ecology has been reviewed in van Vliet (2000) and Lagerloef *et al.* (1989).

Microarthropods (phylum Arthropoda: Chelicerata, Myriapoda, Insecta)

Chelicerata

Representatives of this group in the soil include diverse orders such as the spiders (Araneae), Pseudoscorpiones, Opiliones and the acarids (mites). We will deal with the soil mites, as they are the most abundant and present in all soils. More than 30,000 species of Acari are known, and most species encountered in soils belong to the Oribatida. Recent systematic discussions of these taxa are provided by Walter and Proctor (1999) and Wheeler and Hayashi (1998), and keys to various taxa can be found in monographs such as *A Manual of Acarology* (Krantz, 1978), and the Acarology homepage www.nhm.ac.uk/hosted_sites/acarology For information on the other orders, the student is referred to the *Soil Biology Guide* (Dindal, 1990).

The Acari, or mites

There is an abundant literature on soil mites, and several texts deal with aspects of their biology and ecology, to which the reader is referred to for more in-depth study (André, 1949; Evans, 1992; Bruin *et al.*, 1999; Walter and Proctor, 1999). These microarthropods inhabit soils from every region, including cold Antarctic deserts and hot sandy deserts, and they are functionally irreplaceable in decomposition. In temperate regions, the top 10 cm of forest soil can harbour typically 5–25 × 10^4 mites/m^2, representing 50–100 species. However, they are also significantly abundant in deeper soil, especially along root micro-tunnels, even to 2 m depth (Schubart, 1973; Price and Benham, 1976; Coineau *et al.*, 1978). Most interstitial species fall in the 300–700 μm range in adult body length. Very small species can be found in mineral soil (families: Nematolycidae and Pomerantziidae), as well as in hot or cold deserts (families: Brachychtoniidae, Nanorchestidae, Scutacaridae, Tarsonemidae, Tepnacaridae and Tydeidae) where species are <250 μm. According to Norton (1990), the narrow and elongate species of the Epilohmanniidae, Eulohmanniidae and certain Oppiidae are probably obligate inhabitants of the B-horizon. Several Endeostigmata are even smaller at 150–160 μm, such as *Alycosmesis corallium* and *Nanorchestes memelensis*. The soil mites can be organized into several functional groups based on their feeding preferences. The following categories are proposed by Walter and Proctor (1999): detritivores (or comminution grazers), piercing–sucking, filter feeding, nematophagy, predatory on other microinvertebrates and fungivores.

The most common mites in the soil are the oribatids. This group of organisms is very diverse, and it is difficult to make generalizations, as there are always species which will be exceptions to the rule. However, thus forewarned, a general outline of the biology of species implicated in decomposition is provided below.

The overall body plan consists of a podosoma (the anterior section) and opisthosoma (posterior section), as in other spider-like species (the Chelicerata). The podosoma holds four pairs of legs and an anterior section called the **gnathosoma** (jaw-body) which holds the oral appendages and pedipalps (Fig. 1.11B–H). The anterior-most appendages, the **pedipalps**, are modified to participate in feeding and sometimes also for reproduction. The ends are sometimes shaped like claws or pincers for tearing, grasping or moving objects. The gnathosoma consists of a forward protrusion of the podosoma, with a movable upper jaw (or labrum) and fixed lower plate, and a pair of **chelicerae** on each side of the upper jaw. The chelicerae are also shaped like a claw or pincer and function in obtaining food. Their shape varies with species and the mode of feeding. They can be moved in and out to pierce, saw, scrape, bite and move food. Together with the pedipalps, they gather food into the mouth. The opisthosoma holds most of the somatic and reproductive organs.

The body is covered by a chitinous cuticle as in previous invertebrates we have described. The cuticle is secreted by the epidermis and can be subdivided into several layers, depending on species. The outermost layer is only 50–150 nm and is covered by a waxy layer, which is impermeable to water and prevents desiccation. Between the epidermis and the outer layer, the pro-cuticle consists of chitin microfibrils in a protein matrix and varies in thickness between species from 0.25 to 2 μm. The outer procuticle becomes partially or entirely hardened in some species (sclerotized). It may also become reinforced further with calcium salts, such as calcite or calcium oxalate. The cuticle is not flexible and provides a rigid support for the musculature and a protective casing for the body. Juveniles must therefore moult and shed the old cuticle to grow into adults. Epithelial gland cells with canals extending into the cuticle maintain the cuticle and the waxy waterproof layer.

There are numerous bundles of striated muscles and several thin muscles (consisting of small groups of cells) to operate the pedipalps, chelicerae and gnathosoma. The legs and the jointed oral structures are operated by antagonistic pairs of muscles. These are coordinated by a brain, which consists of fused ganglia organized into lobes, with peripheral nerves into the body. There is an extensive nerve net throughout the body, the axons of which reach the brain through a ventral nerve cord. Sensory organs, **setae** or **sensillae**, are numerous along the dorsal surface, legs and especially on the pedipalps and about the gnathosoma. They can be very elaborate, and are useful in species designation. Setae can respond to vibrations, mechanical contact, moisture content and slight air disturbance, and some are supposed to respond to high-frequency ultrasound. Chemosensory detection in identifying food is an important function for the pedipalps and anterior setae. Photoreception is absent in most interstitial species but, when present, 1–5 eyes can be present anteriorly. These are modified pigmented epithelium cells with an axonal extension, and the lens is modified cuticle.

Respiration is by trachea tubules with opening to the outside by spiracles. However, the trachea are missing in many of the smaller species, and only appear in the adult or later juvenile stages of many other species. When they are missing, gas exchange is possible through the cuticle only. The body cavity consists of a spongy parenchyma filled with a haemolymph that receives nutrients from the digestive tract and participates in gas exchange. The haemolymph contains various leukocytes and amoebocytes of 4–8 μm, similar to coelomocytes in the taxa described above. The haemolymph is stirred and distributed during locomotion by somatic muscle contractions, and tissues depend on diffusion for nutrient absorption. A vasculature is often absent, or it is reduced to a dorsal blood vessel. The dorsal blood vessel beats and forces the haemolymph out through ostioles; it seems that the return is also through the ostioles. Neurosecretion into the haemolymph may regulate life history, behaviour or some physiological changes (Evans, 1992). Oribatids can survive freezing and winter by accumulating glycerol in the haemolymph, and individuals stop feeding and moving until more suitable temperatures return (Norton, 1990). They can also survive long spells in anaerobic conditions.

In many species, the food brought into the gnathosoma cavity is first liquidized, by partial external digestion with salivary gland secretions. However, species of Oribatida and Astigmata may form a solid bolus from microdetritus. The gnathosoma cavity leads to the digestive tract. The digestive tract consists of a pharynx that leads to an oesophagus where digestive enzymes are mixed with the food. The oesophagus leads to a sac with up to ten caecum extensions where the food is digested further with gastric secretions. The number of caecae is variable with the amount of food in the diet and space available in the body cavity, especially in females with eggs. All cells along the intestinal tract seem to function in enzyme secretion, nutrient absorption and nutrient processing. Undigested material is excreted through the anal pore. Interestingly, some species do not have an anus and the intestinal tract empties into the haemolymph cavity. As these excreta are usually liquid, they are disposed of through the uropore. If they are more solid, they may break out through the posterior cuticle, leaving a wound. Osmoregulation, ion balance and nitrogenous waste elimination are achieved by paired Malpighian tubules that empty into a single excretory pouch, which empties out through the uropore. Nitrogenous waste is eliminated as crystalline guanine but, in species with reduced or missing tubules, uric acid is eliminated.

For reproduction, the sexes are separate, with smaller males in some species. Each has a pair of ovaries or testicles with the associated organs. Mechanisms of fertilization and sperm transfer are very varied and will not be described here (Schuster and Murphy, 1991; Walter and Proctor, 1999). It often is an indirect copulation, with externally deposited sperm packets being transferred to the female genital tract with pedipalps. Much remains to be learnt from mites in understanding sexual selection, sperm competition or female choice (Norton *et al.*, 1993). The fertilized eggs are placed in the detritus with or without a secreted silt thread protective covering. Some species are known to remain in the vicinity of the eggs to protect them. However, these examples of parenting in mites are rare compared with other chelicerates, such as the spiders. The eggs vary in number from one to several dozens, but parasitic forms can lay upwards of 10,000. For soil mites, there seems to be a positive correlation between number of eggs, egg volume and female adult body size. There is usually not more than one egg-lay annually. Even with species that develop rapidly, conditions during many days of the year are not suitable for feeding and growth, because of either inadequate moisture or temperature, or lack of food resources. The conditions suitable for egg-laying can be even more restricted. Development of the egg into a small immature individual requires several hundred hours at ambient temperatures. As a guide, the development times from eggs to reproductive adults at 20–30°C, for Oribatids in temperate forest litter, vary between 3 and 50 weeks. Some species, for instance in the Astigmata, depend on ephemeral resources and are adapted for more rapid development (O'Connor, 1994). Parthenogenesis is common in many species. In some species, parthenogenesis alternates with fertilization to produce males or females according to chromosome ploidy. In many species, a limited capacity to regenerate lost limbs has been observed in adults. One particularity that must be kept in mind, however, from an ecological functional perspective, is that adult morphology or mouth structures may differ from the juvenile stages. In fact, not only do some adults have a different diet, but they may also have a different habitat. For instance, the Parasitengona (~7000 species) have parasitic larval stages, but predatory adults that can be >1 cm long. Other species may alternate between the vegetation and soil litter.

Insecta

This subdivision of arthropods contains numerous large species mostly several millimetres or centimetres in length. It would not be appropriate here to elaborate on the insects because most do not participate directly in soil decomposition. For a description of the interplay between larger organisms, such as insects, and decomposition, the reader is directed to Swift *et al.* (1979) and Coleman and Crossley (1996). Although many insect species live on the ground and spend part of their life history in soils, adults are not usually interstitial. Insects, whether in the litter or in the canopy, contribute to decomposition in many ways. Due to their abundance, insects contribute to soil litter through moulting, defecation and dead tissues. Many insect larvae, especially those that hatch from eggs in the soil or in litter, contribute to the decomposition of fruits, carcasses and other litter. Insects in general are important to comminution of litter into smaller fragments which then become available for decomposition by interstitial species. Predatory insects contribute to the accumulation of animal tissues in litter, and grazers to that of frass and plant tissue. **Frass** is the term given to the canopy fallout from insect grazing. Several species, for example several beetles, feed on decomposing litter or on animal excreta. Here only the Collembola are treated at any length because they are found in abundance in the soil and litter.

Collembola, or springtails

The Collembola belong to the subclass Apterigota, together with Diplura, Protura and Thysanura. They are primitively wingless insects, often 1–2 mm long, though a few species can be 7–8 mm long and others, such as the Neelidae, are only 0.25 mm long. Seven families occur in soils: Entomobryidae, Hypogastruridae, Isotomidae, Neelidae, Onychiuridae, Poduridae and Sminthuridae. Their abundance in the top 5 cm of soil is usually 10^4–10^5/m^2 but they are also found deeper in the profile, along roots or microtunnels. Like other small organisms, they prefer humid and moist environments. The **cephalothorax** holds a pair of segmented (four) antennae for olfaction and chemosensory detection, three pairs of legs and the oral structures (see Fig. 1.11I–K). The mouth parts consist of a tongue, sharp mandibles and jaws. These are arranged for tearing and chewing. The abdomen has six segments and holds most of the internal organs. The first abdominal segment has the ventral tube opening which allows water and air intake. Below the abdomen, there is a **furculum** which hooks into the tenaculum in the third segment. It is unhooked and released by strong muscular contraction of the furculum which makes it spring back and outwards. This propels the individual to a new random location several centimetres away. The purpose and use of the furculum is unclear, but it may help them get out of litter as their legs can stick to water films, or escape predators when disturbed. The furculum is vestigial or absent in the Isotomidae, Hypogastruridae and Onychiuridae. It is well developed in the Entomobryidae and Symphypleona. The cuticle may have spines, scales and sensillae. Respiration is tracheal in Symphypleona and Actaletes, or through the cuticle. The intestine is almost linear, from the mouth to pharynx, oesophagus and middle intestine, where digestive glands empty. There may be two caecal lobes from the intestine. The posterior intestine ends with the anus. The intestinal tract is lined by longitudinal and circular muscles, and there are sphincters at the junctions between the middle and posterior intestines. This provides more control over duration of digestion and food retention. Nephridial glands participate in nitrogen waste elimination. Insoluble urates (mostly uric acid) are formed and conserved in special cells (not adipocytes) in the adipose tissue in the abdomen and under the epidermis. The beating aorta functions as a heart, forcing the fluid out through ostioles and circulating the fluid. The overall arrangement of the nervous system is typical of Insecta, with a lobed brain and segmental ganglia, linked by a ventral nerve cord. The antennae and sensillae are the main sensory organs. When present, very simple compound eye cells are located on each side of the head, seen as a pigmented patch. Many species do not have eyes, but it is speculated that the cuticle is photosensitive (Denis, 1949). Even species with eyes do not see well, and respond only to strong light. Behaviourally, most species avoid light, though some are known to move towards a light source. Collembola tend to aggregate even after being disaggregated. The mechanism or benefit is unknown, but they are also known to spend time covering themselves with saliva, even rolling in it to reach difficult parts of the body. Christiansen *et al.* (1992) implicated pheromones in this response. There are two sexes, with indirect internal fertilization. Males deposit sperm packets which are picked up by females. Relatively large eggs are deposited in small packets between hairs, trichomes or in the litter macrodetritus cavities. Species have temperature optima for feeding, copulation and egg development. Females can lay three times or more annually, depending on conditions. Although some adults are known to hibernate, the eggs are the principal means of surviving through the cold, heat or dry seasons. They can partially dehydrate and remain viable. The young will hatch after 1–4 weeks of development with all the somites present. They will grow by successive moults into the adult body size. The life expectancy is assumed to be about 1 year for many species. Feeding preferences in Collembola are varied and more diverse between species than normally accounted for in the literature. Most species are omnivorous opportunists and have a range of prey preferences. They will feed on pollen, hyphae, decomposing litter and even nematodes. Some species are more specialized and have a narrow range of food items. *Archisotoma basselsi* and *Isotomurus palustris* are specialists that lick bacteria films primarily. *Deuterosminthurus repandus*, *Folsonia fimetaria*, *Lepidocyrtus cyaneus* and *Sminthurides virides* are pollen feeders. The *Hypogastrura*, *Sminthurinus* and *Tomocerus* are fungivores. Certain species of Collembola can devastate mushroom cultivations or are plant parasites. Some species feed on decomposing animal tissues or, like *Isotoma macnamarei*, prey on other Collembola.

The Bacteria (Prokaryote: Bacteria and Archea)

The prokaryotes consist of two subdivisions which are biochemically and physiologically distinct. The archea represent species which are mostly known from extreme environments. In the anaerobic environment of submerged rice fields, the methanogens are significant contributors to methane emissions. Archea are also found near deep-sea thermal vents or in other hot environments where other prokaryote and eukaryote species cannot tolerate the temperatures. They also occur in extreme acid or saline environments. There are about 40 genera of archea, but we will not discuss these taxa further. Most prokaryotes belong to the bacteria subdivision, which contains >480 genera, plus the Cyanobacteria, in 17 main lineages, to which we will refer as phyla (Table 1.5). For general reference on the biology of bacteria, the ecologist is directed to Schlegel (1993) which contains sufficient natural history, and to Lengeler *et al.* (1999) for their cellular biochemistry. Two further texts deal with soil bacteriology specifically and in great detail (Paul and Clark 1998; Tate, 2000).

Table 1.5. Organization of the kingdom bacteria according to Cavalier-Smith (1998).

Infra-kingdom	Subphylum	Phylum	Infra-phylum	Classes	Selected genera	No. of genera
Lipobacteria	Heliobacteria					
	Hadobacteria	Chlorobacteria		Chloroflexibacteria, Eochlorea	*Chloroflexus*, *Heliothrix*	6
		Deinobacteria		Deinobacteria, Eotherma	*Deinococcus*, *Thermus*	3
	Spirochaetae	Euspirochaetae		Spirochaetea	*Treponema*	13
		Leptospirae		Leptospirea	*Leptospira*	
Glycobacteria	Sphingobacteria	Chlorobibacteria		Chlorobea	*Chlorobium*	24
		Flavobacteria		Flavobacteria	*Cytophaga*, *Flavobacterium*	
	Eurybacteria	Selenobacteria		Selenomonadea, Sporomusea	*Selenomonas*, *Sporomusa*	4
		Fusobacteria			*Fusobacterium*, *Leptotrichia*	
		Fibrobacteria			*Fibrobacter*	1
	Cyanobacteria	Gloeobacteria		Gloeobacteria	*Gloeobacter*	56
		Phycobacteria		Myxophycea	*Anabaena*, *Nostoc*, *Prochloron*	
	Proteobacteria	Rhodobacteria	Alphabacteria	Alphabacteria, Betabacteria,		
			Chromatibacteria	Gammabacteria, Deltabacteria	*Agrobacterium*, *Rickettsia*,	
					Rhodospirillum	~270
					Chromatium, *Escherichia*,	
					Spirillum	
		Thiobacteria			*'Myxobacteria'*, *Desulfovibrio*,	
					Thiovulum	
	Planctobacteria				Chlamydiae, Planctobacteria,	5
Posibacteria	Posibacteria	Teichobacteria	Endobacteria	Clostridea, Mollicutes	*Bacillus*, *Clostridea*, *Mycoplasma*	~172
			Actinobacteria	Actinobacteria	*Corynebacterium*, *Streptomyces*	
		Togobacteria			*Aquifex*, Thermotogales	3

Data compiled from Cavalier-Smith (1998), Lengeler *et al.* (1999) and other sources.

Bacteria are simpler in their structure and physiology than eukaryotes. The cell is composed of a relatively small number of molecules (Table 1.6). Most bacteria have a cross-sectional diameter of 0.6–1.2 μm, with length varying between 1 and 2 μm. The length is more variable, with some genera having species of about 10 μm long. Many are characterized morphologically as cocci (spherical), rods (box shaped) or filamentous (cells remain attached) (Fig. 1.29). Growth of bacteria by successive cell divisions results in aggregated cells, called **colonies**. Many species naturally secrete a layer of slime or gelatinous sheath. In some species, it provides an environment for gliding, or for keeping the colony together. In other species, it is secreted as protection against desiccation or during physiological stress. The slime or **capsule** helps absorb moisture from the habitat, and provides a protective covering after desiccation. In general, the more motile the species, the less aggregated the colony. Motility is the result of gliding on the substrate, or propulsion by one or more flagella. The **flagellum** consists

Table 1.6. Number of macromolecules estimated in one cell of the bacterium *Escherichia coli.*

	No. of molecule types	No. of molecules
Outer membrane	>50 proteins	10^6 molecules
	5 phospholipids	10^6 molecules
	1 lipopolysaccharide	10^7 molecules
Cell membrane	>200 proteins	10^5 molecules
Ribosome	55 proteins	One set for 10^4–10^5 ribosomes
	3 rRNA	One set/ribosome
	>10^3 mRNA	Variable
	60 tRNA	10^5 molecules
Chromosome	One	4.2 kb

Data from Lengeler *et al.*, 1999.

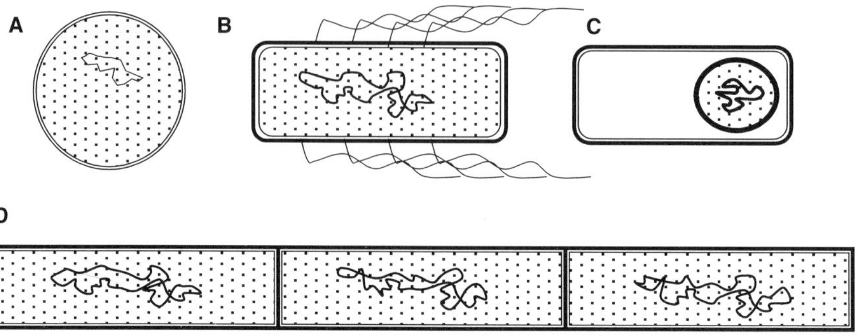

Fig. 1.29. Representation of several forms of bacteria. (A) Coccus cell, (B) rod cell with flagella, (C) rod cell with an endospore, (D) rod cells in a filament.

of a hollow cylindrical tube, composed of polymerized flagellin monomers. The flagellin protein monomers pass into the hollow flagella and polymerize at the tip. The flagellum is anchored on to the cell by a hook and motor complex which consists of about 100 polypeptides. Rotation of the flagellum is caused by proton pump proteins, which establish an electromotive gradient. Normally, the positions of the flagella are polar or medial in species that swim in free water. They are positioned laterally on those that swim or glide in thin water films on solid substrates. Motile forms can move towards nutrients or away from chemicals by chemotaxis. Similarly, filamentous species can grow towards or away from chemicals through the direction of elongation and branching. Many genera do not have the capacity to move on their own, and depend on Brownian movement and external disturbance to be carried. In soils, bacteria normally occur as more or less aggregated colonies, so that species patchiness is expected at the 10–2000 μm scale. They are disaggregated during sampling, and during preparative steps for enumeration or culture. Bacterial abundances can be as high as 10^{10}/g dry soil in litter and surface soils, representing dozens or hundreds of species. They have been reported at great depth, 200–400 m below the surface, at 10^4–10^6/g soil and were demonstrated to be active by ^{14}C uptake (Lengeler *et al.*, 1999, p. 784). However, it is most probable that they are carried by water flow-through rather than being normal residents of that habitat.

Bacteria consist of a cytoplasm bounded by a cell membrane and an outer cell wall. The cytoplasm of bacteria is not compartmentalized by membranes, and the single circular chromosome is free in the cytoplasm. The cytoplasm contains prokaryotic ribosomes for protein translation, and sometimes other inclusions depending on physiological conditions. There are no endomembranes, no membrane-bound vesicles or vacuoles for endo-/exocytosis, and no cytoskeleton. However, phototrophic genera such as the Cyanobacteria, have invaginations of the inner cell membrane to increase the surface area of bound photosynthesis enzyme complex. The cell membrane is chemically different from that of eukaryotes, and is bound by an external cell wall. This confers to bacteria physical and physiological properties different from those of eukaryotes.

The **cell wall** is the barrier between the habitat and the cytoplasm. Its composition and function are central to bacteriology. Bacterial systematics traditionally were based on cell wall chemistry and metabolic biochemistry. More recently, it has been complemented by comparative DNA sequence phylogenies. The cell wall is responsible for keeping the cytoplasm molecules in, selectively transporting nutrient molecules into the cytoplasm, maintaining a favourable physiological environment inside the cytoplasm and providing a chemical and mechanical barrier to the habitat (Fig. 1.30). The two principal

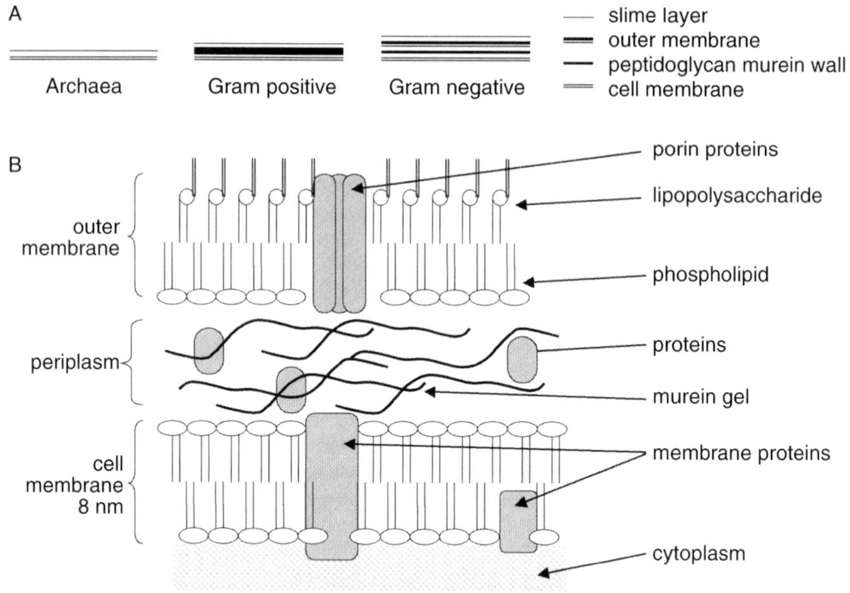

Fig. 1.30. Molecular structure of a prokaryote cell wall. (A) Cell wall structure of archea, a Gram-positive cell and a Gram-negative cell. The chemical composition of a Gram-negative cell wall. (B) Details of a Gram-negative cell wall.

cell wall subdivisions are between bacteria that stain with the Gram stain (**Gram positive**) and those that do not (**Gram negative**). Gram-positive bacteria have a **murein cell wall** (0.015 μm) outside the cell membrane which consists of peptidoglycans with embedded teichoic acid derivatives. It has an external surface layer of proteins or polysaccharides. The murein wall can be enriched with lipidic molecules to increase hydrophobicity, as in *Corynobacteria*, *Mycobacteria* and *Nocardia*. In the *Planctobacteria*, the murein layer is replaced by a proteinaceous cell wall. The second type of cell wall is found in Gram-negative bacteria. They have a loose murein layer (~0.015–0.020 μm thick), but with an additional outer membrane. The loose murein gel of Gram-negative bacteria is interlinked by lipoproteins and contains proteins from the cytoplasm. The **outer membrane** is chemically different from the cell membrane and contains lipopolysaccharide extensions, structural proteins and porins. The **porin** proteins allow the passage of molecules smaller than 600 Da molecular weight into and out of the murein gel. Differences between the exact molecular composition of the cell wall, between families and higher taxa, are useful in species identification. The third type of cell wall is found in the archea, where the outer membrane and the murein wall are absent. In some genera, there is a wall outside the cell membrane, but it is not composed of murein.

The cell membrane of bacteria consists of a lipid bilayer with inserted proteins. The lipids provide a hydrophobic barrier to hold the aqueous cytoplasm in and keep the habitat molecules out. In general, polar molecules (carbohydrates) and charged molecules (ions, amino acids and side chains of larger molecules) do not pass through, except by very slow passive diffusion. Certain molecules do pass through the cell membrane because they are non-polar, hydrophobic or membrane soluble, such as glycerol and fatty acids, sterols, aromatic compounds and some amino acids (phenylalanine). Other molecules can pass through simply because they are small, such as water, short alcohols, ammonia, urea and dissolved gases (mostly O_2 and CO_2). The mechanisms of obtaining nutrient molecules and ions from the environment are regulated by the cell membrane chemistry and its proteins. These are **diffusion**, **facilitated diffusion** by binding to membrane receptors, and **active transport** through specific protein channels or transporters. The latter mechanism requires energy, such as an established electrochemical gradient across the membrane, or using high-energy bonds such as ATP dephosphorylation. The active transport mechanisms are more effective at translocation of nutrients into the cell, especially at low concentrations or against the concentration gradient. The proteins involved are under physiological regulation. They are translated from the genes into proteins only when they are required, then inserted into the cell membrane and activated. *Only molecules that are small enough to pass through the membrane and its proteins can be used as nutrients.* Therefore, cells are limited to monomers or short oligomers of amino acids, nucleic acids, saccharides, short lipids or other small molecules. Most species have several substrate transport mechanisms that allow the cell to use a variety of nutrients from the habitat, depending on what is available or missing outside. The ability and efficiency of each species or strain to import particular nutrients for growth are very useful in species identification.

Many genera of bacteria, particularly from certain taxa such as the Actinobacteria, release small molecules into their environment which prevent cell growth or cause lysis of susceptible cells. These **secondary metabolites** are known as **antibiotics** when they are effective at low concentrations. Antibiotics are usually produced when nutrients are rare, as in conditions when competition for nutrients is high. Secondary metabolites fall into three categories, that are derived from: (i) acetate–malonate condensation; (ii) condensation of carbohydrates or aminocyclitol; and (iii) oligopeptides, rarely long enough to be synthesized on ribosomes. The resulting molecules tend to be membrane permeable and pass through the cell membrane. The mechanisms of function are very varied, and can interfere with one or more aspects of nutrient intake, anabolic or catabolic metabolism or synthesis of structural molecules. In *Myxobacteria*, where an estimated 60–80% of isolates produce antibiotics, they are bactericidal and cause lysis of susceptible bacteria. Other antibiotics target eukaryotic predators such as amoebae (Lancini and Demain, 1999). For example, *Serratia marcescens* (prodigiosin), *Chromobacterium violaceum* (violacein), *Pseudomonas pyocyanea* (pyocyanine) and *P. aeruginosa* (phenazines) cause the encystment or lysis of amoebae that would ingest them. Other secondary metabolites are scavengers for cations (**ionophores**), or for iron (**siderophores**). These are sufficiently effective at removing specific ions or minerals from the habitat, especially where the ions are rare, so as to starve out species that do not have them. They are found in all strains of *Nocardia*, *Streptomyces* and *Micromonosporea*.

When bacteria are starved of nutrients, or in unfavourable conditions, they become physiologically inactive. Certain genera are able to produce **spores**, which are a differentiated state, with a resistant wall. In Actinobacteria, chains of spores form at the tips of growing filaments, in response to localized starvation. The pattern and shape of spores are important in species identification in the Actinobacteria. Other genera form a spore inside the parent cell, which is called an **endospore**. These include *Bacillus*, *Clostridium*, *Sporosarcina* and *Thermoactinomyces*. All species have preferences for optimum temperature, pH and nutrients. However, as the temperature, moisture, nutrient availability and other chemical changes in the habitat are continuously fluctuating, it is important for cells to tolerate a certain degree of fluctuation. It is important for living cells to adapt physiologically or protect themselves from threatening conditions, faster than it would take to cause death. Several mechanisms are available to bacteria, as indicated below.

To resist depletion of nutrients, cells can switch to higher affinity cell membrane transporters, increase the membrane surface area to cytoplasm ratio by decreasing the bulk cytoplasm, modify their metabolism, and divide at smaller cell sizes. These changes occur over several hours as they require changes in gene expression. On the same time scale, cells adapt to changes in temperature by modifying membrane lipid compositions. At higher temperatures, lipid membranes would be too fluid and more permeable, whereas they become too stiff at cooler temperatures. In order to maintain the fluidity of the membrane relatively constant, longer fatty acids and branched fatty acids are incorporated as the temperature rises, and shorter unsaturated fatty acids dominate membrane composition in colder temperatures. When the temperature rises above an optimal range for the species, heat-shock response proteins are synthesized to protect proteins from denaturing, while cells become inactivated.

In contrast, when soils freeze, ice formation outside the cell excludes some salts so that the osmotic pressure in the remaining water films increases, and mimics desiccation as water becomes less accessible to cells. Furthermore, ice crystals forming inside and outside the cytoplasm can break the cell. To protect themselves, most bacteria accumulate solutes which prevent ice crystal formation and freezing. Some species, such as *Pseudomonas syringae*, are able to insert a protein in the outer membrane that promotes more regular ice crystals to reduce the incidence of cell lysis. Lastly, cells can partially dehydrate and bind the remaining water to prevent ice formation in the cytoplasm.

The response to desiccation involves accumulating solutes that remain soluble at high concentration without interfering with enzyme reactions, and that are non-toxic to the cell (Potts, 1994). These are molecules that are metabolic end-points, i.e. molecules that are not metabolic intermediates and will not be required for other reactions. Commonly used molecules in bacteria are betaine, ectoine, L-proline and trehalose. Others are also found, such as glycerol, sucrose, D-glucitol, D-manitol, L-taurine and small peptides.

We will consider briefly several bacteria that are common in soils (Fig. 1.31). Anaerobic species tend to be rare in surface-aerated soils, but can be found in deeper soils, or in saturated fields such as rice paddies or riparian zones. **Mycobacteria** are Gram-positive non-motile aerobic species, that form irregular, slightly branched cells. They contain very hydrophobic wax-like mycolic acids in the cell membrane. This group includes pathogenic bacteria that cause leprosy and tuberculosis. The **Myxobacteria** are Gram negative, strictly aerobic, gliding species (Dawid, 2000). They form extensive spreading colonies. Many species secrete enzymes that lyse bacteria on which they feed, whilst the *Polyangum* and *Cytophaga* genera have cellulolytic enzymes. When nutrients are exhausted, cells aggregate into a mound (fruiting body) with characteristic shapes and differentiate into dispersal spores that are resistant to desiccation and adverse conditions. These fruiting bodies can be 0.1–0.5 mm and visible under the dissecting microscope (with the exception of *Cytophaga* which are missing the fruiting body stage). A large number of secondary metabolites are known from Myxobacteria, including bioactive molecules and coloured pigments derived from carotenoids. Other bacteria are able to hydrolyse and use cellulose, such as the well-studied ***Cellulomonas*** and the genus *Erwinia*. The **Arthrobacter** genus is common in soils and grows on humus, where readily soluble and metabolizable molecules have been exhausted. They grow slowly as cocci, or faster as branching semi-filaments. The **Actinobacteria** grow as filaments (or mycelia) and are found exclusively in soils. They are Gram positive and aerobic, but there are several anaerobic genera. They can be grown on agar media and identified based on their spore and sporangia structures. Their spores are all resistant to desiccation, but not to heat (except *Thermoactinomyces vulgarum*). The *Streptomycetes* (and the Myxobacterium *Nannocystis excedens*) produce **geosmin** which is the characteristic earthy odour of moist soil. As for the Arthrobacter, Actinobacteria grow on 'difficult' substrates that are not degraded by most species, such as cellulose and chitin. Most known antibiotics were first isolated from Actinobacteria. The polyphyletic **Pseudomonads** are aerobic Gram-negative bacteria which are able to grow on very diverse substances. They are

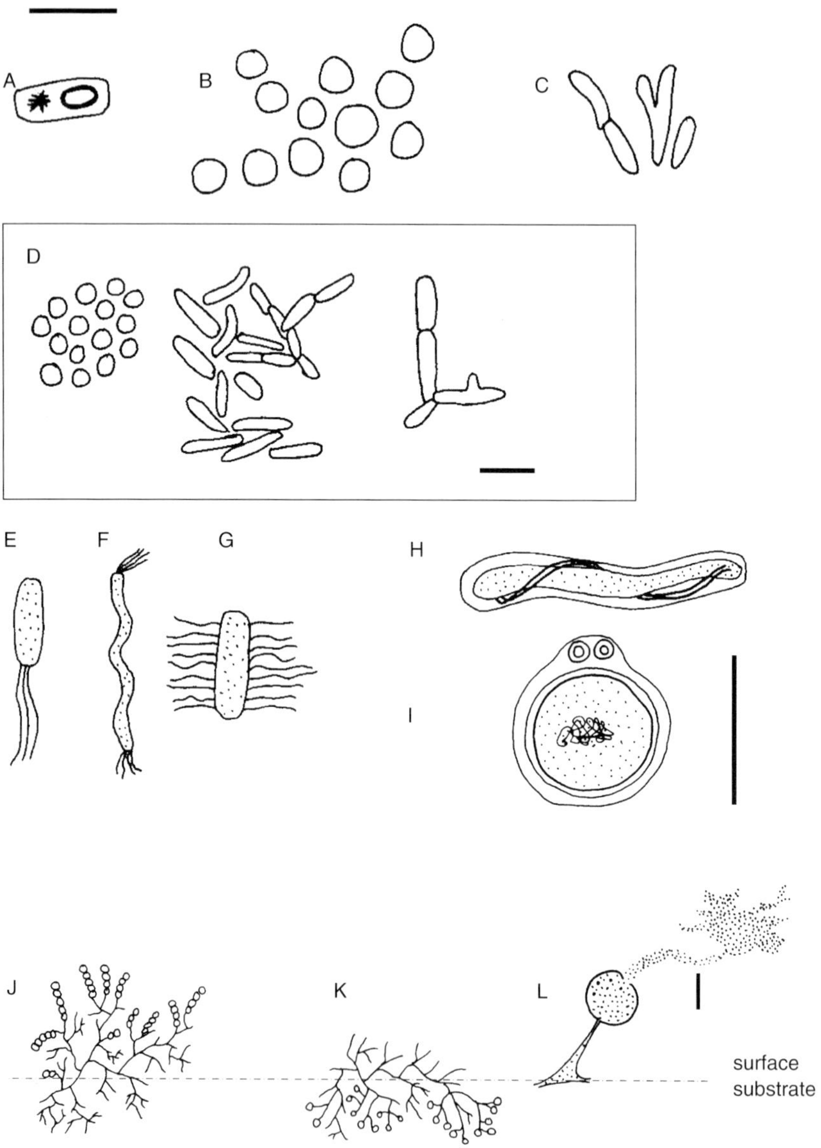

Fig. 1.31. Growth forms of various bacteria. (A) Inactive *Bacillus* rod-shaped cell with an endospore and protein crystal inclusion (e.g. *Bacillus thuringiensus*). (B) Coccus cell (e.g. *Acinetobacter*). (C) Mycobacterium growth forms. (D) *Arthrobacter* coccus, rod and branching cell growth. (E) *Pseudomonas* with apical flagella. (F) *Spirillum* with polar flagella tufts. (G) *Proteus* with lateral flagella. (H) Typical spirochete (Spirochaetae) with two spiral flagella between both membranes. (I) The same as (H) in cross-section. (J) *Streptomyces* (Actinobacteria) filamentous growth with aerospores. (K) *Micromonosporea* (Actinobacteria) filamentous growth with spores forming in the substrate. (L) *Myxococcus* (Myxobacteria) cells aggregated in a mound, releasing exospore cells. Scale bar (A–I) 1 μm, (J–L) 10 μm.

important in soils because some are able to utilize heterocyclic and aromatic molecules (such as lignins and its by-products). Similarly, the **Clostridia** are able to grow on a variety of nutrients, such as proteins, polysaccharides, purines and small soluble molecules. They are anaerobic endospore-forming bacteria which release malodorous metabolic by-products, such as acetate, butyrate, butanol, acetone and many other molecules. More common in soils are genera of the endospore-forming aerobic rods (e.g. *Bacillus*) and Gram-negative facultative anaerobes such as *Klebsiella* and *Aeromonas*.

Bacteria are important symbionts of other organisms in decomposition food webs. Numerous anaerobic or anaero-tolerant species occur in the intestinal tract of animals. They are common in the intestinal tract of soil invertebrates (such as termites) where they are assumed to contribute to digestion, as intracellular or intranuclear symbionts of some protozoa, or as extracellular symbionts or commensals of ciliates and other protozoa. For example, the surface ridges of several Hypermastigea or the surface of the sand ciliate *Kentrophoros* are covered in symbiotic bacteria. There are also species-specific associations between some entomophagic nematodes (Heterorhabditidae and Steinernematidae) and bacteria of the genus *Xenorhabdus*. In this example, the bacteria are released from the nematode when it is ingested by an insect. The released bacteria produce an antibiotic in the insect, that kills the insect host and prevents colonization and growth in the insect of other bacterial species. The nematode then feeds and reproduces on the dead and decomposing insect. This interaction perpetuates a mutual interdependence between the two species.

Roots, Fine Roots and Root Hair Cells

The most significant biomass in most soils, with the exception of certain arid soils, is plant roots. They are a major source of litter input into the soil, as dead roots and from root exudates (Coleman, 1976; Fogel, 1985). Although they do not contribute directly to decomposition of litter and soil organic matter, roots influence the soil physical, chemical and biological environment. Roots play a significant role in holding soil peds together, and in removing gravitational and capillary water and soluble minerals and nutrients. The roles of roots are several: (i) to provide a mechanical anchor and support to the plant aerial organs; (ii) for survival of the plant after aerial organs are senesced or damaged; (iii) to obtain dissolved oxygen and soluble nutrients for the plant; and (iv) to obtain water for maintaining plant cell osmotic pressure, which is necessary for vascular tissue transport and mechanical support of the aerial organs. It is primarily at the level of root interactions and scavenging for soil nutrients that plant species competition occurs (Casper and Jackson, 1997).

The root cap of a growing tip consists of a thick layer of pioneer cells (or root border cells) which secrete substances including mucopolysaccharides into the surrounding soil. These serve the double purpose of mechanically and chemically protecting the growing tip, as well as baiting soil organisms away from the apical dividing cells, by providing a rich source of nutrients (Hawes, 1998). The release of exudates from roots is particularly significant near the growing tip from these pioneer cells. Exudates and nutrient loss also occur from the body of the root network (Merbach *et al.*, 1999; Rowan *et al.*, 2000), although with probably a different composition and origin. The release of organic compounds into the soil adjacent to roots creates an enriched habitat which is directly influenced by the root network and its symbiotic organisms. This habitat is called the **rhizosphere**. The rhizosphere is believed to interact with the soil saprotrophs, both in competing for soluble nutrients and in providing a habitat modified from the bulk soil (Alphei *et al.*, 1996; Grayston *et al.*, 1996; Bonkowski *et al.*, 2000a).

Growth of roots is guided by the fine roots and root hair cells which explore the soil for moisture and nutrients. Root hair cells extend laterally from the root into soil pore spaces and contribute to guiding root growth and in holding together peds, as well as increasing the surface area for soil solution absorption. The abundance of root networks through the soil profile varies for plant species and responds within days to changes in soil water (Hendrick and Pregitzer, 1992; Lou *et al.*, 1995; Bauhus and Messier, 1999). In times of drought, deeper roots may proliferate where deeper soil moisture is retained, whereas near-surface networks develop in wetter surface conditions (Espeleta and Eissenstat, 1998; Lopez *et al.*, 1998). The fine roots of the network are continuously exploring the soil for better conditions. Fine roots in less favourable parts of the habitat are abandoned to senesce and decompose, in favour of growth in another part of the root network. Fine root production and turnover exceed biomass production in the aerial organs during certain seasonal growth periods of plants or in forest trees (Rytter, 1999). Besides nutrient and water availability, temperature, depth and age of the roots affect the longevity and turnover rate of fine roots (Espeleta and Eissenstat, 1998; Lopez *et al.*, 1998; Pregitzer *et al.*, 2000). The same principles seem to apply to mycorrhizal roots (Majdi *et al.*, 2001). It is unclear whether fine root growth and root biomass build-up need to be tightly linked with leaf biomass accumulation (Nadelhoffer and Raich, 1992).

During normal plant development and growth, both morphological and physiological changes occur in the tissues. These changes are reflected in the distribution of biomass in plants, so that over time there are shifts in where new growth occurs. These can be measured in shifts in root biomass, carbohydrate storage in different tissues, rate of carbon fixation, root versus shoot biomass, and total leaf area versus plant biomass (Coleman *et al.*, 1994). Superimposed on these trends, there are shifts in growth allocation between tissues and organs, in response to environmental conditions (McConnaughay and Coleman, 1999). The extent of the developmental plasticity in biomass allocation varies between plant species, and it is well known in agricultural crops. It would appear that plasticity in biomass allocation is more responsive to nutrient limitations from the soil solution than to water, CO_2 and light availability (see McConnaughay and Coleman, 1999).

The effect of plant roots on soil water balance directly affects water availability in the soil habitat for other organisms. Roots are a major sink for soil water, which can exert great osmotic pressure in removing water from the soil solution (Leuschner, 1998). The passage of the water solution into plant root cells and vacuoles occurs through and interacts with the **apoplast**, and is directed into the vascular tissues (Sattelmacher, 2001). Both diurnal and seasonal changes in soil water potential caused by root absorption (Lou *et al.*, 1995; Salisbury and Ross, 1999; Ishikawa and Bledsoe, 2000) will affect the activity patterns of the edaphic community. In this process, one needs to understand the role of plant species physiology interacting with the environment.

Summary

Understanding the functioning of cells is fundamental to the biology of unicellular species which predominate in the soil. Most of the soil invertebrate species are small enough to be affected by the same constraints as the protists that share the habitat. The biology of some organisms that participate in decomposition food webs was described. These included taxonomic groups that are symbionts with plants or animals which participate in decomposition. In protozoa, we described the symbiotic groups in Trepomonadea, Retortomonadea, Oxymonadea, Trichomastigea and Hypermastigea; the free-living Percolozoa, Euglenids, Bodonids and Ciliates; and the very abundant and diversified Amoebae and Cercozoa. In chromista, only the Oomycetes were considered. The fungi are functionally diverse despite the simplicity of their morphology. The Chytridiomycetes, Zygomycetes, Glomales, Ascomycetes and Basidiomycetes were considered. A selection of the most abundant interstitial invertebrates included the Nematoda, Rotifera and Gastrotricha, Tardigrada and Onycophora, Oligochaeta, Collembola and the soil mites. The prokaryote cell is distinct from eukaryotic cells, and examples of bacterial genera in the soil were given. Lastly, the interaction of soil organisms with plant roots, especially through exudates from fine roots, should be remembered as an important part of the soil. An appreciation of the diversity and complexity of the biology of soil species is needed to grasp the many ways in which these species interact together in the soil.

Suggested Further Reading

Carlile, M.J., Watkinson, S.C. and Gooday, G.W. (2001) *The Fungi*, 2nd edn. Academic Press, New York.

Cavalier-Smith, T. (1998) A revised six-kingdom system of life. *Biology Reviews (Cambridge)* 73, 203–266.

Dindal, D.L. (1990) *Soil Biology Guide*. John Wiley & Sons, New York.

Fuller, M.S. and Jaworski, A. (1987) *Zoosporic Fungi in Teaching and Research*. Southeastern Publishing Corp., Athens, Georgia.

Grassé, P.-P. (1949–1995) *Traité de Zoologie: Anatomie, Systématique, Biologie*. Masson et Co., Paris.

Hawes, M.C., Brigham, L.A., Wen, F., Woo, H.H. and Zhu, Y. (1998) Function of root border cells in plant health: pioneers in the rhizosphere. *Annual Review of Phytopathology* 36, 311–327.

Hopkins, S.P. (1997) *Biology of the Springtails*. Oxford University Press, New York.

Lee, J.J., Leedale, G.F. and Bradbury, P. (2001) *The Illustrated Guide to the Protozoa*, 2nd edn. Society of Protozoologists, Allen Press, Lawrence, Kansas.

Moore, D. (1998) *Fungal Morphogenesis*. Cambridge University Press, Cambridge.

Ruppert, E.E. and Barnes, R.D. (1994) *Invertebrate Zoology*, 6th edn. Saunders College Publishing, New York.

Schlegel, H.G. (1993) *General Microbiology*, 2nd edn. Cambridge University Press, Cambridge.

Walter, D.E. and Proctor, H.C. (1999) *Mites: Ecology Evolution and Behaviour*. CAB International, Wallingford, UK.

The Habitat

2

'Through a Ped, Darkly'[1]

If one has never observed soil closely before, it is a useful exercise to place a pea size portion of surface soil under a dissecting microscope and observe its structure. It will probably appear dark, somewhat coloured, with small rootlets and filaments extending out. The uneven surface suggests an intricate labyrinth inside, which holds the secrets to soil processes and harbours the organisms responsible. One can then add a little water to disaggregate the soil and observe the mineral composition. Soil mineral particles will appear as microscopic rocks and crystals. Interspaced between are amorphous organic matter, rootlets, filaments and possibly invertebrates. At higher magnifications (×200) on a compound microscope with phase contrast, one will observe countless moving organisms mostly between 100 and 5 μm. At ×400, the bacteria can be distinguished as small objects of various sizes, some mobile, others not. This chapter discusses the mineral and organic composition of the soil and the many habitats it provides for biological organisms involved in decomposition.

The soil is responsible for **decomposition** of dead organisms and material derived from living tissues, that releases nutrients for roots and growth of soil organisms. Roots constitute the bulk of living plant biomass and provide plants with water, oxygen and other essential nutrients from the soil. Roots also need soil to anchor the aerial portion of

[1]Coleman, D.C. (1985) Through a ped darkly – an ecological assessment of root–soil–microbial–faunal interactions. In: *Plants, Microbes and Animals.* Blackwells, Oxford, pp. 1–21.

plants. Soil interstitial species responsible for decomposition are adapted to this particular habitat. Their trophic interactions release complex organic matter into simpler more soluble molecules, which are accessible to plant roots and their symbionts. One by-product of decomposition accumulates as chemically resistant **humus**. Another by-product of their respiration accumulates in the atmosphere as carbon dioxide, which is required for photosynthesis. The production of biologically useful inorganic molecules from organic compounds, as a result of biological activity, is termed **biomineralization** (Fig. 2.1).

The soil is responsible for irreplaceable ecosystem services, such as water filtration, food production, recycling of nutrients through decomposition, and detoxification of chemicals. However, soils can be variously abused by agricultural overexploitation, chemical pollution or poor management. As our global human population density increases, our demands from the soil and the impact on soil ecosystems are exacerbated. The complexity of the soil, and of decomposition, is illustrated by the hundreds of species of bacteria, protozoa, fungi and invertebrates which can be found in just a few grams of most soils. Soil processes in nutrient cycling, carbon storage and the return of C as CO_2 to the atmosphere sustain primary production upon which organisms, including humans, depend. The challenge before us is to understand soil ecology sufficiently, so as to manage these processes sustainably for the future.

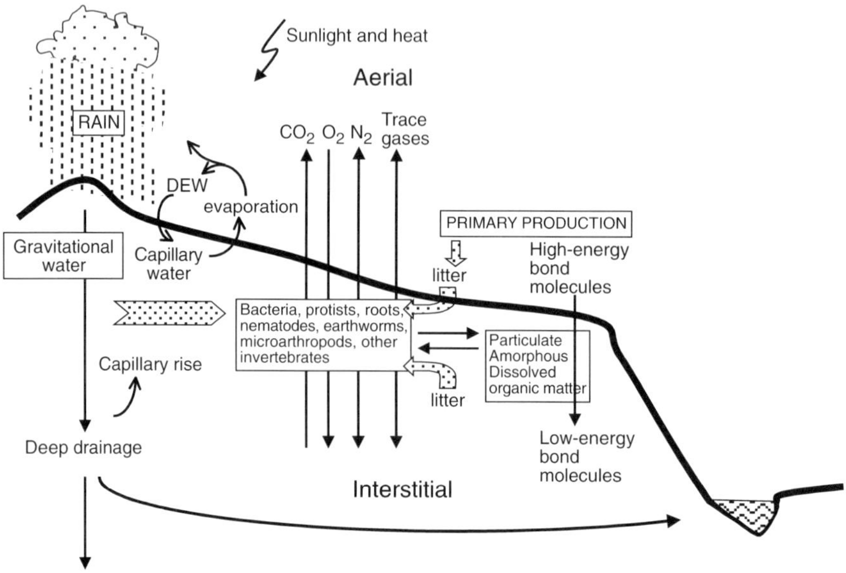

Fig. 2.1. Schema of the ecosystem and flow of water, air and nutrients through soil, across a hill slope to a riparian zone.

The soil habitat

For our purpose, **soil** is the result of a combined mixture of organic matter (derived from living organisms) and unconsolidated minerals (composed of varying proportions of sand, silt and clay) and it provides habitats for thousands of soil-specific species. This definition is more restrictive than the one proposed by the *Glossary of Soil Science Terms* (see http://dmsylvia.ifas.ufl.edu). The **habitat** is the direct soil environment with which a species is interacting. The soil habitat has **edaphic** properties, i.e. pertaining to the soil, or determined by factors inherent to the soil. These properties are the result of interactions between the soil mineral composition, living organisms and their decomposition. The soil provides a porous three-dimensional matrix, with a large surface area, for soil-dwelling species. Thus, species that reside within the soil matrix are **interstitial** species.

Soil Mineral Composition

This is the solid **inorganic matrix**, which consists of **clay** (crystalline mineral particles <2 μm in size), **silt** (soil mineral particles of 2–50 μm), **sand** (soil mineral particles of 50 μm to 2 mm) and **gravel** (soil mineral particles >2 mm). Depending on the proportions of each mineral fraction, a soil is classified into textural types. This changes both the physical and chemical properties in the soil and consequently affects its biological properties. The physical and chemical properties of the mineral component are closely linked to soil **texture** and **structure**, two key factors in decomposition, nutrient release and fertility for plant growth. Texture refers to the percentage of clay–silt–sand proportions in the soil (Fig. 2.2). Various proportions have been given names, referred to as a **soil texture types**. Knowing the soil type is a useful indicator of how workable or malleable the soil is. Sandy soils are friable and clay soils are heavy and difficult to work, especially when wet. The structure of the soil refers to how these soil mineral components aggregate into larger units. The natural aggregates, called **peds**, vary in size and shape with depth and a variety of parameters (mineral soil chemical composition, mechanical and physical conditions, climatic regime and organic composition). The extent of plant rootlets, fungal hyphae, cell filaments and secretions from living organisms all affect the size of peds and soil structure (Fig. 2.3). Secretions and organic matter tend to make mineral soil components clump more and increase mean ped size and stability. The rootlets, hyphae and other filaments form a mesh through the peds and hold the peds together.

The properties of clays are different from those of larger mineral components. Rocks, stones, gravel, sand and silt are just incrementally smaller size fractions of **primary minerals**. They are the result of

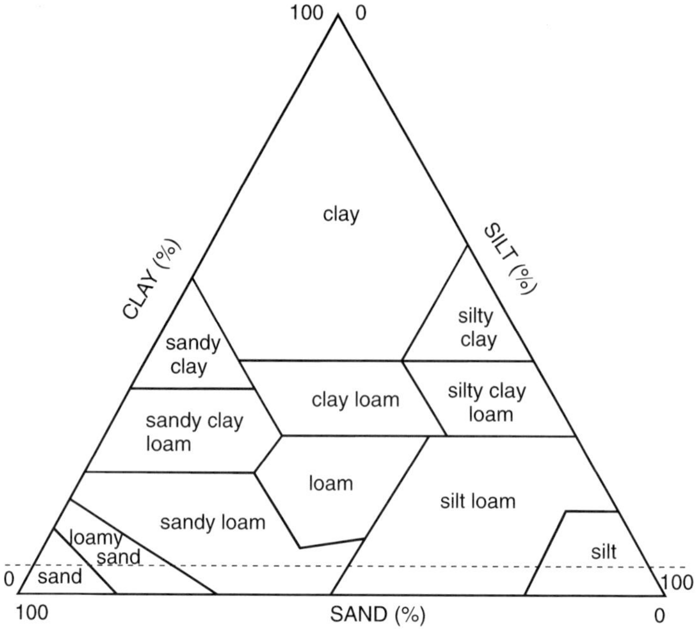

Fig. 2.2. Soil mineral particle composition and resulting soil texture. The soil texture triangle places soils into categories based on the percentage composition of clay, silt and sand. To read the triangle, find the intersection of any two components. An example is illustrated (dotted line) with a soil that was composed of 65% silt, 30% sand and 5% clay. This triangle was proposed by the Soil Survey Staff (USA) and has since come into general use (see Soil Survey Staff, 1998).

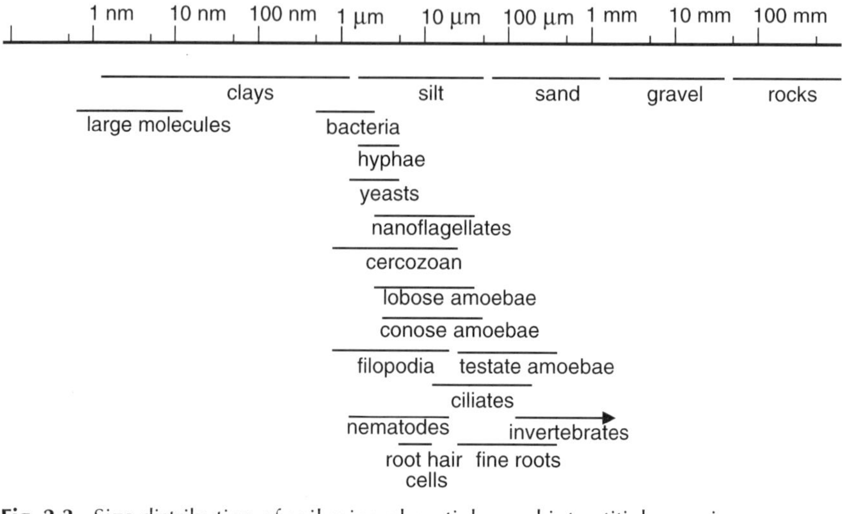

Fig. 2.3. Size distribution of soil mineral particles and interstitial organisms (measurements along a logarithmic scale).

mechanical erosion (breaking and fragmenting) of the parent primary minerals. When primary minerals (such as quartz, feldspars, micas and ferromagnesians) are chemically weathered, they produce **secondary minerals**. This erosion of primary minerals is referred to as **weathering**. Secondary minerals mostly result from chemical reactions of the primary minerals with water and dissolved ions (hydrolysis, oxidation, hydration and dissolution) (Fig. 2.4). The secondary minerals which form are **clays**, and consist of crystalline aluminosilicates, other crystalline minerals and various free oxides, such as precipitates of the soluble crystals of monosilicic acid (H_4SiO_4). These minerals have crystalline or amorphous forms with chemical and physical properties different from those of the parent primary minerals. Clays have negatively charged sites with hydroxyl ions or oxygen and will adsorb cations such as K^+, Mg^{2+}, Ca^{2+}, Fe^{3+} and Al^{3+} in their hydrated forms. The charge of some clays depends on the pH of the soil solution. In tropical soils and those derived from volcanic rock, the charges of clays are more variable. The hydrated cations that adhere on to the surface of clay particles are strongly held, especially if the clay anion charges are high. The hydrated cations further from the clay surface are held less strongly. Thus the former are not available to the soil water solution, but the more weakly held cations are exchangeable with soil water. Only the exchangeable cations which can be released to the soil water are available to soil organisms. This difference between firmly held and exchangeable cations on clay surfaces is described by the Gouy–Stern or Gouy–Chapman double layer model. The quantity of exchangeable cations held by clays is important to determine what proportion are available to living organisms. Clays also interact with the nucleic acids, amino acids, glycoproteins and polysaccharides from the litter (Cheshire *et al.*, 2000).

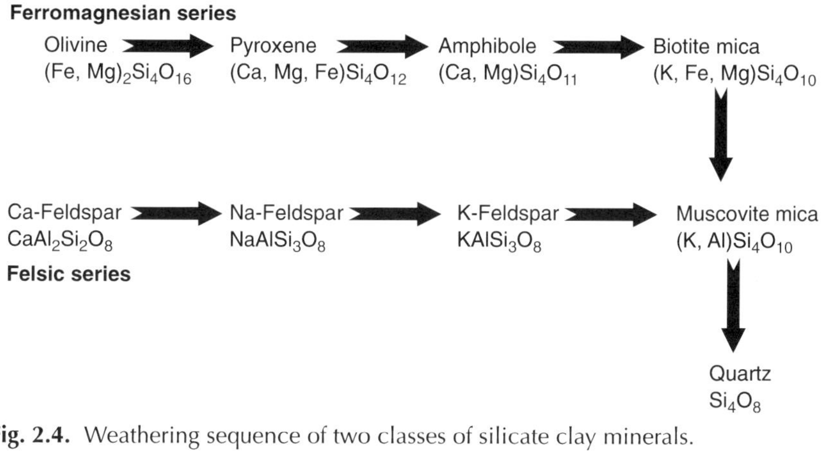

Fig. 2.4. Weathering sequence of two classes of silicate clay minerals.

The types of clay minerals and their proportions impart the soil with many of its chemical properties and impact the biological habitat. One reason is their very high surface area due to their small size. Thus, even when clays are not a major component of the soil type, they provide a large surface area for binding organic molecules and ions, and affect soil chemical properties. Carbonate salts are important soil components in some areas. These mineral deposits often have a biological origin. They were produced by protists (such as prymnesiophytes and foraminifers) as protective covering for the cells or for skeletal support. These deposits form the chalky soils and affect soil chemistry and decomposition rates. In arid and semi-arid areas, $CaCO_3$ can deposit as a surface crust. In wet soil, they are in equilibrium as follows:

$$CaCO_{3(s)} \rightleftharpoons Ca(HCO_3)_{2(aq)} \rightleftharpoons H2CO_{3(aq)} \rightleftharpoons CO_{2(aq)} \rightleftharpoons CO_{2(g)}$$

The $Ca(HCO_3)_2$ form is lime and it is a buffer used to maintain soil at a neutral pH (6.5–7.5). When soil CO_2 levels are high, for instance in some water-saturated fields or deeper in the soil, the pH can be a little more acidic as more CO_2(g) dissolves.

Ped structure

The overall architecture of the ped determines the amount of space occupied by air and water spaces, and that filled by organic detritus and mineral particles. The spaces that can be occupied by air and water are called the soil **pore space**. Soil texture and structure affect the amount of pore spaces in the soil and the size of pores. **Tortuosity** of soil pores refers to the semi-continuous network, or reticulum, of these spaces. Soil organisms live inside the pore space reticulum and on the particles. The **pedosphere** refers to this habitat in the peds. However, not all pore spaces are large enough to hold living organisms. At any one time, a substantial fraction of the spaces are not accessible to aeration, nutrient flow or entry by living organisms. The **habitable pore space** refers to the fraction of pore spaces that is accessible to soil organisms (Fig. 2.4). It is possible to estimate the volume of some of these spaces.

If soil only consisted of solid material, it would have the density of its solid constituents. However, because it consists of solids and pore spaces, the weight of a volume of soil is much less than the density of the solid matrix alone. The **bulk density** is an estimate of how tightly packed the soil structure is. It is the dried weight of a known volume of soil expressed in g/cm^3, or megagrams/m^3 (Mg/m^3). Bulk soil density ($S\rho$) is obtained as $S\rho = W_{Dry}/V_{wet}$, where W_{Dry} is the weight of soil dried to constant weight at 105°C and V_{wet} is the **bulk volume** of the sample before drying. The lower the bulk density, the more loosely packed the soil is, and the more pore spaces there are. The bulk volume is an estimate of total volume of soil solid and pore spaces. This gives no indication of the mean size distribution of pore spaces, just an indication of how compact the soil is. **Porosity** is the ratio of pore space to bulk volume. In general, soil bulk density ranges between 0.9 and 1.7 g/cm^3 with corresponding air porosity of 35–5%. The **particle density** ($P\rho$) is the mass of a volume of soil excluding the pore spaces, i.e. the mass of only the solid matrix. The difference between the bulk density and particle density is an estimate of the percentage pore space (A) in the soil: $A = (P\rho - S\rho)/P\rho \times 100$. An average value of particle density is 2.65 g/cm^3, with a usual range of 2.5–2.8 g/cm^3.

Soil pore spaces are occupied by air and water. The more moisture in the soil, the more space is occupied by water (Fig. 2.5). The **air space** volume can be obtained from the change in volume of a water column. When a known volume of soil is added to a known volume of water, the change in volume is the amount of water displaced by the soil solids. This difference is the sum of the soil solid particles (mineral and organic) as well as its water content. The more wet a soil, the less air volume in the pore space. Thus, dry soils which have more air spaces cause less volume change. By obtaining a series of measurements at different moisture contents, one can estimate the fraction of pore space occupied by air and water. When all pore spaces are filled with water, no air spaces remain and the soil is water saturated. Therefore, water saturation interferes with gas exchange between soil and the atmosphere. **Air porosity** depends on soil water content as well as soil structure. The presence of macroinvertebrates (such as earthworms) and burrowing small mammals can greatly affect soil porosity to air and the quality of soil air, by creating large gas exchange spaces.

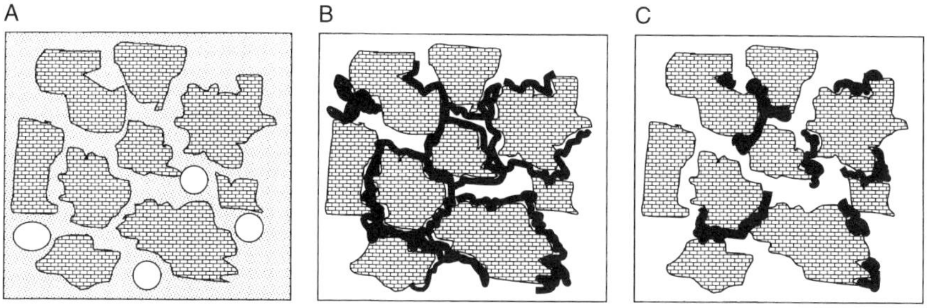

Fig. 2.5. Soil pore spaces. Water moisture content of soil during drainage. (A) Soil pore spaces flooded with gravimetric water after heavy rain; air bubbles remain. (B) Most gravimetric water has drained, with air pockets reforming; capillary water remains. (C) Drying soil, with some capillary water remaining and mostly air spaces; relative humidity will decrease and soil temperature will be closer to air temperature.

Soil Air

This is the gaseous phase within pore spaces. Its composition depends on above-ground conditions as well as soil conditions. The outside humidity and temperature affect soil moisture content and air space volume. Soil biological activity affects how much oxygen and nitrogen is consumed and how much carbon dioxide and other gases are produced. **Aeration** of the soil is the gas exchange between atmospheric air and soil air (Fig. 2.1). Aeration helps to eliminate respiration gases and returns oxygen to maintain an aerobic environment. The **biological oxygen demand** of soil is the amount of oxygen consumed by living organisms over time in a known amount of soil. Similarly, one can estimate the biological demand for any chemical and nutrient under specified conditions. It is a measure of the extent of biological activity and of demand for a nutrient or chemical under specified standard conditions. With depth, as aeration becomes limited, oxygen decreases in concentration, and at greater depths it is absent. Thus, aerobic respiration is no longer possible and other molecules are used as electron acceptors in cellular respiration (see Chapter 1). Therefore, species composition varies with aeration and oxygen availability.

The atmosphere consists of about 78.1% nitrogen (N_2), 20.9% oxygen (O_2), 0.92% argon (Ar), with a small amount of carbon dioxide (0.033% CO_2) and trace quantities of other gases (0.003%). In the soil, the concentration of oxygen is slightly less because of demand by living organisms, and concentrations of CO_2 are slightly higher because of biological respiration. The soil CO_2 level in active soil reaches 0.15% at 15 cm, but can reach 5% and affect the pedosphere chemistry. Carbon dioxide is more soluble than other principal gases in the air. It dissolves in water and forms H^+ and HCO_3^- which is a weak acid and thus increases the acidity. However, it also functions as a buffer and can help to stabilize the soil pH, especially in the presence of charged organic matter. Other gases are present in the pore spaces, in the air or dissolved in soil water. These also accumulate mostly from digestion and respiration and consist primarily of ammonia, carbon monoxide, hydrogen, sulphurous gases and oxides of nitrogen. Organic gases can also be present in trace amounts, such as ethane or volatile derivatives from decomposition and degradation of organic matter. It is unclear to what extent abiotic reactions contribute to mineralization in natural soils. Different gases have different solubilities in water. The more soluble gases can be retained in soil water, whereas the insoluble ones can be flushed out or retained as gas bubbles, by increased water in pore spaces. Thus the solubility constant of a gas in water, and the solutes in soil water affect how fast gas exchange will occur. The rate of diffusion of gases into and out of the soil air spaces also depends on texture and structure, which change porosity and vary with depth. The overall aeration of soils is important to pedosphere health.

With depth, as aeration becomes slower and soil air composition differs from that of the atmosphere, aerobic respiration becomes impossible periodically, and at greater depth it is never possible. With depth, more organisms are anaerobic or facultative anaerobes. Some species are even intolerant to oxygen, which is a very reactive oxidant. Although some protists are tolerant to very low oxygen levels for extended periods, and others are anaerobic, the most important soil anaerobes are considered to be amongst the bacteria. This may or may not be true, as no one has measured the contribution of anaerobic and anaerotolerant protozoa and fungi to soil nutrient transfer and in decomposition. The yeasts can be important due to their abundance in deeper or anaerobic soil. These have not been studied.

One way to estimate the soil oxygen content is to measure the non-equilibrium electrical potential against a platinum electrode, or redox potential (E_{pt}). Between 450 and 800 mV, oxidative aerobic reactions are possible; at 450–0 mV, oxygen availability is limiting to some species. For example, at these levels, nitrification reactions are no longer possible because the bacteria responsible require more oxygen. Overall decomposition is slow, and iron oxides are reduced if the soil pH is acidic. From 0 to –300 mV, the pedosphere is anaerobic, no longer oxidative but a reducing milieu; iron oxides are reduced even at neutral pH. Redox measurements are more important in anaerobic soils and in submerged soils of wetlands or riparian zones in predicting biological activity.

The concentration of gases in soil air is determined with specially devised thin, cylindrical *gas/vapour probes* inserted at required depths. Air samples are drawn out into gas pouches for laboratory analysis. To monitor gas fluxes at the surface, permanent or portable chambers can be installed tightly on the surface, against a collar. There are two types of measurements possible. One is to allow gases to accumulate for a specific time, then to remove a gas sample for laboratory analysis with a gas chromatograph. This is a non-steady-state sample, usually obtained from *static vented chambers*. A steady-state measurement can be obtained by maintaining a continuous air flow through the chamber. Some portable devices include an infrared gas absorption (IRGA) detector for CO_2. The usefulness of any of these instruments depends on accurate calibration of the protocol against known quantities of gas standards.

Water Content

Cells need to be in water in order to function. Water is a liquid medium for cells to remain hydrated, and for nutrients to dissolve in. The biological activity of soil depends on aeration as well as adequate water in pore spaces. The source of water is from snow melt, rain and dew, which depend on temperature fluctuations and air humidity. In some locations, subterranean lakes or streams also contribute significantly to soil moisture. Most soils are hygroscopic and will absorb water vapour from the air.

Water is held on to organic matter, cell membranes and minerals by **hygroscopic forces**. These consist of hydrogen bonds, van der Waals forces, ion hydration spheres, molecular dipole forces and osmotic potential. Part of the water is bound chemically to soil particles. It is referred to as **adsorbed** water, having very strong negative potential, i.e. it is not naturally lost from soil and is held strongly. It is captured from air space humidity or retained by soil organic matter (SOM) and soil mineral matter. It provides a thin film 0.2 μm thick which covers all particle surfaces. **Film water** is not available to roots and cells because it is too strongly held, but it guarantees a wet and humid layer for living cells in the soil. It is retained even in most arid soils. **Capillary** water is drawn into pore spaces by hygroscopic forces, but is held less strongly. It is retained even in dry conditions, especially in smaller pore spaces <10 μm diameter. It is the main source of humidity in air spaces, so that pore space humidity is rarely below 98%. **Gravitational water** is free to drain through soil under the influence of gravitational pull. Vertical and lateral displacement of gravitational water occurs between 0.01 and –0.03 MPa soil water tension, and contributes to replenishing underground and above-ground water reserves. It is an important factor in loss of nutrients to leaching and erosion of the soil profile. Water pressure differences exist due to relative changes in elevation and atmospheric air pressure. These are exerted on the gravitational water. Total gravitational and capillary water content (%W) is measured as the percentage of weight loss due to oven drying to constant weight at 105°C. The percentage water content is obtained from the weight loss. It is expressed as $(S_w - S_d)/S_d \times 100$, where S_w is the weight of the soil sample before drying; and S_d is the weight of oven-dried soil. **Soil water potential** is the sum of gravitational, matric (capillary and adsorbed water) and osmotic potentials. The force required to remove water from soil can be used to estimate how much water is available to living organisms. The soil water potential is affected by SOM content, but also soil texture and structure. The amount of water removed over a range of pressure differentials, by creating suction, is used to measure how strongly water is held by the soil. Alternatively, faster analysis is carried out with a dewpoint potentiometer (Decagon, Pullman, Washington, USA). Graphs of soil water content against the soil **matric potential** (i.e. the force applied in N/m^2) are called 'soil water characteristic curves'. The shape and position of curves vary with soil texture and provide a measure of how difficult it is for the remaining water to be removed by living organisms (Fig. 2.6).

The **osmotic potential** of soil water indicates how easily dissolved nutrients can be removed by cells. It is the amount of pressure that needs to be applied to a solution of known molarity to raise the partial

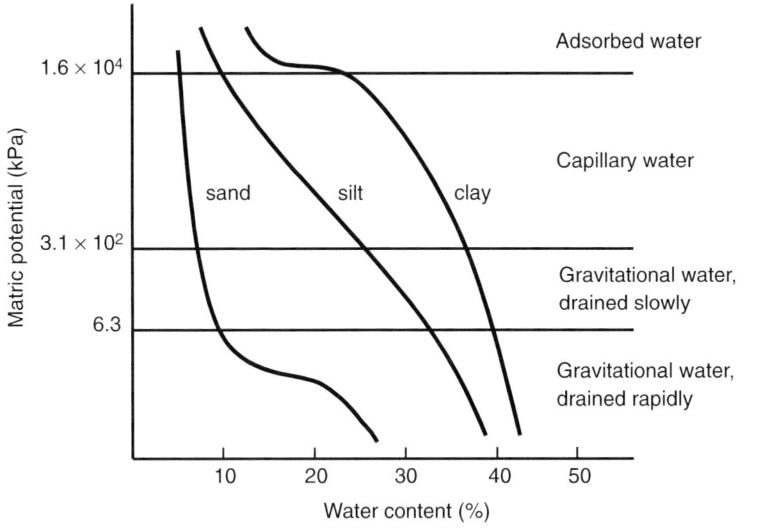

Fig. 2.6. Soil water release curves. Matric potential curves for three arbitrary types of soil. As the suction increases (the matric potential in kPa), the amount of water remaining in the soil decreases. At low suction, silty and sandy soils retain less water than clay soils. Clay soils in general require more suction (higher matric potential of clay soil) to remove the soil water.

pressure of water vapour above the solution to that of pure water. The matric potential can be measured experimentally from water pressure changes between a soil solution and pure water. The less water, the more tightly the remaining water is held by charged ions and organic matter. In very dry soils, the remaining water is no longer available to cells, including plant root cells, once the water tension is too great. For dilute solutions, the osmotic pressure can be calculated as $\psi_s = RTC$, where ψ_s is the osmotic pressure in kilopascals (kPa), R is the gas constant (8.31×10^{-3} kPa m^3/mol K), T is the absolute temperature in Kelvins (K) and C is the molarity in mol/m^3.

One can calculate threshold osmotic potentials for each species. It is the potential above which that species (be it plant root, hyphae, protozoa or bacteria) is no longer able to acquire water from its environment. This value varies between species, and generalization are not possible (Harris, 1981). Certain species are better adapted for moist conditions, while others are only active in low soil moisture conditions. A consequence of changes in soil water content is to affect the rate of diffusion of solutes and gases. **Diffusion** is the movement of solutes (or particles in general) along a concentration gradient, from high to low. When there is sufficient capillary and gravimetric water, solutes can diffuse along concentration gradients. Diffusion of dissolved organic matter through the soil solution and its flow with gravitational water distributes nutrients through the pedosphere.

Natural loss of water

Loss of water from soil is due to **evaporation** into the air, and draining through the soil. Gravitational water near the surface and capillary water contribute to evaporation. A significant portion of soil capillary and gravitational water is removed by plant roots. **Evapo-transpiration** is the total quantity of soil water lost by evaporation to the atmosphere and plant transpiration over a time period. It is expressed in grams of water lost per unit area per unit time. Some gravitational water will also drain into terrestrial and subterranean streams and lakes. **Leaching** is the loss of soil soluble matter with gravitational water drainage. When large volumes of rain are drained, certain physical aspects of water displacement strongly alter the ecology of an area. **Hydrodynamic dispersion** is the change in solute concentration caused by such a flow of water. As gravitational water flows, it removes soluble molecules and particles along pore space surfaces. **Mass flow** of nutrients or organisms can occur with a net directional movement of water. **Darcy's law**, the rate of viscous flow of water in isotropic porous media (such as soil), is proportional to, and in the direction of the hydraulic gradient, i.e. in the direction of water flow. It is stated as $F = (H_s + H_w)/H_s \times \text{area} \times K$; where F is the flow rate in ml/s, H_s and H_w are the height in cm of the soil and water column, area is the surface area of the column in cm^2, and K is the rate of water delivery to the column in cm/s. **Flux density** is the rate of transport or movement of something perpendicular to a unit area. It can be used to calculate the amount of water, solutes or cells being moved by the flow of gravitational water, and is expressed as g/s/m^2 or cells h/m^2. One consequence of mass flow is the deposition of clays and silt from one area to another, and contributes to soil erosion. It is also a factor in leaching soil nutrients and dispersing soil organisms. Figure 2.7 represents drainage patterns for two hypothetical soils, one with higher sand content and rapid drainage, the other with more compacted pore structure and poorer drainage.

Higher temperatures and lower air humidity both increase the rate of evapo-transpiration. Upon desiccation, the osmotic pressure of soil water increases and it becomes unavailable to cells for growth and nutrient uptake. However, cells can remain in a wet film protected from desiccation, especially in the smaller pore spaces <15 μm diameter. Most soils are able to retain sufficient capillary and film water. One way to measure soil water retention is to compare air-dried soil with oven-dried soil. **Air-dry** soil is in equilibrium with air humidity at a given temperature and pressure, and retains film water and some capillary water. Once **oven dried** to constant weight, by evaporating water in a dry oven at 105°C, only adsorbed water is retained. This is the value normally used to express the amount of soil (soil dry weight) in a sample.

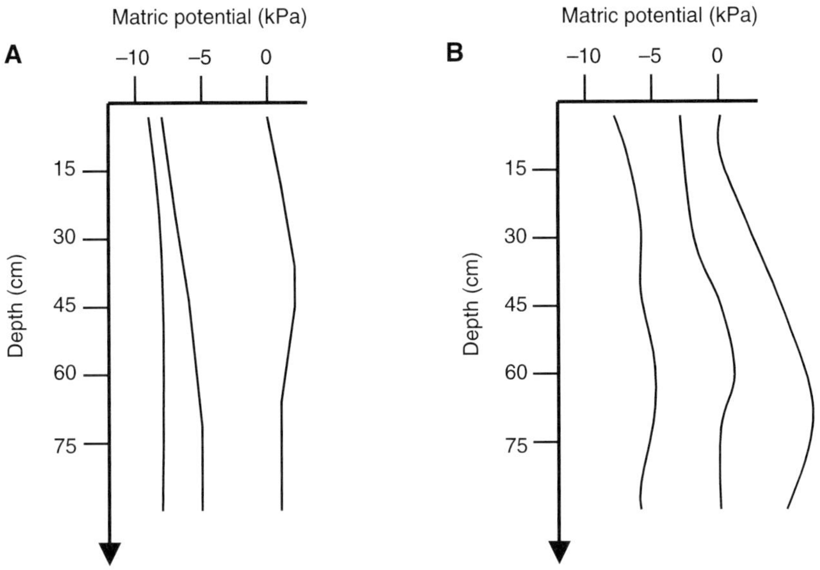

Fig. 2.7. Soil drainage over time with depth for two hypothetical soil types. (A) Well-drained soil high in sand content, such as a sandy loam. (B) Soil with higher silty–clay content and slower infiltration, especially at about 60 cm depth. Lines represent 3-day intervals.

Different methods and instruments exist to obtain estimates of soil water and soil solution content. The correct procedure depends on the amount and type of soil to be analysed and whether field values or laboratory values are required. Sources of information for choosing the adequate protocol are provided in *Methods of Soil Analysis*, *Handbook of Soil Science* and *Standard Soil Methods for Long Term Ecological Research*, referenced at the end of this chapter. Field instrumentation to gather data about processes at the scale of an ecosystem or watershed can be inadequate to provide data on smaller study sites. Small plots may need to be measured individually due to microclimate variations in rainfall or temperature.

Electronic devices are commercially available to measure soil water content directly, by time domain reflectometry (TDR). The apparatus uses microwave frequencies to obtain a measure of the apparent dielectric constant of the surrounding soil (see review in Noborio, 2001). These are available as portable probes for spot measurements or permanent probes for long-term continuous data acquisition. The measured values need to be calibrated, *in situ* and in the laboratory, against soils with known water content. The method is relatively insensitive to mineral soil composition, but is affected by SOM content and can be affected by some clays. Measurements of soil water potential are made with tensiometers in wet soil (0 to –0.48 MPa) and with resistance blocks in drier soils (–0.01 to –3.0 MPa). Tensiometers are water-filled tubes with a porous ceramic cup at one end. The top end of the tube is sealed with a vacuum valve, used to create a negative pressure (or tension) in the water tube. The tensiometer is equilibrated against water to zero tension. The porous cup is inserted at the required depth in the soil. The soil draws water out of the cylinder, creating a negative tension which is read from the instrument. The device only functions if there is a continuous water contact from the tube to the soil. If soil dries too much, the device will empty and it will have to be reset. In drier conditions, resistant block devices are used. These consist of a gypsum or fibreglass block buried at the required depth. The block is connected to a recording device on the surface by wires. The device measures electrical conductance and is therefore sensitive to changes in salinity and will not function in some soils with high salt content.

Soil temperature

Energy loss from the sun is the principal source of heat into the atmosphere and on to the planet surface. Much of the radiation is reflected off the surface of the planet and contributes to atmospheric temperature. Some radiates slowly into the soil and into the planet. It is estimated that 100 years are required for energy on the surface to reach a depth of 150 m, and 1000 years to reach 500 m (Huang *et al.*, 2000). Locally, two physical parameters determine temperature changes in the top soil. One is the organic matter content and litter covering of the soil surface which shields it from direct heat. The other is the soil water content which absorbs a large portion of the radiant energy. Both organic matter and water are efficient buffers against large changes in surface temperature. Water has a relatively high specific heat capacity so that, above 15–20% water content, water is more significant in slowing the rate of heat transfer. Soil minerals and air have low conductivity compared with water and, thus, at lower water content, thermal conductivity increases with water moisture. This complicated relationship is expressed as: $K = D \times C_v$, where K is the thermal conductivity; D is the thermal diffusivity constant, and C_v is the volumetric heat capacity. Both D and C_v need to be calculated separately for the soil. C_v is obtained from measurements of the density of dry soil (ρ_s), the specific heat of dry soil (c_s) and the specific heat of water (c_w), as follows: $C_v = \rho_s (c_s + c_w)$.

Field temperature readings are obtained by inserting a hand-held *thermocouple thermometer probe* to known depths. It is best to obtain a reading just below the surface (2 cm) and at least one more at a greater depth (12 cm). This provides an estimate of the temperature gradient at the time of sampling. The gradient will change with solar exposure at dawn and dusk, or when sunlight first appears or disappears from the site. It is best to avoid these times when making temperature measurements. Permanent probes (*data loggers*) can be buried at the required depths, and programmed to make a measurement at specified times regularly. These are retrieved to download the data to a computer.

Soil Organic Matter

The SOM and soil mineral component together provide the structural matrix and the chemical environment for living organisms in the soil. The source of organic matter in the soil is litter. **Litter** is broadly defined as all that was recently living. This includes fallen leaves, woody debris and fallen bark, dead roots, cadavers, animal dung, insect frass and cuticles. It also includes secretions and excretions from living organisms, such as root exudates. The decay of fresh litter accumulates on the surface and within the soil, as **SOM**. Fresh litter and the more decayed organic matter are sources of food for many species. They support the trophic interactions between interstitial species. The process of litter chemical decay caused by these trophic interactions is called **decomposition**. This is in contrast to the chemical **degradation** of litter and SOM not caused by living organisms. Degradation is caused by chemical reactions between SOM molecules, and of SOM with clays, that lead to new molecular forms and to the aggregation of molecules into colloids and polymers. Thus degradation is caused by **abiontic** reactions.

Mechanical fragmentation of litter and its chemical digestion contribute to its breaking and shredding into smaller more amorphous particles. **Macrodetritus** refers to fragments visible by eye, and **microdetritus** refers to the microscopic size fractions. (Other terms have been proposed in the literature. Terms similar to macrodetritus include 'light fraction', 'particulate organic matter', 'macro-organic matter', 'coarse organic matter' and 'organic matter in macro-aggregates'. Terms similar to microdetritus, when not included in the above terms, include 'fine organic matter', 'micro-organic matter' and 'fine particulate organic matter'. As none of these refer to a consistent degree of decomposition or chemical composition, but to size or density, the pedantic debates are unnecessary.) With time, fresh litter is broken up into a range of increasingly small fragment sizes, down to microscopic debris and organic colloids and flocculants, as it is decomposed and degraded. A fraction of the SOM accumulates because it is difficult to digest or it is indigestible. This SOM can be referred to as **humus**. It is usually defined as all SOM exclusive of living biomass, litter and macrodetritus, i.e. the smaller and more decayed fractions of SOM.

Fractionation of SOM can provide an idea of its quality as food for decomposition. Size fractions are defined by how large the litter particles are, or by how quickly they sediment in a standard heavy liquid of known density (such as sucrose, sodium polytungstate or sodium iodide). The latter is called **density fractionation** and is very useful in estimating the amount of labile litter in the soil. It is based on differences between the density of the minerals (more dense) and the organic matter (least dense). The more biologically accessible fraction (for decomposition) is the **light fraction**. The light fraction is that which floats on top of the liquid and consists mainly of fine litter fragments, macrodetritus and microdetritus, that are decomposed to varying degrees. The intermediate fraction consists of organic matter covering and attached to minerals. This SOM is more decayed than the light fraction and is mostly humus. The heavy fraction consists of mineral particles, with little or no SOM attached. In these assays, the light fraction is usually $<1.6–2.0$ g/cm^3. The organic matter in the light fraction is the most **labile** because it is eaten and digested continuously by living organisms. The SOM associated with the intermediate fraction is less digestible, and thus more **recalcitrant**.

A chemically heterogeneous and poorly defined group of SOM is known as humic acids, fulvic acids and humin. These are obtained by chemical extraction from the soil and defined as follows: humic acids are insoluble in acid, extracted with base; fulvic acids are acid soluble and extracted with base; and humin is insoluble in base (Orlov, 1999; Harper *et al.*, 2000). Base extracts (NaOH) can be partitioned further into aromatic 'humic' and non-humic (mostly polysaccharide) fractions with polyvinylpolypyrrolidone (PVPP), which is used frequently to disrupt hydrogen-bonded complexes (such as tannin–protein complexes). There is a debate on the significance of these extracts, because it is unclear to what extent they are by-products of chemical reactions of SOM with the strong acids and bases used to obtain them (see Stevenson, 1994). From an organic chemistry perspective, the extraction conditions provide conditions for polyphenol condensations or Maillard reaction artefacts, and the hydrolysis (saponification) of the more labile bonds of glycosidic or fatty acid esters. (A less frequently used extractant is sodium pyrophosphate, which solubilizes humic SOM by chelating polyvalent cations (Al, Ca, Fe and Mg) that link otherwise soluble humic constituents into an organo-mineral complex.) Furthermore, chemical extractions of soils do not discriminate between living organisms and SOM. The soil living organisms are lysed and homogenized with the particulate organic matter and SOM in these procedures. Then the homogenate is chemically modified with extracting solutions, so that organic molecules in the soil are no longer all the same as those in the final extracts. These traditional methods are no longer recommended in ecology. Instead, and especially for ecology, the most useful method of analysing SOM composition is the 'cold water wash' extraction. This provides an unmodified extract of the most soluble and extractable fraction, which is biologically available as nutrient.

One important source of organic matter in the soil is released from soil living organisms as secretions. The sources of these secretions are numerous. They include bacterial cell wall components, protective (or defensive) mucopolysaccharides from protists, mucus from earthworms and secretions from root cap border cells. The plant root cell secretions are called root **exudates** (Hawes *et al.*, 1998). These consist of various sugars, amino acids, peptides and defensive molecules. The purpose of root tip exudates is to bait potentially infective organisms, or consumers of root cells, away from the growing root tip. Root exudates are responsible for the coagulation of peds along the root. The region adjacent to roots is thus greatly affected by the plant secretions as exudates. This thin layer of soil along the roots is called the **rhizosphere**, and is supposed to harbour an environment slightly different from the surrounding bulk soil.

Soil horizons

The soil **profile** indicates the changes in ped structure and chemical composition with depth. With depth, there are changes in physical, chemical and biological properties. These changes are caused by several factors. One most noticeable change in the top soil is the amount of litter, its particle size and the extent of litter decomposition. The uppermost layer consists of least decomposed litter, which has fallen to earth more recently. With depth, the litter is increasingly decomposed, degraded and smaller in size. Also with depth, the amount of total organic matter, of root colonization and the extent of fungal hyphae network change. The effect of organic coagulants from SOM, from exudates and from other soil organisms is to increase the natural clumping size of peds. Growth of fine roots, fungal hyphae and Actinobacteria filaments through peds forms a meshwork which further holds peds together. Thus the soil layer most active with organisms and labile litter can have large ped sizes. Therefore, with increasing depth, soils display a gradation of chemical content and physical structure. Together with changes in the amount of water retained, soil air composition, soil water acidity and redox potential, these gradients along the soil depth profiles (over several centimetres) strongly affect species composition. The thickness of each layer along the profile varies for each field, forest or region, depending on biogeography and the ecosystem. Some of the soil characteristics that can be seen in the profile are shown in Figs 2.8 and 2.9. These gradients cause changes in the structure, texture and colour of the soil along the profile. Soil scientists have devised a terminology to describe the soil profile layers, based on appearance and composition. The layers along the profile are called **horizons** (Table 2.1). There are three main horizons, the top organic horizon (A); followed by the inter-

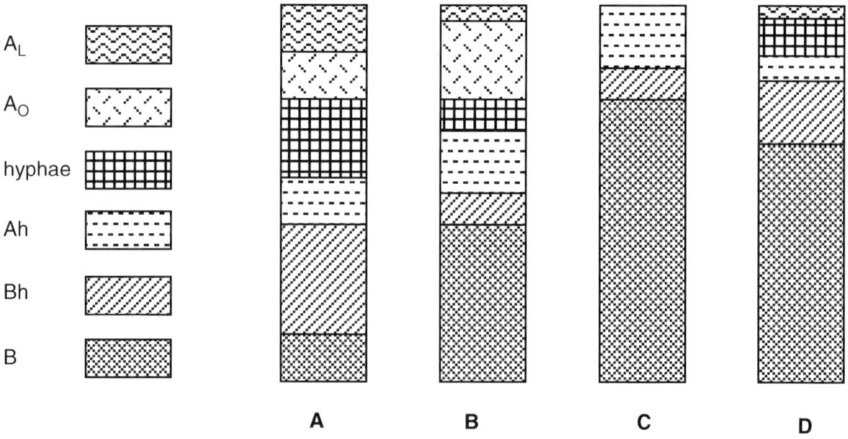

Fig. 2.8. Soil profile and horizons illustrating organic matter distribution in arbitrary soils. (A) Temperate forest site with fresh litter deposit, partly decomposed litter continuing into a densely meshed hyphal net, with organic accumulation below. (B) Forest site with mostly decomposed litter, with a thin mat of hyphal net structures, above the humic layers. (C) An agricultural field with no litter accumulation, but with some humic organic matter. (D) An agricultural field under conservation practice, showing some surface litter, with a hyphal mesh structuring a decomposed litter layer and build-up of organic matter below it.

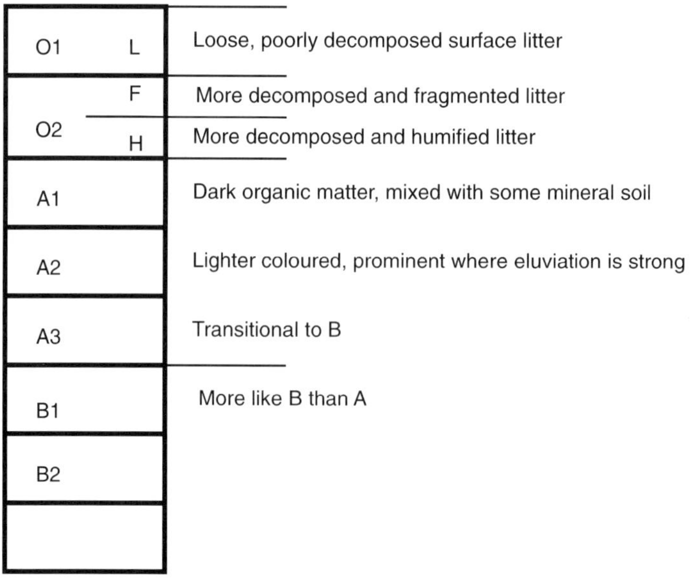

Fig. 2.9. Alternative designation for top soil organic layers. Not all horizons may be present in a profile. The O and A1 horizons contain most of the biological species, plus the rhizosphere, which is not included in these representations.

Table 2.1. International designation of the main soil horizons, with selected subhorizons.

Code	Description
A	Surface horizon with organic matter, often impoverished
(B)	More weathered than C, different structure from that of the surface horizon
B	Below A when present, enriched in fine or amorphous elements: clay, iron and aluminium oxides; may contain humus
C	Parent material from which A, B and (B) are formed
H	Peat organic horizon
G	Rich in ferrous iron, greenish-grey with rust-coloured mottles, within or at the upper limit of ground water
R	Underlying hard rock
Subdivisions (examples)	
A_L	Litter layer with indentifiable cellular tissues
A_O	Cellular tissues modified or destroyed, often >30% organic matter
Ah	Mixture of organic matter (<30%) and mineral soil
Ap	Ploughed, homogeneous humic organic matter, with clear lower boundary
Ae	Poor in organic matter, light colour, leached of clay and sesquioxides
A/B	Transitional layer between A/B, with gradient towards B
Bh	Humus-enriched B horizon

mediate B horizon, which contains much less labile organic matter and less biological activity; and the bottom C horizon which consists mostly of parent mineral particles, with virtually no biological life and the most recalcitrant SOM, if any. Each horizon is subdivided into subhorizons to reflect better the composition and layering. For ecologists, the A horizon subdivisions are the main compartments of biological activity, with little in the deeper B horizon. For a complete review of terminology and methodology in describing soil profiles and horizons, one should refer to the *Soil Survey Manual* (Soil Survey Staff, 1988).

Soil nutrient composition

Litter from above-ground and below-ground is the source of nutrients to soil organisms. A certain amount of minerals is also contributed from the weathering of clays. The nutrient composition of SOM reflects that of cells of the living organisms from which it came. Cells lyse soon after death of the organism or tissue. When a cell wall is present, the lysate is not released into the soil immediately. Digestion of the wall by extracellular enzymes secreted from some bacteria or protists is necessary to release the nutrients. Alternatively, the litter is fragmented and digested internally. These and soil trophic interactions will be dealt with in Chapter 4. However, it is worth remembering that the proteins, lipids, nucleic acids, polysaccharides and derivatives of these macromolecules are the initial source of SOM. Some protective or defensive macromolecules in living cells are digestible only by specialized species. These include sporopollenin of plant pollen; chitin of invertebrate exoskeleton, fungal cell wall and some protozoan cysts and tests; lignin, tannins and cellulosic polymers of plant cell walls; as well as cutin and suberin which waterproof plant cell walls.

As litter is fragmented and digested, it releases nutrients which are used for growth and reproduction of interstitial species. In turn, these species are consumed by their predators. The successive rounds of litter digestion and excretion and predation in the soil are responsible for decomposition. The fragments are eaten by organisms that ingest different particle sizes. Some of the ingested matter is excreted and becomes food for other species. In turn, these species are prey for predators and grazers which digest part of the food vacuoles and excrete the remains. The microscopic digested and excreted organic matter forms the amorphous SOM or humus.

The top organic horizons provide most of the labile nutrients. With depth, the organic matter becomes more recalcitrant as the more labile nutrients are removed and less soluble, less nutritious molecules remain. The labile SOM consists of molecules readily removed from the soil by living organisms and readily soluble molecules. These have a short half-life in soils as they are removed easily by living cells, but consequently they have a higher turnover rate. This pool is reduced over several years when soil is used for agriculture. As much as half the labile organic matter, or the light fraction, is used in the first year of growing a crop on a field. Therefore, it is essential that nutrients be replenished regularly with plant debris, manure, compost or chemical fertilizers.

Nutrients released from litter provide the **essential chemicals** required for growth and division of cells in the soil. These nutrients include species-specific vitamins, dissolved organic matter and mineral elements (Table 2.2). As SOM is decomposed, the amount of non-carbon atoms (especially nitrogen) decreases. They are accumulated in living organisms and released as soluble molecules and ions. The abundance of essential elements is critical to sustain cell growth. **Liebig's law of the minimum** states that the growth of a cell, or an organism, is limited by the 'most limiting nutrient', i.e. the required chemical which is least abundant. Therefore, in deeper soil horizons where the litter is most decomposed, the SOM may not support cell growth because one or more essential chemical is limiting, even in the presence of suitable carbon sources. The abundance of various nutrients or elements can be expressed in moles per soil dry weight, or in μg/g dry soil (Tyler and Olsson, 2001). They can also be expressed as ratios of total organic carbon to particular atoms, ions or molecules, such as nitrogen, sulphur or phosphate content of soils. The ratios (e.g. the C:N ratio) can be used as an index of SOM quality or extent of its decomposition. A useful illus-

Table 2.2. Typical concentration of selected ions and compounds in soil solutions. These values change with depth and with soil types.

Category	Molarity (10^{-2}–10^{-4})	Molarity (10^{-4}–10^{-6})	Molarity ($<10^{-6}$)
Cations	Ca^{2+}, Mg^{2+}, Na^{+}, K^{+}	Fe^{2+}, Mn^{2+}, Zn^{2+}, Cu^{2+}, NH_4^{+}, Al^{3+}	Cr^{3+}, Ni^{2+}, $Cd2^{+}$, Pb^{2+}
Anions	HCO^{3-},Cl^{-}, SO_4^{2-}	$H_2PO_4^{-}$, F^{-}, HS^{-}	CrO_4^{2-}, $HMoO_4^{-}$
No charge	$Si(OH)_4$	$B(OH)_3$	
Natural organic molecules in organic horizon	Amino acids and peptides, simple sugars, nucleic acids, carboxylic acids	Carbohydrates, proteins, phenolic molecules, alcohols, sulphydryls	
Pollutants	Only if accidental spillage (many organic compounds)	Herbicides, fungicides, pesticides, PCB, solvents, detergents, petroleum products	

Data compiled from various sources.

tration of these soil properties is to plot the main variables on a multi-axis graph (Fig. 2.10). Such graphs are useful visual aids when comparing several soils together (Alvim and Cabala, 1974; Alvim, 1978).

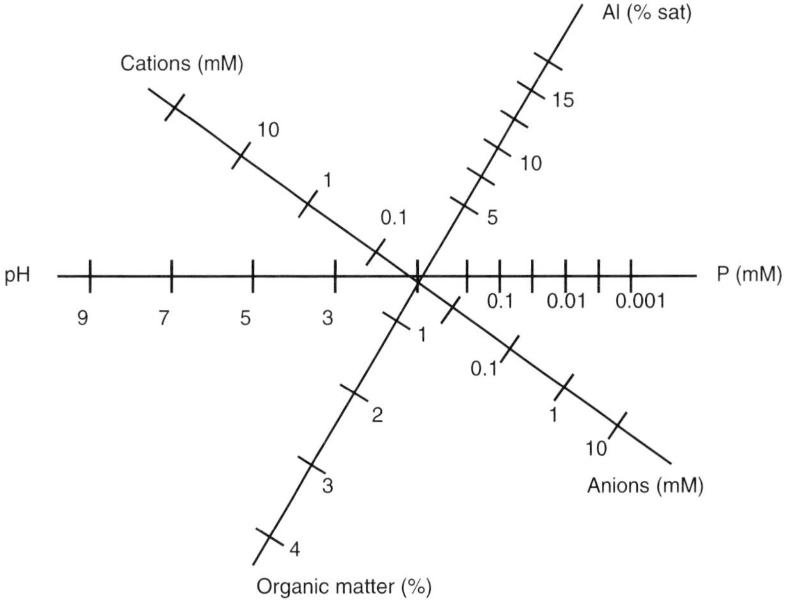

Fig. 2.10. Graphic for representation of the chemical properties of the soil and soil solution (see Alvim and Cabala, 1974; Alvim, 1978).

Charged molecules and ions released from litter are often held tightly in the SOM as **chelates** (i.e. a strong chemical bond forms between an organic 'ligand' molecule and the inorganic 'prosthetic' group). Unlike cations which can bind to clays irreversibly, those bound to organic molecules are still accessible as nutrient. That is because the organic molecules can be ingested and digested to release the nutrients. This is important because organic molecules hold far more chemical charges per mass than clays, to bind ions. Some ions are held reversibly by dipole interactions with organic molecules. One effect of charged organic molecules reversibly binding ions is to contribute to the soil **buffering capacity**. Soils with higher organic matter content are more resistant to small changes in pH. Adhesion or accumulation of molecules, atoms or ions on soil particle surfaces (to organic matter, clays and water), by chemical or physical forces, is called **adsorption**. **Desorption** into soil water depends on the **solubility** of molecules, as determined by their solubility constant K_s, or K_{sp} for sparingly soluble substances. The intake of soluble nutrients into cells is called **absorption**. Once nutrients are inside cells, they are no longer part of the SOM and are said to be **immobilized**. Thus, the ratio of immobilized nutrients in the biomass to that in the SOM and soil water is an indication of nutrient distribution in the soil. A more useful assay is the rate of biomineralization (transfer of SOM through cells) which reflects how active the biological community is.

The composition of soil solution can be obtained from commercial lysimeters. A discussion of differences in solution chemistry is provided by Lajtha *et al.* (1995, 1999), Marques *et al.* (1996) and Haines *et al.* (1982). These studies do not discriminate between minerals present in the soil moisture and those released from lysis of cells in the collected solution. Ion exchange resins and membranes have been used to estimate the amount of soil water available ions *in situ*. The buried resins absorb soil available anions or cations which are then extracted in the laboratory for quantitation. They are thought to mimic uptake of minerals by plant roots, and thus provide some quantitation of mineral availability over time. However, there is concern over interpretation of the quantities as they do not always correlate well with other measures of mineral availability (Lajtha, 1988; Giblin *et al.*, 1994).

Other procedures involve analysis of the wash filtrate from a soil sample, or ultracentrifugation of soil to separate the heavy mineral soil from most of the water (see Lajtha *et al.*, 1990). A modified technique involves centrifugation of soil in a liquid immiscible with water and denser than water, such as tetrachloroethylene (C_2Cl_4). During centrifugation, the mineral soil sediments and water is displaced to the top by the immiscible liquid. Losses of minerals and ions in the immiscible liquid need to be accounted for. In all these techniques, living cells are lysed into the soil solution. The contribution of lysed cells to the solution must be subtracted from the data. This can only be done in a general approximate way, by using mean values for cell composition.

The composition of the soil solution is important to plant nutrition, as it reflects the availability of nutrients and minerals to the roots (Smethurst, 2000) and regulates the rate of primary production. There is a triangular association between the role of the interstitial soil community in decomposing the litter, the influence of mineral soil particles and the overarching role of climate in controlling decomposition and the interstitial species composition. These interactions have been the subject of many reviews and discussions (Kalbitz *et al.*, 2000; Qualls, 2000; Neff and Asner, 2001).

Dynamics of Soil Physical Structure

So far, our description of the soil habitat summarized physical and chemical parameters that interact with soil biology. Although it may be obvious that soil temperature, water content, organic matter content and quality change over time, it is less obvious that the physical structure itself is not static. Fluctuations in water content change the size of pore spaces and displace peds. Flow of gravitational water, for instance after a heavy rain, can carry clays, silt and fine sand, and alter pore sizes and structure. Freeze–thaw cycles in cold weather also physically displace peds. Thus the pore reticulum is continuously changing, simply by physical forces. The interstitial living organisms also disturb peds and mineral particles. The cilia of motile protists is strong enough to displace small particles. The passage of larger species, such as nematodes and microarthropods, creates tunnels (micropores) and contributes to continually changing the reticulum and maintaining tortuosity. Larger species obviously create even larger disturbances (macropores), for instance the burrowing of an animal or elongation of large roots. In soils that are biologically inactive, over time and with gravity, the fraction of pore spaces decreases with compaction, and the bulk density increases. The overall effect of the biological and physical disturbance causes a mixing of the soil, which contributes to moving microdetritus and SOM into deeper soil horizons, maintains the pore spaces, and increases drainage and aeration.

Summary

The soil consists of an inorganic matrix and an organic component, which together provide a habitat for soil organisms. The inorganic component consists of primary minerals and secondary minerals (clays). The chemical properties of clays are important in defining soil chemistry. The organic matter consists of litter in varying stages of decomposition. Its chemistry depends on that of the species from which it originates.

The soil profile reflects changes in composition with depth, of both its organic and inorganic components. The profile can be defined as a series of horizons. Successive horizons differ in compaction (and porosity), particle density, aeration and organic matter content. These changes affect ped size, structure, chemistry and species composition. The overall soil composition also affects water retention, temperature changes and the composition of the soil solution. The reticulum of pores is dynamic and continuously changing in composition under the influence of physical forces and biological action.

Suggested Further Reading

Black, C.A., Evans, D.D., White, J.I., Ensminger, L.E. and Clark, F.E. (1982) *Methods of Soil Analysis. Part I.* American Society of Agronomy and Soil Science Society of America, Madison, Wisconsin.

Duchaufour, Ph. (1998) *Handbook of Pedology*. AABalkema, Rotterdam.

Gregorich, E.G. and Carter, M.R. (1997) *Soil Quality for Crop Production and Ecosystem Health.* Elsevier, Amsterdam.

Harris, R.F. (1981) Effect of water potential on microbial growth and activity. In: Parr, J.F., Gardner, W.R. and Elliott, L.F. (eds) *Water Potential Relations in Soil Microbiology: Proceedings of a Symposium.* SSSA special publication, Soil Science Society of America, Madison, Wisconsin.

Kalbitz, K., Solinger, S., Park, J.-H. and Matzner, E. (2000) Controls on the dynamics of dissolved organic matter in soils: a review. *Soil Science* 165, 277–304.

Page, A.L., Miller, R.H. and Keeney, D.R. (1982) *Methods of Soil Analysis. Part II.* American Society of Agronomy and Soil Science Society of America, Madison, Wisconsin.

Robertson, G.P., Coleman, D.C., Bledsoe, C.S. and Sollins, P. (1999) *Standard Soil Methods for Long Term Ecological Research.* Oxford University Press, Oxford.

Sparks, D.L. (1999) *Soil Physical Chemistry*, 2nd edn. CRC Press, Boca Raton, Florida.

Sumner, M. (2000) *Handbook of Soil Science.* CRC Press, Boca Raton, Florida.

Tate, R.L. (1987) *Soil Organic Matter: Biological and Ecological Effects.* Wiley Interscience Press, New York.

Sampling and Enumeration 3

The soil is a habitat for numerous organisms from diverse phyla, as we have seen in Chapter 1. These organisms are distributed across the soil profile, and affected by the soil physical and chemical properties, as described in Chapter 2. They are not evenly distributed through the soil, nor are they all active at the same time, as we shall see in Chapter 5. There are many different microhabitats with depth and across the surface, over a few millimetres to centimetres of distance, but also between the pore spaces between particles over several micrometres. The aim of soil collection is to sample a representative portion of the biological diversity. To achieve this, one must sample a representative portion of the microhabitats in the area under investigation. Once sampled, conditions in the soil sample begin to change. Species may become inactive or may be stimulated to grow after sampling. Therefore, one tries to differentiate between active and inactive species at the time of sampling. In this chapter, we discuss approaches for: (i) estimating the total number of species in samples; (ii) extracting and enumerating active species at the time of sampling; and (iii) obtaining enriched samples and cultures for certain organisms. The chapter is not intended to be a description of methods, but a discussion of common approaches. Useful practical guides to methodology and laboratory procedures are provided at the end of the chapter and referenced in the text.

Soil Collection

Soil can be collected with a variety of hand-held shovels, soil core samplers and augers, spoons or spatulas. The preferred tool will depend on the quantity of soil required and the type of soil being sampled. To prevent cross-**contamination of samples** with bacteria, protist cysts, spores or active cells between samples, it is necessary to clean sampling tools with a volatile alcohol which disrupts cell membranes (such as 5% ethanol or 5% methanol). A small number of active cells is sufficient to colonize an entire sample in 2 or 3 days, if cells disperse through the sample or grow through several divisions. Therefore, adequate sampling protocols require attention to clean field techniques. Soil augers with an inner sleeve are commercially available and reduce the risk of cross-contamination. The collected **soil sample** must also be protected from **crushing** and **desiccation**. Once a **soil sample** is removed from the field, it will undergo some mixing or crushing, aeration, desiccation and temperature change. One should strive to minimize the changes before samples are analysed. Several precautionary measures are necessary. Excessive friction of soil particles will damage or kill protists, including hyphae, and the microinvertebrates such as nematodes, enchytraeids and microarthropods. A decrease in moisture content will initiate encystment or inactivation of many species. Thus, soil samples normally are placed in zip-lock plastic bags with care taken to avoid crushing the soil. This permits some gas exchange with air in the bag and reduces loss of moisture. A safer method, especially with protists, is to use screw-cap plastic tubes that protect the sample from desiccation, crushing and friction. Only a few centigrams of soil are required to extract and enumerate for protists and bacteria, but up to 100 g may be necessary for microarthropods. It is necessary to compromise between **spatial distribution heterogeneity** and the number of samples to analyse in the laboratory. For efficiency, it is best to obtain several small samples that can be fully analysed rather than fewer larger samples. For transport and temporary storage, soil samples are kept away from direct heat or cold in a cooler, at about the soil temperature.

The distribution of organisms varies with depth through the litter, organic horizon and B horizon, and along the surface. The diversities of conditions and food types available over several micrometres to several metres are called **microhabitats**. Surface soil microhabitats that influence species distribution can be found inside and under a fallen tree trunk or coarse woody debris, in decomposing leaves, or animal debris and carcasses. Litter organisms require sampling of coarse woody debris (twigs, branches and bark), fallen leaves and other debris that may be present. Falling fruit, nuts, catkins or pollen deposition can also impact the litter and soil organisms. To determine **species stratification** with depth, it is more important to preserve the profile. A soil corer, large diameter cork borer or curved spatula are useful because the soil can be subsampled or the profile collected intact. Alternatively, one can dig to expose the profile before sampling at various depth intervals. With depth, the quality of the organic nutrients decreases, soil bulk density or compaction increases, and oxygen availability becomes limiting. These determine species composition. To preserve the **anaerobic species**, it is necessary to fill the container completely with soil (without crushing) and to use a tight lid. Alternatively, anaerobic cores can be sealed in plastic, or commercially available sampling sleeves.

While in the field, it is important to remember the following points before samples are collected: (i) whether the soil profile needs to be preserved or subsampled, or if mixed bulk soil will do; (ii) whether the microhabitats in the litter can be amalgamated or if they need to be kept separate; (iii) how deep in the soil horizons one needs to sample; and (iv) how many replicates per sample are required. Once the soil is collected, the heterogeneity of the study site needs to have been sampled adequately for the project. The **number of samples** collected needs to be manageable in the laboratory, by the number of people and amount of time dedicated to sample analysis. The idealized number of samples required statistically can be impractical or simply superfluous once in the field. The correct balance between site heterogeneity and number of samples must be adjusted to what can be analysed (see below, this chapter). This may require narrowing the scope of the project. It is more informative to have fewer well-analysed samples than too many poorly studied samples.

Soil handling

The soil matrix contains decomposing litter and organic matter that are a source of nutrients to species of bacteria, protists and invertebrate animals. The source of some of the litter is animal dung and corpses, including those of mammals. Since animals carry internal and external parasites, they can release them to the environment through excretions, secretions and cadavers. Therefore, viral, bacterial, protist and invertebrate **parasites** of animals can be found in the soil, where some have intermediate hosts or life cycle stages. To avoid infections, it is necessary to avoid contact with potentially contaminated soil. Some species are not normally parasitic to humans, but can still cause a mild infection. However, non-parasitic soil species can become invasive and infectious, under certain conditions. These are called **opportunistic parasites**. For example, *Acanthamoeba castellanii* and similar species, which are common soil amoebae, can proliferate in the eye and cause blindness. Similarly, many species of soil bacteria are pathogenic, or can become pathogenic if in contact with a wound, or internalized. There are also fungi that are inedible and poisonous, so contact can be dangerous, especially if ingested. Lastly, tissues of certain plant species (roots or leaves) contain poisonous chemicals, poison glands or trichomes (Colgate and Darling, 1994). These may remain in the litter for some time even if partially decomposed. Skin irritations (**contact dermatitis**) can be caused by contact with these molecules in the soil and litter.

Another source of toxicity in soils is **pollution** caused by human activity. These anthropogenic sources can affect large areas (e.g. through atmospheric deposition) or be very localized (an accidental spillage). Deposition of air pollutants from the atmosphere causes accumulation of heavy metals, radioactivity, acid rain and organic poisons. Accidental spillage of petroleum products and hazardous chemicals, at the site of manufacturing or in transport, are realistically unavoidable. Accumulation of toxic chemicals occurs gradually and cumulatively, or in large discharges. Soil pollution can be legal, for instance at a mining site, or illegal, by undeclared and improper disposal of toxic materials. Agricultural and urban applications of pesticides, herbicides and other poisonous chemicals further contribute to soil toxicity. The dispersal and range of a pollution event depend on regional and global biogeochemical cycling, hydrology, soil composition and many other parameters. Standards for collection, storage, handling and analysis of pollutants, and other chemical analysis of soils, are set by various national governmental agencies. These sources must be referred to for guidelines on methodological assays and protocols.

For biological analysis, collected soil must be handled on a clean bench space and with clean or sterile equipment. **Cysts** and **spores** of soil protists and some bacteria are resistant to desiccation, but also to a variety of detergents and disinfectants. Thus, contamination of samples from laboratory surfaces is easy without sufficient care. Moreover, cysts and spores in spilt air-dried soil can be disturbed and carried in the laboratory air. Once airborne, contamination of soil isolates and cultures is more frequent. Cysts of colpodid ciliates are notorious invasive colonizers, and they can be isolated from sinks and floors of most soil laboratories. They are the most common protozoan culture contaminant in cell biology laboratories. Often, the laboratory space is shared with bacteriologists who apply sterile techniques, or with invertebrate or plant biologists who do not. Clean work habits are the best deterrent to permanent contamination problems.

Soil storage

Storage of soil samples can affect the biological and chemical properties of the soil. For example, the sorption and release of dissolved organic carbon (DOC) in soil samples varies with soil storage conditions, which is effectively a pre-treatment step (Kaiser *et al.*, 2001). The least effect is observed in fresh samples, but with an increase in soil DOC release of 23–50% depending on the horizon after 1 month at 3°C. Freezing at −18°C more than doubled the DOC measured, and air drying the samples increased the measured DOC by 4.3–4.7 times. Variations in soil storage and handling are significant in altering the chemical release of material from soils. In part this is due to killing and lysis of organisms and denaturing the chemical–physical properties of the organic matter. For many soil organisms, biological preservation is possible by fixation of soil samples in ethanol or formalin at sampling or soon after. This will preserve the microinvertebrates, fungi and bacteria, but not protozoa. Biological fixation with ethanol is not suitable for protozoa which do not have cell walls, because the alcohol dissolves membranes and cells are lysed. However, the ethanol fixation procedure is suitable for DNA extraction as the nucleic acids are preserved. This procedure has been tested with bacteria for DNA recovery (Harry *et al.*, 2000) and is probably suitable for other organisms with modifications.

Once soil has been collected from the environment, it is no longer part of the open ecosystem. It is a **materially closed system**, in that it is held inside a sealed container. The sample no longer receives nutrient input, or exchanges material with the outside of the container; only energy can be exchanged, in terms of temperature. Thus, organisms in the sample will respond to this new situation, where the only food and moisture available were those present at the time of sampling. Furthermore, since the respired, secreted and excreted materials cannot escape, they accumulate in the sample. Over time, the sample will become depleted of food and oxygen, and will accumulate CO_2 and other by-products of respiration and growth. One can maintain the sample aerated and moist, or even add nutrients; but by-products of respiration and growth accumulate none the less. For this reason, over time, the active species in the collected soil will no longer be representative of the active species at the time of sampling. Some biological activity continues even in the cold, through cold-tolerant species, so that moist stored soil is not completely inactive.

Site Variation and Statistical Patterns

One gram of soil can harbour 10^4–10^{10} protists and bacteria, so that removing some soil will always reveal some living organisms. (These values are greatly reduced in conditions of stress and nutrient deficiency as we shall see in later chapters.) However, species are not evenly distributed through the soil. One must bear in mind that species abundance and species diversity vary along horizontal and vertical gradients. For this reason, it is imperative that one has calculated what scale of resolution is needed before any sampling pattern is adopted. The ideal is to sample in such a way as to have a statistically robust estimate of species abundance or diversity at the correct scale of resolution for the experiment. First, there is variation in time over several days with soil moisture and temperature changes (such as rainfall or sunny days), or over seasonal patterns. Species diversity, composition and abundance also vary over physical distances: (i) along the soil depth profile; (ii) across a distance of an area of field or forest (micrometres to centimetres); (iii) along local abiotic conditions and microclimate (such as shade, temperature, rainfall, along slopes with altitude and soil moisture gradients from riparian zones (stream-side) to hilltops); and (iv) across climatic regions, from lowlands to hilly or mountainous, and geographic patterns. Therefore, any single soil sample represents only, and only, the exact location and depth of the soil at the time it was obtained. To obtain a fair estimate of species abundance and diversity across scales of time and space requires previous planning in the experimental design.

Spatial distribution patterns

Multicellular organisms tend to dominate ecological studies of above-ground processes. The individuals of these species can be viewed as scattered packets of attached cells, where the number of cells in each 'packet' or individual varies (i.e. the size of the individuals within and between species varies). The important distinction with interstitial species is the significant distances between each individual or packet of cells, above-ground. The dominant organisms below-ground are unicellular or filamentous (bacteria and protists), vermiform (nematodes and enchytraeids) or burrowing (mites and Collembola). These individuals are more evenly distributed spatially, in a more continuous array. One can think of an ocean of cells and microorganisms between soil particles. This difference in spatial distribution makes sampling from soil more comparable with sampling plankton from water than sampling above-ground species.

Before collecting soil samples from an area (field, forest or stream bank), it is necessary to have clearly defined and delineated the area and microclimates under investigation. Depending on the size of the area, the number of samples required to describe the area will vary. Furthermore, depending on the type of study being conducted, the resolution required from the data will vary. If general estimates of total protist abundances are required, few samples may be sufficient. If the aim of the sampling is to describe species of a taxon from a defined area, it is best to obtain many small samples. The data are then analysed according to various *species diversity measures*. If a description of the dynamics of particular species or functional groups is desired, then many more samples would be necessary. In the following outline, a procedure is described that can be easily scaled-up as required, for larger areas, or for more detailed sampling analysis. As a general guide, protists and nematodes are sampled from 2–3 cm diameter cores, and microarthropods with 5 cm cores. For larger or rarer species, 25 cm cores may be necessary. For macroarthropods, pit-fall trap diameters of 2 cm are sufficient for ants and similar sized species, but normally wider traps <25 cm diameter are used that will collect a broader array of species.

The basic unit of study from which samples are removed is the **pedon**. It is similar to a quadrat, with the addition of a depth dimension. The pedon must have a stated area, depth and shape. Each of these parameters strongly affects the data collected and the accuracy of the results (Elliott, 1977; Lal *et al.*, 1981). The depth of sampling needs to be pre-determined for each project requirement. The area of the pedon, number of pedons and distribution of pedons need to be optimized and established at the beginning of the study. This is to ensure that the sampling protocol reflects the field spatial distribution. Theoretically, the **spatial distribution** of individuals in the pedon can have three identifiable patterns. The individuals can be **randomly distributed** in the soil volume; they can be found in randomly arranged patches or **clusters**; or they can have a **uniform pattern** of clusters. Motile species can disperse and will tend to have a spatially random distribution. Immobile species or those with restricted dispersal ability will tend to be clustered randomly. Individuals aggregate where litter is present, or where prey is clustered. Because of heterogeneous litter distribution on the surface and in the soil, and varying motility of species, the overall distribution will tend to be a combination of clusters and more dispersed individuals between clusters, from all species. A more uniform (periodic) pattern can be obtained in managed forests or agriculture fields, where plants are regularly spaced, if there is an association between plant roots and interstitial species. This would also affect the distribution of symbionts such as mycorrhizae, or root pathogens.

Pedon sampling and enumeration

One cannot enumerate all the individuals in a pedon. Instead, several samples must be removed to the laboratory for extraction and enumeration. The number of samples removed must reflect the abundance, diversity and heterogeneity within the pedon. Guidelines to calculate the number of samples and pedons to use, in order to obtain adequate confidence intervals, can be found in most standard statistic text books (see *quadrats*), using *Hendricks' method* or *Wiegert's method*, and at http://www.exetersoftaware.com The latter can be particularly useful to estimate optimal quadrat sizes and the numbers of quadrats to use, if it cannot be done quickly on a calculator or with spreadsheets. Figure 3.1 illustrates the effect of sample size and number of samples on a particular area with the same spatial pattern of a species. The variation is demonstrated further in Table 3.1. It is possible to use **nested quadrats** to obtain an experimental estimate of the best pedon size and sample volumes for the site under investigation. The idea is to define pedons of incremental areas (20 × 20 cm, 50 × 50 cm, 100 × 100 cm) and to sample each with five 2 cm cores, and five 5 cm cores. Each core sample is

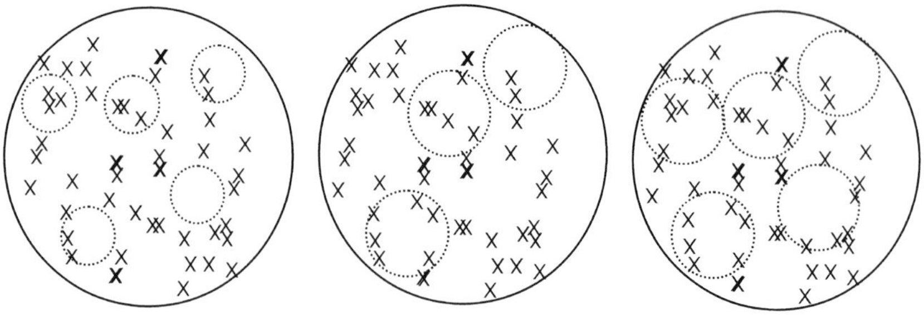

Fig. 3.1. The effect of soil core area on sampling of population abundance from a hypothetical pedon. The same pedon is shown sampled three or five times with two different soil cores, at approximately the same locations. The crosses represent individuals in the pedon. The results are summarized in Table 3.1.

Table 3.1. Summary of data from different sampling choice for the same hypothetical pedon, shown in Fig. 3.1.

Sample diameter (cm)	n	Enumerated	Mean	SD	Abundance (no. cm^{-2})
2.50	5	2, 3, 3, 0, 2	2.00	1.23	0.41
5.00	3	2, 6, 5	4.33	2.08	0.22
5.00	5	2, 6, 5, 5, 4	4.40	1.52	0.22

then extracted and enumerated in the laboratory. One needs to determine from these data the minimum number of samples from which area provides an adequate estimate of field variance, in a timely or feasible manner. It may be that sufficient resolution is obtained from enumerating fewer than five samples from each pedon.

Wiegert's method is useful in choosing a compromise between how much time one is willing to invest in enumeration and how much resolution is required. An estimate can be obtained as follows: P_c = (relative cost) × (relative variance of data), where P_c is the *cost for a pedon area*, in terms of hours spent in experiment set-up, extraction–enumeration, and data variability obtained; the relative cost is obtained from (time/time$_{min}$), where time is the time to enumerate all samples from one pedon, and time$_{min}$ is the minimum time required to enumerate all samples from one pedon; the *relative variance* is obtained from (standard deviation (SD)^2/minimum SD^2), where the SD for the data from a pedon of a given area is divided by the minimum SD obtained for a pedon.

A hypothetical situation is described in Table 3.2. In this case, the optimum situation is to obtain five 5 cm cores from the 100 × 100 cm pedon. However, it may be decided from the table that using 2 cm cores from a 50 × 50 cm pedon is adequate resolution.

Table 3.2. Wiegert's estimate of optimal sampling procedure for a hypothetical research plot. Pedon area sampled five times (*n* = 5), with different sized cores (sample area) and total duration of sample examination in the laboratory required (enumeration) are used to obtain Wiegert's estimate (P_c) of time required versus data resolution.

Pedon area (cm)	Sample area (cm^2)	*n*	Enumeration (h)	Samples SD	SD^2/SD_{min}^2	Time/time$_{min}$	P_c
20 × 20	5	5	7.5	2.3	2.7	1.0	2.7
20 × 20	20	5	10	3.1	4.9	1.3	6.5
50 × 50	5	5	10	1.8	1.7	1.3	2.2
50 × 50	20	5	12.5	1.7	1.5	1.6	2.5
100 × 100	5	5	10	2.4	2.9	1.3	3.9
100 × 100	20	5	12.5	1.4	1.0	1.6	1.7

In the description above, sampling one pedon of different sizes was considered. These calculations are particularly useful in field investigations, where a large number of pedons are sampled. In this case, the total time that each sampling design would require must be calculated, against the variability of data between pedons, or between samples within a pedon. Considerable time can be saved if an estimate of these parameters is known in advance, to avoid oversampling of the study area, or undersampling that fails to resolve differences over time or between treatments. Two studies with nematodes have shown that 16 × 2 cm cores, removed at random from a forest site, were sufficient to obtain 96% of species diversity (Johnson *et al.*, 1972) and that ten random 2 cm cores were sufficient to determine abundances in a 7 ha agricultural field (Prot and Ferris, 1992). These experimental values are far below those estimated theoretically without field data and those recommended by the Society of Nematology in 1978 (Barker, 1978).

Since several samples are removed from each pedon, and each sample is then subsampled in the laboratory, the correct statistical procedure for error analysis is described by *multistage sampling* analysis. The number of subsamples should be sufficient to obtain a robust mean, with about 10% coefficient of variance about the mean. An inescapable fact whilst enumerating samples is that different microscopists have different efficiency and skills. It is particularly important for beginners that samples are enumerated consistently. Thus, recounts of the same samples by the same person can be analysed by *one-way ANOVA* (analysis of variance), followed by a measure of **repeatability**. It is calculated from $R = s^2_a/s^2_b + s^2_a$, where R, the measure of repeatability, is obtained from s^2_a, the variance between repeats, and s^2_b, the variance between counts in subsamples (for further details, see Krebs, 2000). To obtain useful data requires that enumeration be reproducible. These will improve with experience and with the skill of the microscopist.

Experimental design and field sampling

The defined study area normally will require several pedons to be sampled adequately. There are a number of ways that additional pedons can be positioned on the study area. Since the soil is spatially heterogeneous and patchiness in abundances is expected, two methods are recommended. One is to sample a small number of pedons at random across the area, the other is to sample 3–5 pedons along 2–5 transects of fixed length. The latter resemble long and narrow rectangular quadrats which are usually best for clustered or heterogeneous distributions. They differ in that they are discontinuous and several samples are removed from each pedon within the transect (Fig. 3.2). Although applying equally spaced pedons along a transect is acceptable in most situations, there are

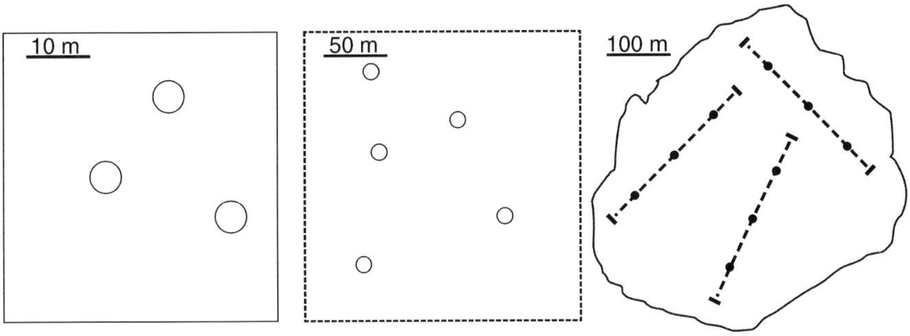

Fig. 3.2. Random sampling of research areas of different sizes, showing circular pedons and transects (see text for explanation).

two situations where caution is warranted. One is in agricultural fields where there is a periodicity to the plough line or seeded line, and the other is in managed forests or replanted forests. In these study areas, it is best to avoid a systematic pedon distribution (same spacing between pedons) in favour of unequal (or randomized) pedon position, or to ensure the sampling period does not correspond to the site periodicity. However, in most situations, a systematic pedon distribution, imposed on a heterogeneous landscape, provides a random sampling.

The frequency of individuals counted from samples is then used to obtain the **statistical distribution** of species. If the frequency data fit a **Poisson distribution**, then the individuals must be assumed to be randomly distributed in geographical space. In this case, it is best to obtain an estimate of the mean number of individuals per sample (or in each pedon) that is statistically robust. Soil species rarely have a completely random pattern. Instead, there is a tendency to find clusters of higher abundance of one or more species or functional groups. If the spatial clustering of individuals is statistically significant, as it could be in a heterogeneous spatial space such as a forest floor, then a **negative binomial distribution** is often a better statistical descriptor of the spatial pattern. It is important to distinguish between these two mathematical frequency distributions, as they determine the correct methods to chose for calculating confidence limits for the data. The spatial scale at which the distribution of populations is being described should be tested. Since the investigator samples from within a pedon (<1 m^2), to several pedons across a landscape (>100 m^2), spatial clustering or non-random distribution of species can occur within the pedons or between pedons on a larger scale. The data can be analysed further with *indices of dispersion for quadrat counts*.

Normally, the soil samples removed from each pedon can be mixed into a single **composite sample**, prior to subsampling and extraction. The resolution of this sampling design can be improved using a **stratified random sampling** procedure. The procedure is particularly useful when sampling soil, because of the variability in habitat over short distances and the diversity of microhabitats. The idea is to group data from similar soil samples (each being one stratum). For instance, when sampling a pedon in a crop field, one could remove, from within the same pedon, samples from the base of the plants (stratum 1, enriched in rhizosphere soil), and from between the plants (stratum 2). During data analysis, all the samples from the base of the plant from all the pedons can be compared together, and those from between the plants together. Another situation would be to sample surface litter in forest quadrats, but to subsample coarse woody debris, fresh leaf fall or insect frass separately. Lastly, it may be useful to compare the distribution of species and abundances through the profile. In this case, each sample would be subsampled into depths along the profile (0–3 cm, 3–8 cm, 8–15 cm). This procedure provides improved resolution from the sampling design. It is more precise than simple random sampling within the pedon, and should be applied wherever possible.

Extraction and Enumeration

There are two different purposes for **extracting** and **enumerating** soil species. Each requires a different approach. The aim of the first is to determine how many species in total, active and inactive, are present in the samples. The aim of the second is to identify the species active at the time of sampling, and to estimate how many individuals of each species there are. To obtain estimates of the active species requires more care, as bacteria and protists respond to changing conditions immediately, and can become inactive or active within hours once the soil is disturbed by sampling. Microinvertebrate activity is also affected by sampling, and they respond physiologically over several hours to 1–2 days. Samples that are to be analysed for active species cannot be stored and must be analysed within a day. Soil for total species enumeration can be stored under conditions suitable for the species of interest, or stored after chemical fixation.

Before proceeding with microscopy and **enumeration**, it is necessary to have calculated several numbers. The number of cells or individuals counted, in each replicate subsample, depends on the dilution factor applied to the collected soil. Therefore, the weight of soil in subsamples and the amount of dilution must be known in order to calculate the estimate of abundance in the collected soil. To obtain a robust estimate, about 5–50 individuals must be counted in each category. Normally, the standard error between three replicates should be about 10% of the mean value, or better. If the deviation is greater, less dilution or more replicates should be considered. Lastly, the minimum number that can be detected in each subsample should be known, i.e. the **detection limit** for the subsample, at that dilution.

Total potential species diversity

This approach aims to obtain an estimate of all species present in a sample, irrespective of how many were active when sampled and how many were encysted or otherwise inactive. The collected samples can be air dried at room temperature, away from places where they can be contaminated with cysts and spores from other sources in the laboratory. Freezing soil samples will damage many cysts, and fewer species will survive. Even air-dried soil should not be stored at cold temperatures that the soil would not have been exposed to normally. Most soils can be stored at 4°C, or below 12°C once air dried slowly.

Bacteria

Identification of bacterial genera and strains requires description of growth characteristics on various media at different temperatures, membrane chemistry, DNA sequencing of ribosomal genes, or nutrient uptake biochemistry. These procedures assume sufficient cells are obtained to carry out growth assays or molecular extractions. However, many isolates of bacteria are not cultured easily. Therefore, most assays that rely on cultures provide data on a small subset of bacteria, those that were cultivable with the particular laboratory conditions and media. It is recommended that media and incubation conditions be adjusted to isolate functional groups of species or strains. The following need to be adjusted accordingly, to reflect soil conditions at the time of species activity: (i) soil ion concentrations and pH; (ii) PpO_2/CO_2; (iii) temperature; (iv) source of nutrients and metabolizable carbon; and (v) availability of final electron acceptors for metabolism. Often, multiple carbon sources or inhibitors in the media will prevent species activity. Too high or too low concentrations of nutrients will inhibit substrate utilization. Substrate utilization patterns may vary with incubation conditions and medium composition for more versatile species. Moreover, substrate utilization may require nitrification or accompanying metabolic pathways in parallel. The most common mistake is to try to culture too many species under rich aerobic conditions, at inadequate temperature and pH. Many bacteriologists are trained with pathogenic strains and assume glucose is a universal substrate. However, there is not enough soluble glucose in the soil to support growth of organisms. Nutrients for growth must be derived from litter and decomposition of soil organic matter (SOM).

One strategy for identification of bacteria without problems associated with culture or single strain isolation is by selective polymerase chain reaction (PCR) amplification of target DNA sequences. The target DNA from the soil extract is probed with PCR primers of varying specificity to identify bacteria at the appropriate taxonomic resolution (Macrae, 2000). This approach has been the focus of numerous studies recently that strive to optimize soil DNA extraction (Miller *et al.*, 1999; Bürgmann *et al.*, 2001). Caution must be exercised to remove inhibitory substances from the soil that co-purify with DNA (Miller *et al.*, 1999; Alm *et al.*, 2000). These methods can be used to quantify particular species in soils (Johnsen *et al.*, 1999). Targeted PCR amplification from 16S rDNA separated by differential gradient gel electrophoresis (DGGE) is more precise, and the appropriate choice of rDNA segments can provide resolution at various taxonomic levels (Muyzer *et al.*, 1993; Grundmann and Normand, 2000; Nakatsu *et al.*, 2000). Similarly, temperature gradient gel electrophoresis (TGGE) has also been used. A more recent and simpler procedure involves *single strand conformation polymorphism* (Peters *et al.*, 2000). By avoiding plating and culturing of soil samples, which tend to activate or inactivate many species, and using taxa-specific primers, it was possible to obtain a high resolution of the sequential activity of bacterial species.

An alternative approach that does not require cultures is based on whole soil bacterial separation and extraction of membrane lipids. It relies on differences in membrane chemistry between bacterial taxa, and the method can detect changes in bacterial community structure between samples (Zelles, 1999). Whole community phospholipid fatty acid (PLFA) profiles are analysed by multivariate statistical analysis to detect shifts in the profile (White, 2000). Although PLFA does not provide resolution at the genus level, it is fairly sensitive in detecting community changes caused by the environment (Griffiths *et al.*, 1999a; Steinberger *et al.*, 1999). Two other methods based on fatty acid methyl ester (FAME) profiles or the carbon substrate oxidation pattern on single substrate BIOLOG plates (BIOLOG, Hayward, California) seem to misidentify or fail to detect certain genera and should be used with more caution (Oka *et al.*, 2000).

Protozoa

Species identification at the microscopic level is the traditional approach to protozoa species enumeration, and to protists in general (Table 3.3). However, in recent years, with an increasing use of specific DNA probes, phylogenists have come to realize that in many cases morphotypes represent a cluster of species that vary in behaviour, life history and gene sequences. Protist phylogenies are in an advanced state and there are adequate oligonucleotide probes available from the research literature to identify most organisms at least to the family level. These molecular methods should be used in more formal studies of protozoa diversity. Their use in identifying soil protozoa has been timid and limited (Van Hanen *et al.*, 1998; Fredslund, 2001). These methods are discussed further in the next section on fungi, because their use in soil ecology of fungi is more advanced. This section discusses approaches to the enumeration of protozoa by microscopy.

Table 3.3. Size range and microscope observation methods for protists, listed by extraction method.

Taxa	Size (μm)	Microscope	
		Enumeration	Identification
Naked amoebae	5–100	Compound	Compound
Testate amoebae	20–200	Dissecting/compound	Compound
Small flagellates	4–8	Compound	Compound/SEM
Flagellates	8–30	Compound	Compound/SEM
Ciliates	8–150	Dissecting/compound	Compound/SEM
Hyphae	5–10 diameter	Dissecting/compound	Reproductive structures
Reticulate	100 to 1cm	Dissecting/compound	Compound
Yeasts	5–10	Compound	Compound

Most protist species in the soil are inactive at the time of sampling. If the sample has been air dried for storage, species need to be activated first. Therefore, to enumerate and identify species, it is necessary to **excyst** the individuals. Even before the soil sample is stimulated to excyst, changes in storage conditions (a pre-treatment) will affect species excystment. Different protozoa species have different preferences in pre-treatment requirements for **excystment**. For instance, some Gymnamoebae cysts require exposure to slightly elevated temperature as a stimulus. Other cysts require a period of desiccation before they can be stimulated to excyst. Thus, pre-treatment storage conditions (especially moisture and temperature) and duration of storage affect the excystment response. Usually soil is moistened, with or without powdered litter amendment, to provide an excystment stimulus. Cysts and spores of various species will respond differently. Some may not be stimulated to initiate excystment under particular conditions. Some species can abort excystment if the conditions are not adequate. Therefore, one protocol is inadequate to excyst all species. Common stimuli to initiate excystment are shifts in temperature (up or down), or changes in moisture, aeration, or nutrient quality or quantity. Variations in temperature, amount of moisture and aeration, or litter additions can help bring out other species, gradually over several weeks. Once excysted, cells remain active only for a certain time. The duration of activity and sequence of species excystment result in a sequence of species activity in the soil samples, over several weeks. Species which are stimulated to excyst do not represent the active community in the soil, because some species can remain inactive or encysted for several years. Therefore, conclusions about an experiment at the time of sampling cannot be obtained from excysted cultures. However, some conclusions pertaining to the soil in general can be derived from analysis of total species composition.

In general, manipulation of excysted cells is carried out by hand with a micropipette or with a bench-top centrifuge. Micropipettes are made from Pasteur pipettes pulled into a thin tube over a small but hot flame. Known volumes of the suspension can be obtained with a pipettor. The cone of the pipette can be cut to obtain a wider bore, to prevent clogging by the soil suspension. For collection of live ciliates and testate amoebae on to a membrane, gentle suction can be applied. Soil suspensions can be filtered with nylon mesh or cheese cloth, without suction, to prevent cell lysis. Nucleopore membranes or nitrocellulose membranes are useful because they are transparent under the microscope. Nitrocellulose membranes are resistant to the fixatives used and do not stain. They are made transparent at the end of the staining procedures by dehydration in alcohols and clearing in xylene. Protocols for general histology and cell handling are provided in standard texts in protozoology (Kirby, 1950; Sonneborn, 1970; Lee and Soldo, 1992). Use of ocular grids and micrometre scales in the microscope eye-piece objective is needed for estimating abundances or measuring cell dimensions.

Once soil samples have been prepared to stimulate excystment and organisms extracted, the next step is species identification. Most taxa are large enough to identify under a compound microscope. However, staining of cell structures for ciliate identification or scanning electron microscopy (SEM) for smaller species of protozoa are necessary. Many species of nanoflagellates are unknown or indistinguishable from each other unless thin sections are obtained with transmission electron microscopy (TEM). New species require TEM sections to reveal the ultrastructure of organelles, especially of the basal body and cytoskeleton. As a rule, new species require SEM and TEM for formal descriptions and identification. These are accompanied by descriptions (or video footage) of locomotion, sampling site description, food preferences, growth rates, nuclear and somatic division, and other potentially useful observations. Storing a DNA extract in ethanol is a valuable additional reference.

Ciliates are best obtained from the '**non-flooded Petri plate**' technique (see Lee and Soldo, 1992). The soil sample is moistened and, at regular intervals over several days to weeks, more water is added and drained from the soil into a Petri plate. The free gravitational water will carry organisms with it. Many larger flagellated species can also be obtained with this procedure. The extract can be scanned with an appropriate microscope, or fixed and collected on a membrane for staining. The quantitative protargol staining protocol can be modified for other stains or soil suspensions, to obtain quantitative estimates (Montagnes and Lynn, 1987a,b). Species identification of ciliates requires staining of the ciliature and oral structures with protargol (Foissner, 1991). The general antibody against α-tubulin (Amersham, bovine brain α-tubulin) is also useful for immunofluorescence of the basal bodies and cilia. Many of the smaller species can only be identified accurately by SEM.

The **amoeboid** species can be observed from soil droplets on agar, supplemented with 0.1% yeast and malt extracts or other appropriate organic nutrient solutions. The medium provides a source of nutrients for osmotrophy and for bacterial growth. Different species will excyst over several weeks of observations. A variation based on serial dilutions can be used to obtain estimates of abundances (Anderson and Rogerson, 1995). Individuals are identified based on locomotion, overall shape and dimensions, and details of the pseudopodia. These observations can provide identification of the family or genera. However, confirmation requires TEM, and final species identification will require DNA sequencing of selected genes. Several genera require hyphae, yeast or other protozoa as food sources. **Testate species** are not easily cultured or activated. On agar plate preparations, excystment and activation occur after >10 days, and some species will only excyst after 3 weeks or more. However, the tests, which are used for species identification, can be found in the litter and soil. It is best to sift through the organic horizon and surface litter under a dissecting microscope. Two useful protocols are recommended for species in the organic horizon. One provides a preparation on a filtration membrane (Couteaux, 1967), and the other provides a fixed preparation on a microscope slide (Korganova and Geltzer, 1977). The preparations can be stained to highlight test features and to differentiate between living individuals and empty tests. The addition of leaf powder or fragmented macrodetritus can stimulate growth of some species on agar plates.

Many species have more specialized food requirements, and several sources must be tried. This approach is called '**baiting**' with an appropriate selection of organic nutrients or other food sources supplied to agar plates. Although the technique can be applied to many taxa, it is used primarily in obtaining preparations of Labyrinthulea (including Thraustochytrids), Oomycetes and Hyphochytrea (chromista), Chytridiomycetes (fungi) and specialized species. More details about collection and isolation of these species are provided in Fuller and Jaworski (1987). These protists with **filamentous** and reticulate growth forms are obtained in liquid or on agar media supplemented with a source of organic nutrient. Common organic matter 'bait' includes pollen, cellulose, chitin, keratin or powdered leaf litter. The soil subsample spread on the plate is incubated for several weeks on various agar–bait combinations and observed regularly. Reticulate and phyllose forms, such as Vampyrellidae and the Eumyxa (protozoa), are also obtained on agar plates. They can be fed with oats, hyphae or macrodetritus.

Most other species, primarily heterotrophic **nanoflagellates**, can be obtained from soil suspension dilutions. The suspension can be pre-filtered through 50–25 μm mesh and centrifuged to sediment cells. This may be repeated several time to separate cells from soil particles and microdetritus, although with significant loss of cell numbers. The suspension can be observed under a compound microscope or fixed for SEM for species identifications. These small flagellates are mainly osmotrophic and bacterivorous species. The preparations can be supplemented with bacterized soil extract, 0.1% yeast extract or a similar dilute organic medium such as 1% wheat grass powder infusion.

Many species of protists across phyla are tolerant to aerobic and anaerobic conditions. Other species are adapted for growth in **anaerobic conditions** and cannot tolerate long exposures to oxygen. These species include yeasts, several species from various phyla of protozoa, such as free-living Archamoebae, and uncharacterized nanoflagellates. They occur in anaerobic soils, submerged soil or deeper in the profile of aerated soil. In order to isolate or observe these species, the soil must be anaerobic for excystment and enumeration. For isolates kept a few days only, an open tube of soil kept saturated with water may be sufficient. For isolates that need to be kept longer, alternative methods exist. The anaerobic environment can be obtained by keeping the soil or the organisms in a sealed container deprived of oxygen. The oxygen can be removed by placing a small lit candle inside a holding jar sealed tight (Fig. 3.3), or with commercially available apparatus.

Lastly, one final group of anaerobic protists needs to be considered in samples. These are **endosymbionts** of several soil invertebrate taxa. The symbionts live in the intestinal tract of host species, and contribute to the host's digestion and assimilation of nutrient derived from microdetritus. The most common are species of the phyla Trichozoa and Metamonada (both protozoa) that reside in the gut of several wood-feeding termite families, along with specialized bacteria. The protozoan

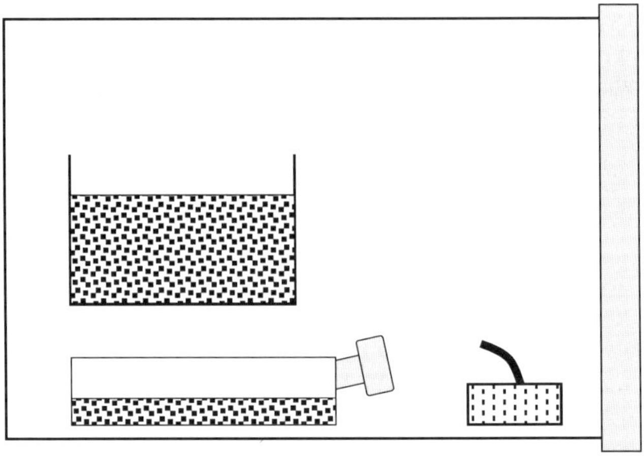

Fig. 3.3. Anaerobic conditions can be recreated easily in the laboratory. The diagram illustrates soil in an open container or tissue culture flask, inside a sealed canning jar. The candle burns until the oxygen is exhausted.

symbionts ingest the wood microdetritus ingested by the termite host. The cellulose is partly digested and some nutrients assimilated for growth of the symbionts. The remainder is excreted back into the gut, where the termite can continue the digestion and further assimilate the wood. The Termitidae family do not contain protozoa, but have bacterial and fungal symbionts. One subclass of ciliates, the Astomatia with 10 families, is specific to the intestinal tract of annelids, especially oligochaete worms. It is likely that many other species of wood- or detritus-feeding invertebrates, such as some Scarabaeidae, depend on bacteria or protozoa endosymbionts. Endosymbionts are obtained from dissection or squashed preparations of gut contents. However, most species are not tolerant of oxygen and will die soon after exposure, so they must be observed immediately under a coverslip.

Estimates of bacterivorous protists

One popular technique is to assay for total protist species from one functional group only, the bacterivorous species. Generally referred to as the '**most probable number**' (MPN) method, it was devised initially to estimate the number of bacteria species that could be cultured under one set of conditions. This approach was abandoned by bacteriologists. Several modifications were proposed to apply the technique to protists (see Coleman *et al.*, 1999). The method relies on serial dilutions of a soil suspension, incubated in a bacterized suspension, and observed at fixed intervals for enumeration. The method does not assay species that do not grow in aerated conditions, in liquid, on the inoculum bacteria or that have been damaged in pre-culture treatments. Furthermore, as explained earlier, excystment conditions vary between species. Soil handling and storage conditions after sample collection affect excystment. Several methodological parameters also affect excystment and growth. These include solution pH (which should be close to that of the collected soil), heat or cold treatments, duration and extent of soil desiccation, and soil dissolved nutrients in the solution. As these conditions vary between laboratories and are not always reported, it is difficult to compare results between laboratories. Moreover, the assumptions of the method are not met when applied to protists. These assumptions are that cells can be evenly distributed in the soil suspension, that all species will grow under the set laboratory conditions on the food provided, and that encysted cells can be differentiated from excysted cells. None are met, and growth of small flagellates can be hidden by growth of predators on them. Among the bacterivorous species, testate amoebae are not cultured, naked amoebae are grossly underestimated, and ciliates are overestimated. Early studies showed that direct counts yielded far more amoebae and nanoflagellate species than culture methods (Sherman, 1914; Martin and Lewin, 1915). The MPN method has poor resolution between collected soils from varied environments, and between trials with the same soil samples, and fails to correlate between variants of the method (Berthold and Palzenberger, 1995). Because the variations in protocols between laboratories are not reported, and because they affect the results, it is impossible to compare results between research groups. This method is not recommended by protozoologists or bacteriologists.

Besides problems in its resolution and repeatability with collected soil, the method is also applied to microcosm and laboratory experiments. This is an erroneous application of the method. The soil for these laboratory treatments is derived from one collected site, and it is then subjected to the same culture conditions for MPN. It is not surprising, therefore, that little statistical difference is obtained between treatments in MPN results from air-dried samples. All the treatments contain the same species diversity, from the same collected soil and, under MPN conditions, the same species are excysted and enumerated each time. The variability is then due to systematic errors, variability inherent to the technique and variations in pre-treatment conditions that affect excystment rates. Attempts at discriminating between encysted and excysted species are also not successful with the MPN approach. For example, HCl-treated subsamples, to kill active species and preserve cysts, were compared with non-treated subsamples. The study showed that almost all cysts were also inactivated by the HCl treatment (Bodenheimer and Reich, 1933). An adequate discussion of problems with this method is provided by Foissner (1987) and Lüftenagger *et al.* (1988).

Fungal hyphae

The mycelium of fungal hyphae must be separated from the soil matrix and collected on a surface prior to examination. It is impossible to release the hyphae intact from the soil, as they are in large part responsible for holding peds and soil–litter aggregates together. Soils rich in hyphae will hold >100 m hyphae/g soil. The soil will also hold yeasts, $>1 \times 10^4$/g in organic horizons and 600–1800/g in subsurface horizons. The yeast forms are more easily separated with bacteriological protocols. Their species identification is by morphology of growth on agar media (if culturable), with consideration for the range of metabolized substrates and ability for anaerobic fermentation (see Slavikova and Vadkertiova, 2000). However, for stringency and accuracy, molecular markers are essential, such as rDNA sequence, mitochondrial DNA restriction fragment length polymorphism (RFLP) patterns or nuclear magnetic resonance (NMR) spectra of cell wall mannans (Spencer and Spencer, 1997).

Separation of the mycelium from the soil can be achieved rapidly by homogenizing a soil–water suspension briefly (~1 min) in a Waring blender. This will shred the hyphae into short fragments, but it is suitable for many molecular and biochemical protocols. More gentle homogenization of the soil by hand in tissue homogenizers provides less damaged preparations. Alternatively, the soil can be disaggregated with a suspension of suitable density, and hyphae separated from the heavy mineral particles by density and collected on a fine mesh or membrane. This procedure is similar to the invertebrate extraction by density sedimentation/decantation protocols (see below). The latter is possible in soil with dense mycelia.

For estimates of abundance, a known volume of the suspension or weighed cellular material is smeared on a microscope slide for visualization of hyphae and slide preparation. Useful stains include calcofluor white M2R (a fluorescence brightener which stains fungal cell walls), and fluorescent DNA stains such as 4',6-diamidino-2-phenylindole (DAPI) for nuclei. Basidiomycete hyphae can be differentiated from the Ascomycetes with diazonium blue stains after a KOH pretreatment, which stains cell walls of Basidiomycetes red–purple. The Ascomycete's simple septate pores between cells in the filaments can be detected by trypan blue and congo red in NH_4 or sodium dodecylsulphate (SDS). The Ascomycetes cell wall is bi-layered, with a thin outer and thick inner component. The fluorescent stain 3,3''-dihexyloxacarbocyanin ($DiOC_6(3)$) is specific for Ascomycetes. Other morphological characters are useful to distinguish between these two classes (Fig. 1.21). Basidiomycetes can be recognized from dolipore septa and clamp connections, with monokaryotic and dikaryotic hyphae, while Ascomycetes have simple septa. However, from most hyphal lengths, it will be impossible to determine any identity. Hyphae inside macrodetritus, leaf litter or in mycorrhizal associations require staining of the host material. Common techniques and detailed procedures are described in Johnson *et al.* (1999) and Norris *et al.* (1994). Trypan blue is the recommended stain, but acid fuchsin, chlorazol black E, cotton blue in lactic acid and rose Bengal can be useful with some root preparations.

Hyphal preparations can be assayed for biomass from estimates of length/g soil. Common microscope procedures use the *line transect* or *variable area transect* method to scan the slide preparation (Paul *et al.*, 1999; Krebs, 2000). The ratio of nuclei/m is a useful index of living hyphae material. However, the presence of nuclei does not indicate that the isolate was active when sampled. Many species are only seasonally active and become transiently inactive when conditions are unsuitable for growth.

A culture-based approach to identification of fungal species involves baiting the soil with appropriate litter or specific plant roots (Johnson *et al.*, 1999). The idea is to stimulate the growth of hyphae from spores, sclerotia or inactive hyphae. Species that are not abundant can be detected under suitable conditions. Relatively intact soil cores are necessary so that the hyphal network is not shredded. The soil is then analysed at regular intervals for the presence of identifiable reproductive structures, spores or mycorrhizae.

In order to identify hyphae, traditionally one relied on descriptions of conidiospores, basidiospores and sporocarps. This procedure requires the visual identification of morphology to a corresponding species (see Johnson *et al.*, 1999). The method does not provide quantitative estimates of species abundance, because some species produce abundant spores, whereas others sporulate infrequently or in small amounts. One study recommends that at least 5 years of repeated observations are required to identify most species in an area (Arnolds, 1992). Therefore, on short sampling periods, even abundant species could be missed if they are not sporulating during that time.

Modern methods based on DNA hybridization with oligonucleotide probes provide an alternative approach. The taxonomic resolution to family or species level is determined by specificity of the oligonucleotide primers chosen. Extracted DNA from soil samples is purified and the DNA can be amplified by PCR techniques. Extraction of DNA requires bead-beating in a lysis buffer. This has been shown to be adequate to break even dehydrated cysts and spores. Lysis buffers such as Trizol allow for sequential extraction of DNA, RNA and proteins from the lysate. Separation of amplified products from nested PCR techniques by DGGE was able to resolve banding patterns between fungal species and can follow species dynamics and succession trends (van Elsas *et al.*, 2000). Detection of known sequences is by Southern hybridization with labelled probes. To quantify known local species of mycorrhizal fungi, RFLP of amplified rDNA has been useful for identification against standards (Kernaghan, 1997; Viaud *et al.*, 2000). The usual chromosomal region for detection of species is the internally transcribed spacer (ITS) region of the nuclear rDNA. This ITS region is flanked by the conserved 18S and 28S genes, which are present in hundreds of copies in each nucleus. It is therefore possible to obtain amplification products from very few nuclei. These PCR-based methods theoretically are more efficient at detecting rare species if the protocol is optimized. However, it is more difficult to quantify species abundances with PCR protocols. Unknown species can be identified from PCR amplification using more general primers (such as one targeting the fungi or the Ascomycetes), separated by DGGE and unknown bands cut out for cloning or direct sequencing. The detection and identification of endomycorrhizal fungi pose a difficulty, in that the hyphae must be identified from within the plant host tissue. A combination of staining, *in situ* enzyme reaction assay and molecular protocols is recommended, as discussed by Dodd *et al.* (2000).

Invertebrates

Interstitial and litter invertebrates are best separated from the soil by distinguishing between microarthropods and other invertebrates. The **microarthropods** are obtained from soil cores placed on to modified

Tullgren or Burlese funnel extractors (Farrar and Crossley, 1983; Coleman *et al.*, 1999). The soil core or litter is heated gently with a low-wattage light bulb, and the organisms migrate down, away from the heat and desiccating surface, and fall into a container of 70% alcohol (Fig. 3.4). The efficiency of these extractions is variable, depending on the soil, and some Collembola families (such as the Onychiuridae) are poorly extracted. Specimens may remain adhered to the collecting funnel and not be counted. The extraction is improved using floatation methods, which are more labour intensive (Walter *et al.*, 1987). These require disaggregation of the soil in an organic solvent or mineral oil. The cuticle of microarthropods has an affinity for the organic molecules, and individuals float to the surface. Alternatively, density gradient sedimentation with salt or sugar solutions can separate the denser mineral particles from the less dense biota. These methods are discussed in detail elsewhere (Coleman *et al.*, 1999). The most abundant individuals are usually the Oribatida, with up to 40 species in a 20 cm diameter core to 5 cm depth, and 150–200 species represented in temperate forest soil and litter.

In a comparative study of extraction procedures, Snider and Snider (1997) examined samples collected over several years. The micro-arthropods were first Tullgren extracted, followed by three extractions in saturated sugar solution floatation-decanting. Very little organic matter was recovered by the final extraction, on 200 mesh sieve, so that it can be assumed that the invertebrates were also fully removed. The efficiency of the Tullgren heat gradient extraction was compared with total recovery after a further three floatation extractions. The results showed that even though Acarina were in general well extracted (>90%) by the

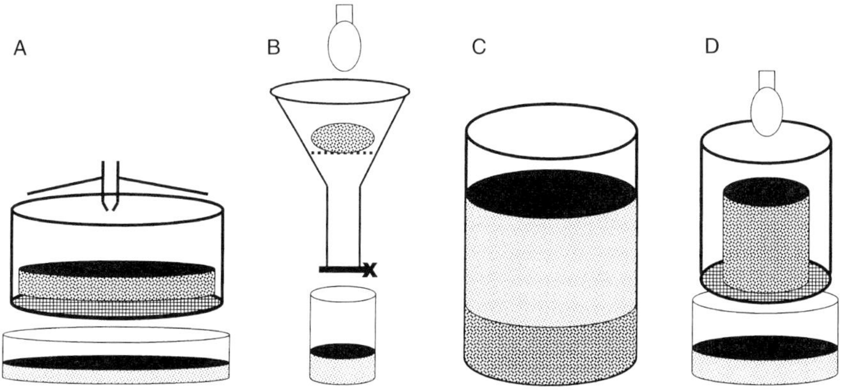

Fig. 3.4. Extraction of invertebrate species from soil and litter samples. (A) Mist chamber above the sample for gravimetric extraction with water flow. (B) Baermann funnel with a light bulb as heat source. (C) Density sedimentation and separation of organisms from soil in a suitable liquid, followed by decanting. (D) Modified Tullgren and Burlese dry extraction of microarthropods with a light bulb as heat source.

heat gradient, it was more variable for some taxa. For example, a Mesostigmata species was only recovered at 14.4 to 42.1% between samples. The results were more variable for Collembola, with the Onychiuridae being poorly extracted by the Tullgren method. In particular, the authors note that greatest variations between heat extraction efficiencies were obtained between samples on the same sampling date. There was also date-specific variability at the same site on the same soil. These observations argue against using one correction factor for all sampling dates, or all soil samples.

Other small invertebrates, namely the **tardigrades**, **nematodes**, **rotifers**, **gastrotrichs** and **enchytraeids**, are wet extracted. The soil or litter is placed in water in a funnel, and the organisms are collected into a recipient. The set-up is called a Baermann funnel extraction, and relies on organisms migrating down through the sample and sinking into a recipient. For interstitial nematodes, the Baermann funnel is the easiest method, but it is also the least quantitative (see Grassé, 1965; Nagy, 1996). Alternative and efficient methods are also recommended at the Society of Nematology homepage www.ianr.unl.edu/SON/ecology manual. A variety of centrifugation and elutriation devices have been described that are more quantitative, but require an initial investment in set-up (Freckman and Baldwin, 1990; Coleman *et al.*, 1999). The centrifugation method relies on separation of the nematodes from the soil sediment by floatation in a sucrose solution (454 g/l in water), followed by sedimentation of the organisms. It is noteworthy that different families of nematodes are extracted with different efficiency between techniques, so that choice of procedure should reflect the taxa studied (Nagy, 1996). One can expect about 50–100 species represented in a temperate forest, with 30–80 individuals/g of soil.

The wet extraction for the other taxa requires a low-wattage light bulb above the water, to drive individuals away from the heat source. More accurate enumeration requires floatation, differential sedimentation, elutriation or hand sorting, as for the nematodes.

Collection and hand sorting of larger litter invertebrates are described in detail in Sumner (2000). For the **macroarthropods**, pit-fall traps of varying diameter are used. The traps consist of a collecting funnel or conical-shaped container placed into the soil, with the opening below the litter surface. A trap is placed into the soil below the collector to capture organisms that fall in. The trap consists of a receptacle that can be capped tight, partially filled with a fixative. Both 70% ethanol and polyethyleneglycol (PEG) are commonly used. The latter has the advantage of not evaporating or becoming diluted with dew and rain, but it is very toxic. Salt water is also an effective and non-odorous alternative that is non-toxic. The collected samples are hand sorted and enumerated in the laboratory at a dissection microscope or with an inverted microscope.

Lastly, earthworms are best sampled by hand sorting soil. Samples from pedons (25 × 25 × 25 cm and deeper) are hand sifted with some washing and sieving to remove the mineral particles. One can expect to find 10–100 individuals/m^2 in temperate regions, and about six species in a given area of forest or agricultural land. Active individuals can be differentiated from inactive species, cocoons and juveniles. Depth of sampling varies with soil type, and the anecic and deeper burrowing species are always difficult to sample. Absence of a species in samples during certain seasons is probably due to deeper burrowing. A variety of other approaches exist, such as pouring formalin to irritate individuals to the surface, and electro-shock pulses. The efficiency of these methods is usually poor, especially with deeper species, and they cause a sampling bias for certain functional groups and species. Mark recapture has been possible using dyes, fluorescent markers and radioactivity. Species identification can be done visually, but may require dissection, especially in smaller species. Useful keys are provided in Sumner (2000) and references therein.

Active species at time of sampling

The aim is to estimate living and feeding organisms at the time of sampling. This provides a snap-shot of soil activities at a particular moment. Soil species **populations** are **dynamic**, and both the active species and the number of individuals of each species are continually changing. The number of individuals of each species will vary over time as cells grow and divide in favourable conditions, or will encyst or disperse in unfavourable conditions. The soil environment itself is a continuously changing environment. It undergoes diurnal warming and cooling, wetting and drying, variable litter input and localized disturbance of microhabitats. Superimposed on these changes, there are seasonal climatic variations. There is a physiological lag between cells detecting a significant change in their immediate environment and responding. There is a further lag in the numerical adjustment of the active population. Thus, species composition and the number of active individuals reflect past conditions, and individuals are adjusting to current soil conditions. Therefore, soil organisms are permanently in a state of adjusting. To obtain an idea of the rate of change and the direction of change, one must sample at regular intervals. The correct interval depends on the resolution required and the nature of the investigation.

For rapid estimation of general cytoplasmic activity by soil organisms, extraction and quantification of ATP can provide a useful index. This assay does not give any indication of the species type or functional groups which are active. However, because ATP is unstable outside cells, with a short half-life, it is a suitable marker for cellular activity. The levels of ATP in inactive cells, cysts and spores are very low, and both cysts and spores are unlikely to be extracted without bead-beating and lysis buffer. The method needs to be standardized between laboratories, and efficiency of ATP extraction optimized for the soil condition (Martens, 2001).

For ecological studies, one needs to focus on species abundances and estimates of species activity at the time of sampling. Therefore, it is necessary to extract organisms and enumerate the abundance of different species that are interacting. Several precautions are necessary to obtain living specimens and to prevent the inadvertent inactivation or activation of species after sampling soil. Rapid changes in osmotic pressure with desiccation, or in temperature, will lyse cells before they encyst. Gradual desiccation (over several hours) after sampling will affect species composition, as some will no longer be active when the sample is analysed. As protists respond within hours to changes in conditions, to observe species active at the time of sampling, it is necessary to analyse the samples within hours of collection. Some taxa, such as the filamentous hyphae and reticulate species, cannot be isolated in active form. They are either destroyed in collection or are too entangled with the soil matrix and damaged in extraction. These precautions are unnecessary for bacteria, which are too small to be damaged mechanically by sampling, and cannot be lysed by desiccation as they are protected by a cell wall.

Bacteria

When sampling for bacteria, it is often assumed that a composite sample of about 1 g accumulated from multiple milligram quantities of soil adequately represents a sampling area or pedon. Bacteria can be dislodged from particles of a soil suspension in water using a Waring blender, sonication, mechanical shaker and mild surfactant solutions, with minimal mechanical damage. However, if applied to eukaryotic cells, these techniques produce lysed homogenates, but cysts and spores will survive. Once the bacteria are dislodged from the soil mineral particles and organic matter, differential sedimentation by centrifugation with or without filtration steps can yield clean separation of the bacterial assemblage. These protocols vary with the soil types and purpose of isolation. The protocol must seek to balance how many bacteria are extracted and how much damage to cells can be tolerated (see Mayr *et al.*, 1999).

Physiologically, bacteria will responds in minutes to new conditions, so that they can be inactivated or activated rapidly. Therefore, standard extraction procedures will tend to inactivate species through dilution of the soil solution, temporary desiccation, disturbance of the soil pore microhabitats and changes in temperature. It is a challenge to determine which species were active before sampling. Probably the least questionable procedure is to use tracers and assay uptake into cells, before the sample is disturbed. Suitable markers with fluorescence, or stable or radioactive isotopes should be added in solution to the soil core and quantified after extraction. The tracer molecule should be able to enter bacteria by active transport reflecting cytoplasmic activity. These could be amino acids, nucleotides or fluorescent molecules. Common fluorescent markers include fluorescein diacetate (FDA) which is activated by a membrane esterase through cleavage of the acetate, releasing the fluorescent molecule in the cytoplasm (Söderström, 1979). Another molecule is 5-cyano-2,3 ditolyl tetrazolium chloride which is reduced by electron transfer in metabolism (Rodriguez *et al.*, 1992). Another redox activated dye is tetrazolium chloride (INT) which releases formazan in the cytoplasm. The latter is not fluorescent, but is visible under transmitted light microscopy. One must verify that the species under investigation do take up the markers under controlled active conditions.

Enumeration of bacteria from soil samples by microscopy usually requires staining of nucleic acids and cytoplasmic proteins to distinguish between living cells (with nucleic acids and cytoplasm) and dead cells (cell wall without nucleic acids inside). This provides an estimate of the fraction of cells that are potentially active and inactive, or dead. Suitable stains for nucleic acids include acridine orange, DAPI and ethidium bromide. Common protein stains for the cytoplasm include 5-(4,6-dichlorotriazin-2-yl)aminofluorescein (DTAF) and fluorescein isothiocyanate (FITC). A soil suspension extracted for bacteria is smeared and stained on a microscope slide, or observed in a haemocytometer counting chamber. Appropriate microscope settings include a magnification of 60–100 with phase contrast or fluorescence setting.

An exciting approach to identification of bacteria, without prior culture of the organisms, exploits a combination of these methods. This protocol combines the metabolic uptake of radioactively labelled substrates, together with microautoradiography of smears on slides, and direct microscopy for visualization of fluorescence from rRNA probes against specific genera (Lee *et al.*, 1999). The procedure is amenable to modifications to provide microenvironments suitable for the habitat and organisms studied. Substrate utilization profiles, conditions for activity and the identity and abundance of the bacteria can be quantified. These more focused, experimental and functional approaches are the more informative approach, rather than enumeration by taxonomy or biomass estimations.

Motile protists

The protist taxa are usually separated from the soil matrix based on their locomotion. Although the mode of locomotion is not related to ecologically relevant functional groups, it provides a convenient method of separating the cells from the matrix for identification. The following

locomotion categories are proposed: (i) amoeboid with lobose or conose pseudopodia; (ii) amoeboid with filose pseudopodia; (iii) reticulate or phyllose amoeboid species; (iv) ciliated and swimming; (v) with cilia but at least partially amoeboid and surface associated; and (vi) filamentous or hyphal growth. To avoid technical problems associated with sampling fragile cells, soil desiccation and excessive perturbation, the soil portion to be analysed is briefly and gently mixed with a spatula in a watch glass or suitable container. Once evenly distributed, a subsample is removed and weighed for microscopy, and another to obtain the oven-dried weight (Fig. 3.5). Deionized water is added to the subsample for microscopy, to prevent further desiccation, which occurs within minutes if the soil only contains capillary water. Table 3.4 summarizes common methods that have been shown to be suitable to obtain these species.

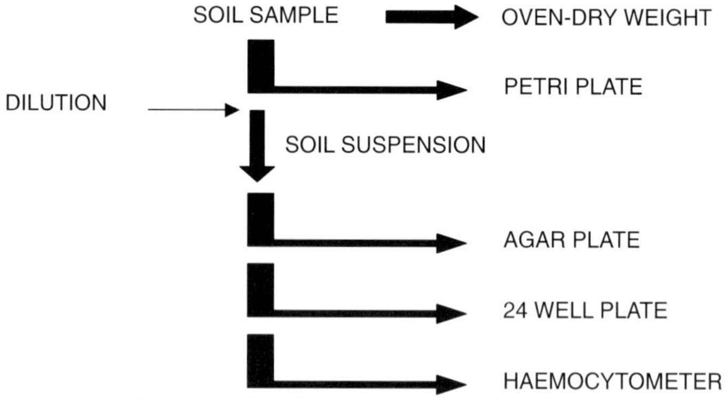

Fig. 3.5. Subsampling flow chart for enumeration of active protists. The moisture content from oven-dry weight for each sample is required. Weighed subsamples are then placed in various containers for enumeration of taxa at the microscope (see text for explanation).

Table 3.4. Typical abundances of active protists and total species in soil samples. Taxa listed by motility group based on extraction method.

	Extraction method		Numbers (g dry soil)	
Taxa	Active species	Total species	Cells/g	Species/g
Naked amoebae	Non-nutrient agar plate	Nutrient agar over time	10^3–10^6	10–100
Testate amoebae	Soil dilution, litter sections	Litter bait agar plate	10–10^4	10–20
Small flagellates	Soil dilution	Dilution, nutrient amendment	10^4–10^6	10–50
Flagellates	Soil dilution	Non-flooded Petri plate over time	1–10^2	<10
Ciliates	Soil dilution	Non-flooded Petri plate over time	0–10^2	10–30
Hyphae	Ergosterol:chitin ratio	Nutrient agar or bait	m/g	<10
Reticulate	Agar plate & membrane	Agar plate bait	?	<10
Yeasts	Radioactivity	Nutrient agar or bait	0–10^4	<5

The **amoeboid** species in the Amoebozoa (protozoa) (except the Testacealobosea) and the Labyrinthulea (Chromista) can be obtained from non-nutrient agar plates. This preparation also reveals Actinobacteria, Myxobacteria and sometimes other bacterial taxa, which can be identified from reproductive structures or colony morphology. Soil droplets, with or without dilution, are placed on a thin layer of 1.5% non-nutrient agar. The plates are incubated overnight and observed with an inverted microscope under phase contrast, at appropriate magnifications (this is not possible if the agar is too thick). The active amoebae can be visualized at the periphery of the soil droplets, as they crawl looking for food. As some cells will remain in the soil, this only provides an estimate of the number of individuals which are active. The number of cells remaining within the soil can be reduced by decreasing the quantity of soil used in each droplet. These eventually will encyst. The advantage of this technique is to provide sufficient moisture for amoeboid species, but not for swimming species to disperse. It also provides a clean preparation of living amoebae required for species identification. Amoeboid organisms are recognized by their overall shape, type of pseudopodial locomotion and details of pseudopodia.

The **Testacealobosea** and most **Filosea** (protozoa) need to be separated from the litter or particulate organic matter. These organisms are best observed in soil smears, collected by filtration on membranes, or by dissecting through litter and detritus. The **filamentous** or **hyphal** species (such as Thraustochytrid, Oomycetes and Hyphochytrea (chromista), the Chytridiomycetes and other fungi) cannot be obtained intact or active from collected soil. These species are not motile, except for the zoospore dispersal stages, or not at all. Similarly, the **reticulate interstitial species** of amoeboid organisms cannot be obtained intact. These are normally damaged during sampling, caused by movement and friction between soil particles. However, some can be isolated from relatively undisturbed soil by placing a wet membrane or tissue of large mesh on a soil sample. The organism will crawl through the pores or mesh, and can be separated from the soil beneath. Although some of these species are known, their abundance cannot be estimated, as they are not usually assayed.

The **Ciliophora** species in the soil consist of species that swim in the gravimetric water, as well as species that crawl along surfaces and on to microdetritus. They can be identified by their characteristic **swimming** and locomotion. Less abundant, larger species of other taxa that swim in the gravimetric water can be observed along with the ciliates in this preparation (such as Sarcomonadea and Euglenoida). This preparation requires the soil subsample to be water saturated, then decanted into a Petri plate several times. The wash water will contain mostly the larger free-swimming species (>20 μm) that tend not to hold on to particles. Alternatively, decigram quantities of soil can be disaggregated in a Petri plate with a large quantity of deionized water and observed. For identification, the preparation is scanned with a compound microscope at appropriate magnifications. Most species will require staining of the ciliature or, for smaller species, SEM observations are necessary.

All other taxa of swimming species (with one or more cilium), in the chromista and protozoa, are best observed in centigrams of soil subsamples in water, with an inverted microscope or compound microscope, and phase contrast (Fig. 3.6). These species tend to be about the size of small soil particles (4–12 μm long and smaller diameter), and tend to hug particles with a cilium. It is best to scan between soil particles, at the bottom of wells, and to look for movement. Transects of known dimensions can be scanned across the wells, or an ocular grid can be used to scan known areas of the well bottom. If the abundance of the nanoflagellates is $>5 \times 10^4$ individuals/ml in the suspension, the Neubauer grid of a **haemocytometer** can be used to count cells in smaller grid areas. (The haemocytometer grid is also useful to enumerate bacterial cells in soil suspensions, as their abundance is sufficiently high.) These small protist species are difficult or impossible to identify without a compound microscope. In general, SEM of fixed preparations is recommended to avoid misidentification and guessing. Many species of the soil '**nanoflagellates**' are still unknown.

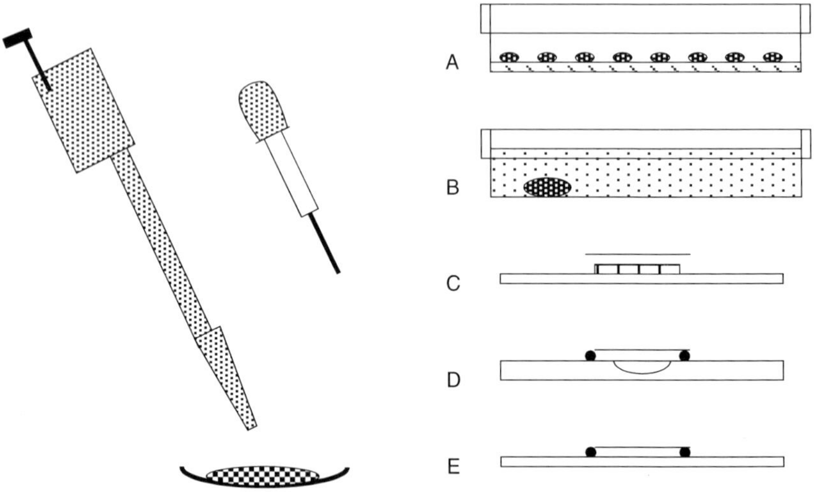

Fig. 3.6. Manipulation of living soil organisms with a calibrated pipette and micropipette, from soil in a watch glass into recipients for enumeration. (A) Soil suspension in droplets on an agar plate. (B) Soil sample in liquid medium on a Petri plate. (C) Soil dilution on a haemocytometer counting chamber. (D) Soil suspension in a depression microscope slide, with Vaseline edges to prevent evaporation. (E) Soil suspension under a coverslip on a microscope slide, with Vaseline edges to prevent drying and squashing of the specimen.

Foissner (1987) provides a discussion of recovery rate of species from soil samples. Known numbers of individuals can be recovered from a subsample to estimate the efficiency of a protocol. He reported generally good success with species of ciliates (55–100%, mean 76%), and testate amoebae (30–100%, mean 60%), and passable rates for other large cells (~50%). However, the diluted soil suspension was inefficient for amoebae and small cells (<10% recovery).

Attempts at staining protists within soil samples have not been successful. There are several problems that cannot be avoided. One is that cells attach to particles, and background staining of the soil matrix hides the cells. Furthermore, some stained cells will remain hidden behind particles, away from the microscope objective. Counts of stained soil protists are 100 times lower than direct counts of live samples, and the nanoflagellates and most amoebae are simply missed. Secondly, methods that would detach bacterial cells from particles will, in general, lyse protists and cannot be applied. Lastly, once stained, cell shape, locomotion and behaviour are no longer discernible. Thus identification becomes impossible. Staining is not recommended for counts of active species. Instead, it is preferable to look for motility under phase contrast, at the correct magnification, in each preparation, as it is a true sign of cellular activity.

Fungal and filamentous species

Approaches to the enumeration of hyphae and spore abundances were described above for total species counts. To distinguish between active and inactive species at the time of sampling, one should rely on fresh samples. Stained preparations are scored for the presence of cytoplasm and nuclei in hyphae, as a ratio of total hyphae. However, as for the bacteria, confirmation of activity at the time of sampling is by uptake of labelled substrate from intact soil cores. The label can be in the form of radioactive atoms (^{14}C, ^{3}H and ^{35}S) or stable but rare isotopes (^{13}C and ^{15}N). Choice of substrate needs to account for the diversity of species function as saprophytes. Monosaccharides or oxygen isotopes are not suitable markers because they tend to be lost through respiration as CO_2. However, they are adequate for total respiration rate data from soil cores or microcosms. Generally, amino acids, thymidine or ergosterol precursors are suitable substrates. Radioactively labelled cells can be identified by autoradiography on microscope slides, or assayed in scintillation counters. Their incorporation into proteins, DNA or cell membranes is an indication of substrate intake and growth. Stable isotopes need to be quantified by accelerator mass spectrometry or using a thermal combustion elemental analyser for stable isotope analysis by gas chromatogram separation (new instrument Delta-plus XL) (Werner *et al.*, 1999). These techniques are suitable for detection of growth in other soil taxa, but need to be exploited further. They are becoming sensitive enough for detection of change and quantification of rates from microcosm amounts of cells.

Invertebrates

Invertebrates do not reproduce and change in abundance within hours or days as protists and bacteria do. They are also more resistant to manipulation. They can be enumerated as for total species counts described in the previous section, from soil samples within 1–2 days after collection. The specimen can also be scored into age categories, from eggs and juvenile stages, to reproductive adults.

Number of Species in Functional Groups

Species isolated for enumeration by motility or by an extraction method do not represent **trophic functional groups**. Instead, they represent those species extracted by a particular method. It is not very useful to report the number of individuals or species in various motility groups, or in broad taxonomic categories. Ecologically useful information is obtained by categorizing the species into functional groups, based on feeding preferences and periods of activity. Consideration for other species that interact together through competition, symbiosis or predation is also necessary. Trophic interactions convey information about the structure of food webs and decomposition processes. Alternatively, one can assay the number of species in each functional group by baiting with particular nutrients or food items. In baiting or other culture methods, it is important to use abundances and conditions that are naturally encountered. The main functional groups that can be identified in decomposition food webs are: osmotrophy, bacterivory, fungivory, detritivory on microdetritus, predation on protists or on microinvertebrates, and saprophytic on litter. Table 3.5 lists common habitats and food preferences of protists that are co-extracted. These are discussed extensively in Chapter 4. Some species are **specialists** that have a narrow range of food alternatives (for instance on bacteria of a specific size range, or on pollen). Other species are more diversified in their food preferences. These less specialized species may interact with one or several functional groups, depending on food availability. **Generalist species** that belong to several trophic groups are **omnivorous**. Recognizing species is useful, because functional groups, seasonal periods of activity and feeding preferences can be determined from what is known of their biology.

Table 3.5. Preferred habitats and common functional groups of protists. Groups do not represent taxonomic categories, but are based on the extraction method.

Taxa	Habitat	Food preferences
Naked amoebae	Interstitial, some in litter	Bacterivory, fungivory, osmotrophy, predatory
Testate amoebae	Litter, organic horizon	Saprotrophic, predatory, detritivory, (fungivory?), (bacterivory?)
Small flagellates	Interstitial, some in litter	Osmotrophy, bacterivory
Flagellates	Interstitial, some in litter	Bacterivory, detritivory, others possible
Ciliates	Interstitial, rare in litter	Bacterivory, fungivory, predatory, detritivory, osmotrophy
Hyphae	Litter, interstitial	Saprotrophic, predatory
Reticulate	Interstitial, some in litter	Bacterivory, osmotrophy, fungivory, others possible
Yeasts	Interstitial, litter	Osmotrophy

Summary

The aim of soil collection is to obtain a representative sample of the species diversity. The sampled soil is delicate and subject to crushing and drying, which damage soil organisms. Storage and handling of collected samples affect species activity and the composition of species that can be enumerated. The number of samples required to describe a site over time completely is usually very large. Soil ecologists must balance the number of samples with the amount of time spent on each sample. In practice, the experimental design tries to limit the breadth and depth of analysis to what can be achieved in a timely manner. The organisms are extracted from the soil for species enumeration and estimation of their abundance. One must be aware of what taxa are not extracted by each procedure. Similarly, there is no one culture medium that is suitable for all bacteria or all protozoa. Different taxonomic groups require different methods that are appropriate for their biology. There are two strategies for extraction of species from soil. One approach aims to identify as many species as possible, and the other aims to identify only those species active at the time of sampling. In general, wet extractions have focused on nematodes, at the expense of all other co-existing wet-extracted species.

Suggested Further Reading

Coleman, D.C. and Crossley, D.A. (1991) Modern techniques in soil ecology. *Agriculture, Ecosystem and Environment* (Special issue) 34, 1–507.

Fuller, M.S. and Jaworski, A. (1987) *Zoosporic Fungi in Teaching and Research.* Southeastern Publishing Corp. Athens, Georgia.

Krebs, C.J. (2000) *Ecological Methodology*, 2nd edn. Benjamin-Cummings, California.

Lee, J.J. and Soldo, A.T. (1992) *Protocols in Protozoology*. Society of Protozoologists, Allen Press, Lawrence, Kansas.

Robertson, G.P., Coleman, D.C., Bledsoe, C.S. and Sollins, P. (1999) *Standard Soil Methods for Long Term Ecological Research*. Oxford University Press, Oxford.

Reconstructing the Soil Food Web

4

The top soil is a skin over the earth's surface which functions as a natural compost. It receives a continuous input of dead tissues from primary production on its surface, and below-ground from roots and soil organisms. The trophic interactions within the soil are responsible for decomposition and biomineralization. Certain soil organisms invade fresh litter, while other species continue to feed on older organic matter within the profile. Once the sampled soil has been extracted for various organisms, the spatial structure is lost and the organisms are outside their natural environment. One needs to determine: (i) what the role of each species is in the soil, in terms of feeding preferences; and (ii) what their spatial preferences are in the soil structure. In this chapter, we discuss the ecological role of species, based on what they feed on. Similarities in feeding preferences and in mechanism are used to group species into trophic functional groups. Using knowledge of the biology of species to place them into functional groups simplifies the task of dealing with many interacting species. It is more convenient to describe interactions between the functional groups. In this chapter, we define interactions between functional groups that are responsible for the primary and secondary decomposition of litter. In Chapter 5, we will discuss species distributions vertically through the soil profile, horizontally across the landscape, and species succession through time.

Functional Categories

The sources of substrate for decomposition food webs are litter from above- and below-ground sources, other organisms in the soil, and the excreted and secreted material from all these trophic interactions. Litter is used as a general term for all that was recently living together with excreta (see Chapter 2). The organic matter derived from tissues in the litter is increasingly digested by repeated cycles of ingestion and excretion. Species participate in the ingestion of food, its partial digestion and absorption, and excretion of the remaining portion. Therefore, repeated cycles of biological processing and chemical modification of organic matter have two effects. One causes an accumulation of increasingly indigestible (or recalcitrant) organic matter. The other supplies a steady source of soluble nutrients into the soil solution. The end result of decomposition trophic interactions is the return of nutrients into primary production food webs, in the form of soil solutes, carbon dioxide and other gases (Figs 2.1 and 4.1).

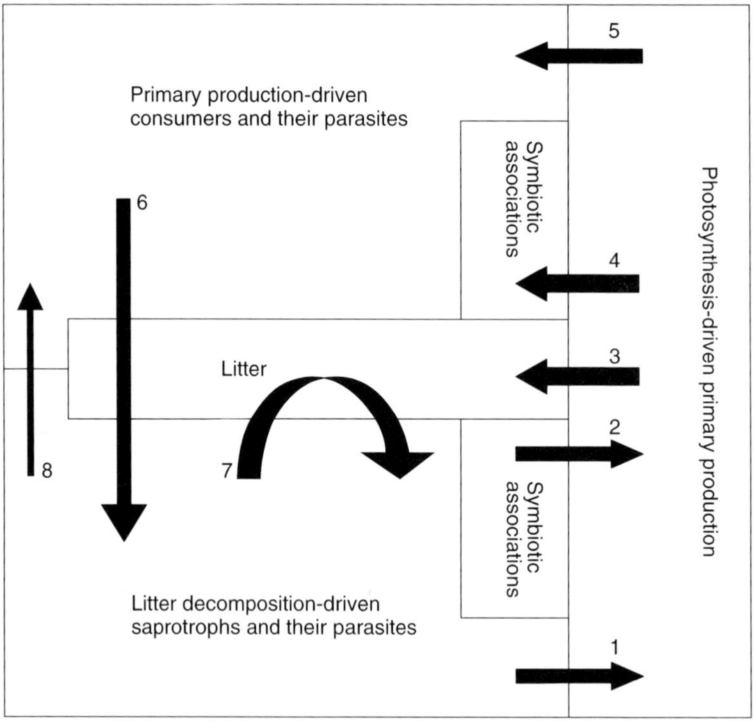

Fig. 4.1. Flow of nutrients between primary production and saprophytic species. The diagram does not indicate the biomass and amount of flow, but shows the direction of flow between living and decomposing biomass. Numbered arrows indicate (1) uptake of nutrients from the soil solution by roots and rhizoids; (2) uptake of nutrients by mycorrhizae; (3) litter from plant-derived tissues; (4) plant tissue digestion by symbionts of consumers of primary producers; (5) consumption of primary producers; (6) litter from consumers in the primary production subsystem; (7) litter from saprotrophs; and (8) consumption of saprotrophs by consumers in the primary production subsystem.

Primary decomposition is initiated in part by organisms called **primary saprotrophs**, in that they release digestive enzymes into their environment to solubilize a source of nutrient, or substrate. It is a form of external digestion that releases soluble nutrients from previously insoluble organic matter. Most primary saprotrophs are found in several classes of the bacteria kingdom and most families of the fungi kingdom, but some are also found in other kingdoms. Decomposition also requires **fragmentation** of litter. It is performed by a broad variety of species that facilitate the access of other primary saprotrophs to the litter and its permeation into the soil. Species that carry out fragmentation are also called shredders, and occur mostly in the Arthropoda classes Insecta, Arachnea and Crustacea, in the Annelida, and very few in some Aves and a small number of Mammalia. Invertebrate shredders often require a symbiotic association in their gut to carry out the digestion. These are specialized anaerobic protozoa which are usually Ciliata but, in some termites, they are Hypermastigota, as well as a variety of bacteria. Further decomposition is achieved by organisms that feed on microscopic particles of organic matter that become released from the litter by fragmentation. These are called **detritivores** and include species in the Nematoda, Oligochaeta, Testacealobosea, Gymnamoeba (Lobosea), some Colpoda and other protozoa. Another functional category includes the **osmotrophs**, which absorb soluble organic matter present in their environment, mostly released from the above interactions. These species differ from primary saprotrophs because they do not secrete enzymes to solubilize their nutrients. Although some classes and species are exclusively osmotrophic, many species throughout the protozoa, chromista, fungi and bacteria are only partially osmotrophic. An important and large category responsible for most biomineralization is **consumers** (secondary saprotrophs), which include species that feed on primary saprotrophs, fragmenters and detritivores. The consumers are subdivided into bacterivores, fungivores and cytotrophs on unicellular species, or predators on invertebrates.

Primary Decomposition

Primary decomposition involves the initial processes that modify the dead tissues and excreta from primary production, which constitute the litter. The litter was defined in Chapter 2, to include shed leaves, fallen trees and branches, animal cadavers and shed body parts, shed reproductive structures, animal excretions and all secretions. These accumulate primarily on the soil surface (Table 4.1), although tree canopies also accumulate litter, particularly in wet tropical forests. A substantial portion of below-ground litter accumulates in the form of dead roots and growing root tip exudates. Primary decomposition involves the initial

Table 4.1. Summary of litter biomass (standing crop) on temperate forest floor. Estimates for cold and warmer climates in temperate forests, indicating residence time on forest floor, and turnover rates for woody and non-woody litter.

	Forest floor biomass (kg/ha)		Mean residence time (year)		Litter fall (kg/ha/year)			
					Non-woody		Woody	
Temperate forest type	Warm	Cold	Warm	Cold	Warm	Cold	Warm	Cold
Broadleaf deciduous	11,480	32,207	2.7	10.2	4236	3854	891	1046
Broadleaf evergreen	19,148	13,900	3.1	3.9	6484	3590	?	?
Needle leaf evergreen	20,026	44,574	4.6	17.9	4432	3144	1107	602

Data from Boddy and Watkinson (1995).

lysis of cells in the dead tissues, release of soluble molecules, physical fragmentation of the tissues, and colonization by primary saprotrophs which further dissolve and partially ingest the organic matter. The organisms grouped together as primary saprotrophs secrete enzymes for the extracellular digestion of litter, and participate in the uptake of soluble nutrients from the soil solution. These primary saprotrophs are in part dependent on litter fragmentation by the macrofauna and other litter invertebrates.

Plant senescence and necrosis

Senescence of leaves involves a massive modification and mobilization of cytoplasmic components, out of the living cells, in preparation for shedding the organ. A similar process occurs in shedding other plant organs, such as roots and reproductive organs, but the biochemistry is less studied. By the time the plant organ is shed, the tissues are low in nutritional value compared with the living cytoplasm. The exception is when viable parts of plants are physically broken by animals or the wind.

The chronosequence of events is coordinated, and involves the dismantling of organelles and translocation of nutrients to growing tissues (Smart, 1994; Biswal and Biswal, 1999). There is a total loss of chlorophyll and degradation of macromolecules (proteins, lipids, nucleic acids and starch). The genes that regulate this process are known as *senescence-associated genes*. More than 30 such genes are known from a variety of plants. Senescence is initiated by environmental cues through plant physiology. It is driven by the accumulation of abscissic acid and involves other plant hormones. It ends with the **abscission** of the organ or its physical breakage from the plant. The fall of incompletely senesced leaves therefore contributes a greater supply of nutrients to the soil.

Translocation of nutrients from tissues during senescence aims to salvage nutrients from the parts to be shed, and use them in the growing parts of the plant. Nitrogen from proteins is removed as NH_4 and used for the *de novo* synthesis of glutamine and asparagine. The two amino acids can be translocated to other parts of the plant. Gluconeogenesis enzymes use lipid substrates (primarily from thylakoid membranes) through the glyoxylate cycle to form sucrose. Sucrose is the primary plant sugar for transport. Nucleic acids are released as purines and pyrimidines. These are degraded further into smaller soluble forms for translocation. The phosphate component is also salvaged, as well as sulphur and other minerals and metal ions. The main plant cell storage compound is starch. It is hydrolysed into monomers and other sugars which can be transported.

A variety of species of bacteria and yeast adhere to living leaves (Beatty and Lindow, 1999). These species are mostly osmotrophs. They benefit from the increase in available soluble nutrients from senescing tissues, prior to leaf fall. It is an adaptation for earlier access to plant soluble nutrients, compared with species that reside in the litter. Many mildews, rusts and other fungi also proliferate in senescing parts, as tissues lose their natural resistance to infections.

Plant species have adopted different physiological thresholds for shedding older or less efficient organs. There are also different mechanisms for survival through stress. These adaptive strategies to variations in seasons and to climate result in a variety of patterns of senescence (Leopold, 1961; Larcher, 1973). Some are evergreen, others are seasonally deciduous, and others are annuals that survive as roots or bulbs. Therefore, the pattern of annual and seasonal contribution to soil litter depends on plant species composition. The quantity and the quality of litter input are determined by plant-driven primary productivity.

Physical degradation of tissues

Several factors contribute to the initial degradation of dead animal and plant tissues in the litter. The first is the continued autolysis of cells that begins with cell death. Autolysis involves an internal digestion of cell organelles and components. In senesced leaves, cell lysis begins before a leaf falls. In a fallen tree, some cells may continue to live for weeks. Cells of animal tissues begin autolysis soon after death of the tissue. Autolysis increases leakage of soluble molecules from tissues, especially when the cell membrane is lysed. However, the cell wall of plant cells is not lysed and prevents larger molecules from being released by leakage. Instead they are trapped inside a cellulose box which must be broken into. Further abiotic degradation of tissues occurs as a result of desiccation. Loss of cell and tissue water causes a clumping of denatured cytoplasmic molecules. It further contributes to the physical breakage of plant cells and tissues as the litter becomes brittle and easily fragmented. A similar effect on the cytoplasm results from freezing. Repeated cycles of drying and wetting, or freezing and thawing, further denature the cytoplasm and fragment the litter, releasing soluble molecules. Storage of leaf litter for decomposition experiments usually involves drying of leaves. It is a pre-treatment step which can alter the initial character of the litter and the survival of organisms that have already colonized it.

Macrofauna invertebrates

A large variety of taxa are represented in the macrofauna. It is necessary to distinguish between those that are predatory on litter organisms (such as spiders), those that are part of the above-ground primary production food webs and those that contribute to decomposition as saprotrophs. Some species have a particularly significant impact on the initial fragmentation and **comminution** of the litter. The macroarthropods, earthworms, onychophorans and gastropods contribute to this. These organisms feed on the litter by removing small pieces and help break it up into smaller fragments. This has the effect of increasing the surface area of exposed tissue. The salivary secretions and excretion of partially digested litter contribute to the initial litter decomposition. They increase leachate from the lysed cells in the litter, and facilitate its colonization by primary saprotrophs. Since the sites of ingestion and of subsequent excretion are not the same, these organisms also contribute to the redistribution of the litter, both vertically into the profile and horizontally across the soil surface.

A comprehensive treatment of species implicated in comminution and initial litter breakdown by the macrofauna can be found in several books (see Swift *et al.*, 1979; Dindal, 1990; Coleman and Crossley, 1996). A few brief points are made below as examples of their contribution. The principal macroarthropods include the Isopoda, Diplopoda (millipedes), Isoptera (termites), Coleoptera (beetles) and the Hymenoptera family Formicidae (ants) (Table 4.2). The earthworms include the epigeic (litter) and anecic (vertical burrows with permanent surface openings) species (see below, this chapter). Several families of Coleoptera have species that contribute to litter and macrodetritus fragmentation. *Anobividea* and *Blattodea* are important on detritus in general, or leaf litter and softened wood. Several genera prefer animal matter and cadavers, in the *Ptinidea*, *Dermesidea* and *Nitidulidea*. Some *Dilphinidea* species also have a strong preference for cadavers. The genus *Necrophorus* buries cadavers by digging under them, and a reproductive couple will then feed on it through time, in the burrow. Several species prefer mammal dung, such as *Scarabaeus sacer*, *Hister quadrimaculatus* and in the genera *Copris* and *Ontophagus*. *Typhoeus typhoeus* hoards dung inside deep tunnels. Some insects in the orders Diptera and Coleoptera are important particularly in the initial decomposition of animal cadavers (Byrd and Castner, 2001). Many ant and termite species are efficient scavengers of detritus which they accumulate in nests. As with the action of some earthworms, these have the effect of mixing litter into the soil, as well as breaking it up and carrying out a first round of digestion–excretion. However, the organic matter may not become accessible to the soil outside the nest, while the walls are maintained by the colonies. Finally, the function of invertebrate larvae in the soil can be very different from that of the adults. Species that may not participate in decomposition as adults may contribute to decomposition as larvae or juveniles.

The macrofauna are also important shredders and saprotrophs on animal cadavers. The impact of animal cadavers, especially those of large animals, affects soil nutrients and plant species composition in the

Table 4.2. Invertebrate taxa with usual habitat preferences and food preferences.

Taxa	Habitat	Food preferences
Rotifera (phylum)	Wet litter, riparian	Water current capture or predation on nematodes, protozoa, bacteria
Nematoda (phylum)	Interstitial, litter	Bacteria, hyphae, roots, predation or omnivorous
Tardigrada (phylum)	Wet litter, bark, moss	Nematodes, detritus
Platyhelminthes (order Tricladida)	Litter	Earthworms
Lumbricina (class)	Litter, interstitial	Litter and log fragmentation, soil
Enchytraeid (family)	Interstitial	Protists, bacteria and invertebrate excreta
Isopoda (order)	Dark and wet	Litter fragmentation, plant consumers
Diplopoda (class)	Litter or interstitial	Litter fragmentation and detritus, Ca
Chilopoda (class)	Litter	All predators, some fragmentation
Pauropoda (class)	Litter	Hyphae, fungi-infested litter, logs and soil
Collembola (order)	Litter, top soil	Hyphae and nematodes, some are live root and plant consumers
Protura (order)	Litter, top soil	?
Diplura (order)	Litter, top soil	Predation on microarthropods, or detritivore
Thysanura (order) (family Machilidae)	Bare rocks	Protists and detritus
Isoptera (order)	Nests	Wood, soil, detritus, or hyphae cultured on detritus
Hymenoptera (order)	Nests	Predators, detritus, other
Coleoptera (order)	Litter	Predation, varied
Scarabaedidae	Litter	Wood, litter, dung
Aranea (order)	Litter	Predation
Pseudoscorpionida	Litter, top soil	Predation on small invertebrates
Acari (order)	Litter, top soil	Nematodes, hyphae, detritus

immediate vicinity. Large carcasses, such as ungulate species, constitute an intense localized disturbance, which releases a concentrated pulse of nutrients into the soil, and changes the vegetation for years after (Towne, 2000). It is unclear to what extent soil nitrogen supply is derived from invertebrate and vertebrate animal cadavers and excreta. The effect of cadavers (or animal detritus) has been much less studied than plant-derived litter, presenting opportunities for further research.

The impact of macroinvertebrates on litter decomposition is usually demonstrated using mesh bags enclosing leaf litter. Leaf litter is placed on the field site, using mesh sizes to bias against size classes of the litter fauna (exclusion of fauna), and against larger mesh sizes that do not (control litter bags) (Fig. 4.2). Using mesh of 1–2 mm, most macrofauna are excluded, but this often increases moisture content in the litter bag

Fig. 4.2. Litter bags of plastic mesh holding decomposing leaf litter. The large mesh allows access to macroarthropods and the larger organisms, while fine mesh bags limit access of macroinvertebrates.

(Vossbrink *et al.*, 1979). The rate of litter mass loss in control and exclusion litter bags is obtained from weight loss through time. This is approximated by a simple decay equation $W_t = W_0 \, e^{-kt}$, with the decay constant k representing the rate of mass loss through time t, and W_t and W_0 representing the final and initial dry weight of litter. The decay constant k can be obtained from the slope of the natural logarithm of the graph, from $\ln W_t - \ln W_0 = -kt \ln e$. However, the actual pattern of mass loss is often multiphasic, with periods of rapid loss and periods of stable or slow mass loss (Fig. 4.3). The effect is caused in part by seasonal variations and by species succession on the litter. The pattern mostly reflects plant-specific differences in litter nutrient composition, timing of leachate of toxins and rate of decomposition of the cell walls. The procedure can overestimate the rate of decomposition, because macroinvertebrates will chew on the litter, but defecate the unassimilated portion at a distance outside the litter bag. Furthermore, a portion of litter fragments falls out of wide mesh bags that permit access to macroinvertebrates. The rate of mass loss can be underestimated by a variable amount, due to an increase in mass of organisms in the litter tissues. This value is normally low, but can be more significant in some cases. For instance, pine needles can be embedded with enchytraeids, Collembola, oribatids and other organisms (Ponge, 1991; Edslung and Hagvar, 1999). The presence of primary and secondary saprotrophs inside the litter does, however, contribute significantly to the litter nutrient content (such as elemental nitrogen).

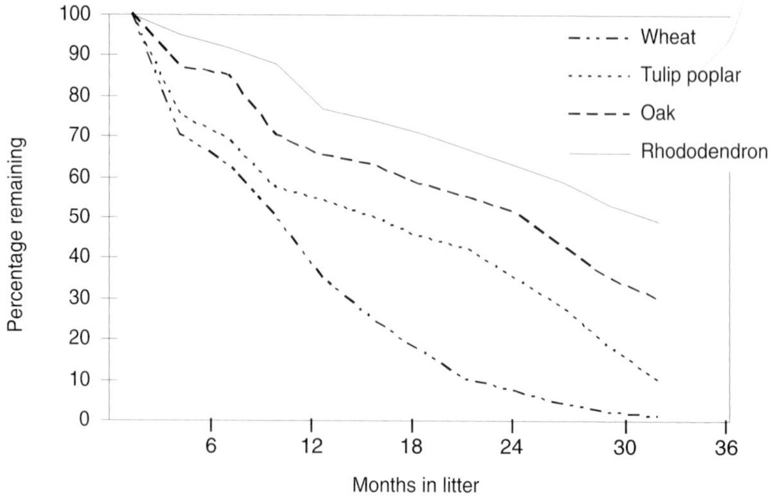

Fig. 4.3. Rate of mass loss over time in several leaf litter species. Representative data from single species of leaf litter placed in large mesh litter bags, in a temperate forest in North Carolina.

Litter mass loss rates

The rate of initial litter decomposition from mass loss studies in litter bags varies with climate (temperature and moisture), litter chemistry and faunal composition (Table 4.1) (Cadish and Giller, 1997). Comparison of leaf mass loss across climatic regions shows that the effect of climate is the principal predictor of the initial decay rate (Swift *et al.*, 1979). This was shown further in a review of leaf litter decomposition from 44 varied geographic locations (Aerts, 1997). The climate was characterized from the *actual evapotranspiration* (AET) from monthly temperature and precipitation data (Thornthwaite and Mather, 1957). This value is superior to temperature or moisture values used alone. It is an index of climatic energy and the availability of capillary water from soil and litter (Meentemeyer, 1978). Several generalities were observed from the comparisons reported by Aerts (1997) (Tables 4.3 and 4.4).

1. After the actual evapotranspiration value (climate effect), leaf litter chemistry was the next most important parameter determining initial decomposition rate. Climate exerts the strongest influence on initial litter decay rate (k) between geographic regions, but litter chemistry (nutrient content, lignin and secondary metabolites) is a better descriptor of variations in decay rate locally.
2. A threefold increase in AET from temperate to tropical regions is accompanied by a sixfold increase in the decay constant.

Table 4.3. Summary of data of 44 leaf litter decomposition studies, from temperate, Mediterranean and tropical sites (modified from Aerts, 1997).

Region	Temperate	Mediterranean	Tropical
Lignin (%)	22.7	15.6*	20.9
Lignin:N ratio	32	29.4	24.2*
Lignin:P ratio	848	660	764
C:N ratio	62.2	78.9	53.8*
Decomposition rate, *k* (per year)	0.36	0.35	2.33*
AET (mm/year)	590	571	1475*

*Significant difference between data along one row.

Table 4.4. Selected most significant regressions (coefficient of regression, r^2) between decomposition rate (k) and a single or two other parameters (litter chemistry and AET), with statistical significance level (P) indicated.

k versus single parameter	r^2	P
Temperate regions		
N (%)	0.05	<0.05
Mediterranean regions		
N (%)	0.08	<0.05
Lignin:N ratio	0.24	<0.005
Tropical regions		
N (%)	0.24	<0.0001
Lignin:N ratio	0.57	<0.0001
C:P ratio	0.39	<0.005
Total data set		
N (%)	0.24	<0.0001
Lignin:N ratio	0.24	<0.0001
AET (mm/year)	0.46	<0.0001
***k* versus two parameters**		
Temperate regions		
No significant multiple regressions		
Mediterranean regions		
AET + lignin:N ratio	0.58	<0.0007
AET + C:N ratio	0.38	<0.0251
Tropical regions		
No significant multiple regressions		
Total data set		
AET + C:N ratio	0.57	<0.0001
AET + lignin:N ratio	0.53	<0.0001
AET + C:P ratio	0.52	<0.0001

Modified from Aerts (1997).

3. Much variation exists between specific leaf litter chemistry and the decay rate. In temperate climates, there was no suitable chemical predictor of mass loss rates. In Mediterranean and tropical climates, the lignin:N ratio was the best chemical predictor.

However, the effect of soil mineralogy on the overall process of decomposition and nutrient recycling must not be neglected (Fig. 4.4). For instance, the more weathered tropical soils (often oxisol or ultisol) are usually poor in phosphorus (P-limited soils) (Vitousek and Sanford, 1986). In contrast, temperate forest soils tend to be nitrogen limited more often (Vitousek and Howarth, 1991).

Litter chemistry and decomposition

The interaction between nitrogen and lignin content of leaf litter was characterized further in Hawaiian tropical forests (Hobbie, 2000). The study used leaves of *Metrosideros polymorpha*, a dominant tree in Hawaiian forests, because its lignin content varies with the environment. Leaves from drier areas have <12% lignin concentration, and those from wetter areas have >18% lignin. The litter bags were constructed with 2 mm mesh, which minimizes macrofauna grazing on the litter, but permits microbial and microarthropod decomposition. Litter mass loss rates are slower with increased lignin content, or higher lignin:N ratio. Since these sites are naturally poor in nitrogen, rates were compared with N-fertilized sites. There was an increase in mass loss rate (*k* value) in fertilized plots, regardless of lignin content (Fig. 4.5). However, the amount of increase was much lower in the high lignin litter. The study concludes that on these N-limited soils, litter decomposition is naturally limited by

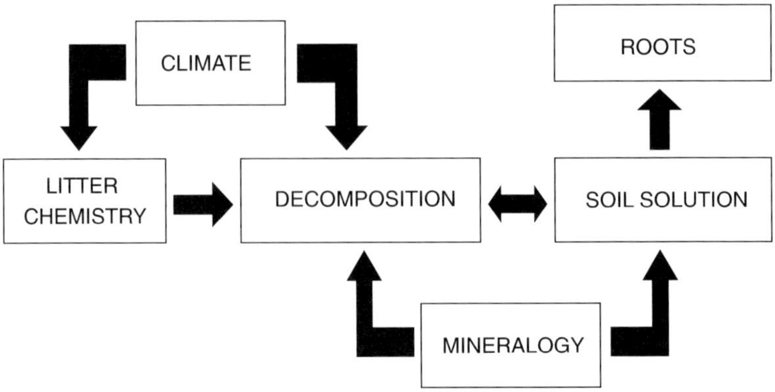

Fig. 4.4. Overarching effect of climate and soil mineralogy on regulating the rate of decomposition. This affects the rate of biomineralization and soil solution composition. In turn, the effect is seen in plant nutrition and subsequent litter chemistry.

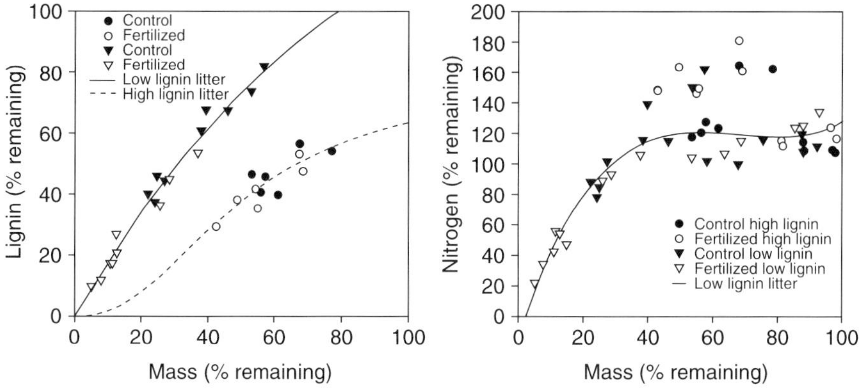

Fig. 4.5. Implied trends of lignin and nitrogen decrease, in high and low lignin litter of *Metrosideros polymorpha*, in a Hawaiian forest. Data for N in high-lignin litter are scattered in the circle (data modified and adapted from Hobbie, 2000).

insufficient N nutrient. (According to **Liebig's law**, organism growth is limited by the most limiting nutrient.) The pattern of lignin loss was linear with litter mass loss. However, for N content, there was little decrease until about 60% of the initial mass was lost. This may reflect the stage at which sufficient cell wall breakdown permits nutrient loss from within the leaf litter, and tissue comminution. However, the cytoplasm of primary saprotrophs, especially active hyphae, will contribute significantly to litter N content. In temperate forests, there were several reports of inhibition of lignin digestion in the presence of sufficient nitrogen (Keyser *et al.*, 1978; Fog, 1988; Berg and Ekbohm, 1991; Magill and Aber, 1998). No such inhibition was observed in the Hawaiian study. It raises the question of whether the inhibition observed in temperate sites can be extended to all regions. There may be species and regional differences. In general, there are no clear trends between studies of fertilization with inorganic nitrogen on litter decomposition rate (see Hobbie, 2000).

Mesofauna diversity and climate effect on initial litter decomposition

The following study was established to consider the added effect of macro- and microarthropods on initial mass loss rate of litter (Gonzalez and Seastedt, 2001). The experiment consisted of litter bags representing two leaf litter qualities, compared at two subtropical sites (wet and dry sites), and at two subalpine sites (north- and south-facing slopes) (Table 4.5). Each was replicated with and without arthropods. Normally, studies and models begin with the assumption that decomposer organ-

Table 4.5. Percentage composition and ratios of two leaf litter species, in a control and arthropod exclusion experiment, at four climates.

Litter species	Site				
Initial content		C:N ratio	Lignin:N ratio	Lignin %	Cellulose %
Quercus gambelii		63.6	26.0	20.0	41.2
Cecropia scheberiana		35.6	26.7	34.6	46.1
After 1.5 years		Control C:N	Exclusion C:N		
Quercus gambelii	Tropical dry	28.7	26.7		
Cecropia scheberiana	Tropical dry	20.7	19.9		
Quercus gambelii	Tropical wet	23.8	27.8		
Cecropia scheberiana	Tropical wet	19.7	21.3		
Quercus gambelii	Subalp. North.	34.4	40.0		
Cecropia scheberiana	Subalp. North.	33.5	29.4		
Quercus gambelii	Subalp. South.	39.0	40.4		
Cecropia scheberiana	Subalp. South.	30.9	29.6		

Data from Gonzalez and Seastedt (2001).

isms are totally constrained by the climate (AET) and litter chemistry (quality). For example, among tropical sites with less variability between seasons, there should be less climate influence on decomposition, and a more important role for the fauna, whereas in more northern or southern latitudes, temperature constrains cellular activity, and thus the rate of decomposition. However, this is not a complete argument, since the abundance and diversity of the decomposer species also vary with climatic region, ecosystem and latitude (Swift *et al.*, 1979; Gonzalez and Seastedt, 2001). This experiment attempted to measure the impact of arthropod functional diversity and abundance on initial litter decomposition between climatic regions. In this study, macroarthropods were removed by sieving the soil, and by preventing recolonization with a physical barrier and chemically with naphthalene (an arthropod repellent). The naphthalene has the added effect of also greatly reducing the microarthropod abundance. The control plots were also sieved, and had a physical barrier. The macroarthropods were not removed and naphthalene was not added, to maintain the arthropod abundance. One indirect effect of sieving soil is to destroy soil stratification, many organisms and the hyphal mesh. In all plots, the 2 mm mesh of litter bags would have reduced macroarthropod access.

The results of this arthropod exclusion study (Table 4.6) show that climate, litter substrate and the biota component each independently influenced the decomposition rate. The abundance of the microarthropods was much greater in the wet tropical site than at the other sites. The functional diversity (per g dry litter) was also greater at the tropical wet site. The effect of arthropod exclusion in slowing the rate of mass

Table 4.6. Abundance of microarthropods (mean, per g of dry litter) and the percentage effect of arthropod exclusion on litter decomposition rate (k, per year).

Site	Treatment	Oribatida	Mesostigmata	Prostigmata	Collembola	Total % effect	Litter species
Tropical	Control	1.45	0.1	0.19	0.01		
dry	Exclusion	0.01	0.03	0.03	0	1.6	*Quercus gambelii*
						23.9	*Cecropia scheberiana*
Tropical	Control	43.17	8.9	0.72	1.48		
wet	Exclusion	1.66	1.92	0.25	1.15	45.2	*Quercus gambelii*
						66.2	*Cecropia scheberiana*
Subalp.	Control	1.45	2.27	1.07	0.92		
North	Exclusion	0.05	0.13	0	0.05	35.6	*Quercus gambelii*
						49.8	*Cecropia scheberiana*
Subalp.	Control	4.97	3.07	0.85	0.82		
South	Exclusion	0.01	0.04	0.01	0.03	39.1	*Quercus gambelii*
						12.4	*Cecropia scheberiana*

Based on the experiment in Table 4.5 (data from Gonzalez and Seastedt, 2001).

loss in litter bags was greatest in the wet tropical site, followed by the subalpine sites, and least affected was the tropical dry site. The climate effect (measured as AET) correlated with the abundance of microarthropods in the litter bags in the control plots. Instead, the climate effect correlated with the decomposition rate (*k*) in the arthropod exclusion plots. There was no difference between the effect of the arthropods on the two litter species chosen in this study (Tables 4.5 and 4.6).

Normally, there will be litter from several plant species accumulating together. One can speculate that in mixed species litter, faster and more nutritious litter types can promote the decomposition rate of slower litter. This can occur through the release of particular nutrients into the soil solution (Seastedt, 1984; Chapman *et al.*, 1988). Alternatively, the overall rate can be slowed through leaching of secondary metabolites, tannins and phenolics from some litter species (Harrison, 1971; Swift *et al.*, 1979; Chapman *et al.*, 1988). In some combinations, there may be no effect at all on the mass loss rate in mixed species litter (Blair *et al.*, 1990; Prescott *et al.*, 2000). Blair *et al.* (1990) also reported a significant effect of mixed litter on the composition of the decomposer community (microarthropods, nematodes, hyphal length and bacteria) and on the pattern of nutrient release (such as nitrogen). Hansen (1999) also observed that some combinations promoted mass loss in litter bags. Prescott *et al.* (2000) further observed that with mixed litter, initial mass loss rates (1–2 years) did not always predict the effect of treatment over several years (5 years).

Secondary Decomposition

Most biological studies of decomposition have focused on primary decomposition, using fresh litter (usually roots or leaves) placed in mesh bags and sampled for 1–2 years. At this point, only 20–80% of most fresh litter is broken down into organic matter. Observations are continued longer more rarely. Animal sources of litter, such as insect parts, excretions and frass, or animal dung and cadavers, are largely neglected. These are an important source of pulsed nitrogen into the soil food webs. With time, the litter is fragmented and decomposed into humus by a succession of species within and between functional groups. As fresh litter accumulates on top, the older litter becomes part of deeper soil horizons. A fraction of the decomposing organic matter is respired by saprotrophs (CO_2), a fraction is metabolised into new cellular biomass and exoskeleton, a portion is excreted by the saprotrophs, and a portion is abandoned as fungal cell wall material. The excreted and abandoned portions become part of the soil humus or organic matter. Secondary decomposition occurs in the soil horizons below the A_L and A_F horizons. It involves the decomposition of organic matter that has already been partly decomposed from fresh litter. Not all of the organic matter below the A_F horizon is in secondary decomposition. The deeper horizons also contain fresh litter from below-ground sources, such as roots and dead organisms. In some ecosystems, such as forests and prairies, the predominant flow of carbon into the soil is from roots, particularly fine roots <2 mm in diameter (Coleman, 1976; Fogel, 1985; Fahey and Hughes, 1994; Wells and Eissenstat, 2001).

Secondary decomposition cannot be studied from litter in mesh bags, and it is more difficult to trace and study *in situ*. The controlling roles of community structure and trophic interactions in secondary decomposition are best studied in the laboratory using intact soil cores and soil microcosms. The main sources of substrate in secondary decomposition are excreted pellets, microdetritus in varying stages of decay, colloids and amorphous organic matter. The faecal pellets can constitute >80% of the organic matter in the forest floor horizons (Pawluk, 1987). The role of the food web in the decomposition of this fraction of organic matter has been detailed insufficiently. The effect of community structure in regulating the rate of secondary decomposition is poorly understood. There are reviews based on chemical extractions of the organic matter in secondary decomposition (Stevenson, 1994; Sollins *et al.*, 1996). These analyses highlight the increasing influence of clays on soil solution composition and dissolved organic matter in the solution. The adsorption of organic matter and ions to clays is greatly affected by the mineralogy and cation exchange capacity (Sposito, 1989; Kalbitz *et al.*, 2000). However, almost none of the studies were conducted *in situ* or taking into consideration the biological species that release the organic matter into the soil solution.

Primary Saprotrophs

Species that first colonize litter and are involved in its digestion are called **primary saprotrophs**. Most of these are prokaryote and fungal species that release extracellular enzymes. The digestion is extracellular, and complex molecules and polymers are hydrolysed to smaller molecules which can be absorbed through cell membranes. The transfer of nutrient molecules into cells through the cell membrane is called **osmotrophy**. Yeasts and testate amoebae are also important early colonizers of litter and contribute to its decomposition and soil nutrient absorption.

Saprotrophic bacteria

Bacteria obtain their nutrients from the soil solution by osmotrophy. They rely on nutrients and minerals already dissolved from litter. Most species secrete small enzymes that can pass through the bacterial membranes, or that assemble on the outer membrane (White, 1995; Van Wely *et al.*, 2001). These enzymes digest specific chemical bonds of substrate compounds in the soil environment. The soluble molecules released into the soil solution become accessible for uptake by the bacterial cell membranes, and other osmotrophs. The secretion of specific enzymes, and the activation of substrate molecule transport mechanisms, depend on chemical detection of the substrate. In the presence of sufficient nutrient molecules, nutrient intake can be sufficient to maintain cell growth to division. This condition also requires adequate pH, salt balance, temperature and moisture. Some species have single substrate requirements for growth. Other species require more than one source of nutrients or may need cofactors in the soil solution. Most species can grow on a variety of substrate molecules, such as the *Pseudomonads* which are particularly versatile in their substrate utilization. In these species, when a preferred substrate is absent or at low concentration in the soil solution, an alternative substrate is obtained for growth. The switch to new substrate usually involves synthesis of new enzymes by the bacterium. Availability of a preferred substrate inhibits the uptake of alternative substrates. In order to understand substrate use and regulation of nutrient intake by bacteria, one must understand the metabolic regulation of their **operons** (Lengeler *et al.*, 1999). The substrate preferences of bacteria are key in assigning their taxonomic species name, genus and family. Their main sources of organic nutrients in the soil are indicated below.

Cellulose

The ability to digest cellulose is significant, because it is about half of the total biomass synthesized by plants, which returns to soil as litter. The plant cellulose cell wall holds the more nutritious cytoplasm remaining inside. Therefore, breaking into these cellulose boxes provides access to a better source of nutrients. It consists of chains of β-D-glucopyranose in 1,4-glycosidic linkage that are polymerized into long fibrils, about 80 units in length (Fig. 4.6). The fibrils are held together in a three-dimensional matrix by inter- and intramolecular hydrogen bonds. The enzyme subunits assemble on the external cell membrane surface into **cellulosomes**. Aerobic digestion of cellulose occurs through the action of several genera, such as *Archangium*, *Cellovibrio*, *Cellulomonas*, *Cytophaga*, *Polyangium*, *Sorangium* (Myxobacteria), *Sporocytophaga*, *Thermomonospora*, the Actinomycetes *Micromonospora* and several *Streptomycetes*. The anaerobic digestion is also possible by species in the genera *Acetovibrio*, *Bacteroides*, *Clostridium* (*C. thermocellum* and *C. cellubioparum*) and animal rumen bacteria such as *Ruminococcus*.

Digestion of the **lignin** component of plant cell walls is discussed below with the fungi. Some bacteria have been implicated in lignin decomposition, but they are poorly studied. Usually, mixed species are

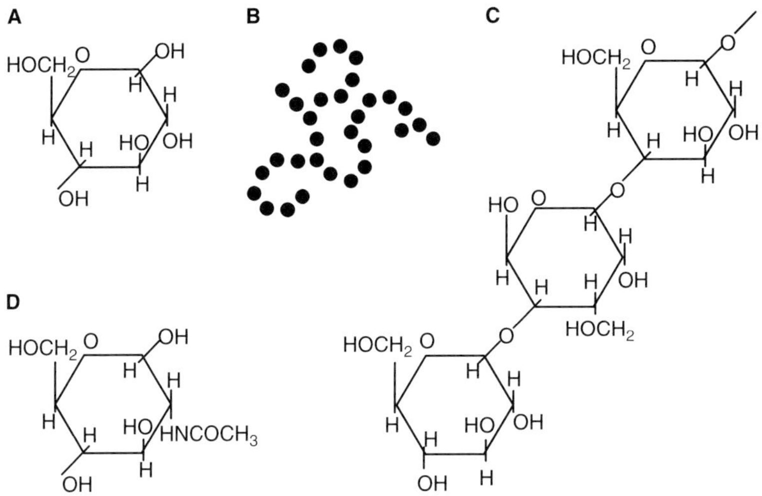

Fig. 4.6. Common saccharide and polysaccharide molecules in litter. (A) Glucose monomer of cellulose, starch and glycogen polymers. (B) Starch or glycogen branching structure of glucose polymer. (C) End of a cellulose chain of glucose monomers. (D) Monomer of chitin polymer found in fungi, other protists and invertebrates.

required to provide the adequate combination of enzymes by **syntrophy** (Blanchette, 1995; Boddy and Watkinson, 1995). Only moderate digestion of the lignin occurs. These species occur more commonly in water-saturated near-anaerobic conditions, or where fungi are inhibited by wood chemicals. They have been recognized in high lignin hardwoods which are difficult for hyphae to access. For example, they occur in *Eusideroxylon zwagei* wood which is very resistant to decomposition because it contains fungal inhibitors. In general, a good working hypothesis is that bacteria will be out-competed by fungal hyphae in lignin decomposition.

Xylan

Xylans are part of the heterogeneous group of molecules known as hemicelluloses. These are the next most abundant after cellulose in plant cell walls (Table 4.7). These fibrils consist of 1,4-glycosidic links of β-D-xylose, but also of a variety of other pentose and hexose sugars. The polymer forms branched structures, 30–100 units long and partially soluble in water. The variety of xylanase enzymes synthesized by species have preferences for different specific bonds of the fibrils. In *Clostridium*, secretion of xylanases is constitutive, although it is under inducible regulation in other bacteria. Similar branched structures of other sugar polymers occur and are collectively known as **hemicelluloses**. Mannans, for instance, contribute <11% of conifer wood dry weight, and are also produced by some yeasts in their cell walls. Fructans occur as storage

Table 4.7. Example of plant and an Ascomycetes cell wall composition for selected molecules (% dry weight) in several litter substrates.

Substrate	Lignin	Cellulose	Hemicelluloses	Protein	Sugars	Chitin
Lucerne stem	6.0–16	13–33	13–33	15–18	0	0
Wheat straw	18–21	27–33	21–26	3	0	0
Beech wood	18–21	45–51	45–51	0.6–1.0	0	0
Aspergillus nidulans	0	0	0	10.5	78.4	19.1

Data from Paul and Clark (1996).

compounds in some plants such as *Phleum pratense* (Timothy grass). They are also excreted by some bacteria when grown on sucrose. In some cases, it was demonstrated that the fructans are then used as substrate when sucrose runs out (in *Azotobacter*, *Bacillus* and *Streptococcus*) (Schlegel, 1993). The genus *Erwinia* belongs to the soft rot (see below) and, when present, is an important contributor to hydrolysis of hemicelluloses, pectins and cellulose.

Starch

This is the primary plant storage compound, and therefore a rich source of energy for organisms that can metabolize it. However, since plants translocate nutrients from senescing organs before shedding them, much less starch becomes litter than one would anticipate. Starch consists of aggregates of two polymers, amylose and amylopectin. Amylose consists of 1,4 α-glycosidic links of D-glucose, forming unbranched, water-soluble chains of hundreds of glucose units in length. Amylopectin consists of the same polymerized units, but with a 1,6-branching approximately every 25 units, with PO_4^{3-}, Mg^{2+} and Ca^{2+} ions incorporated. Numerous bacteria synthesize glycosidases to dissolve amylose and amylopectin. α-Amylases are endoglycosidases and release glucose, maltose and various soluble oligomers. β-Amylases are exoglycosidases and release monomers until the branch points.

Chitin

Chitin is the most abundant polysaccharide in the soil after cellulose. It is one of the main components of the cell wall of fungi and of invertebrate cuticle, and has been reported from numerous protozoa in cyst walls and tests, as well as in the zoosporic fungi and chromista. It consists of the polymer of *N*-acetylglucosamine, linked in 1,4-β-glycosidic linkage (Fig. 4.6). The chitin polymer is usually associated with other fibrils, which hold the matrix together. Therefore, it is important to note that as for plant cell walls, digestion of one wall component alone, although it may loosen the matrix, will not dissolve or remove it. For example, invertebrate cuticles also contain fibrous proteins and fungal cell walls may contain β-glucans. Many bacteria have the ability to digest chitin polymers, such as species in the genera *Bacillus*, *Cytophaga*, *Flavobacterium*, *Micromonospora*, *Nocardia*, *Pseudomonas* and *Streptomyces*.

Murein

This component of bacterial cell wall consists of repeated units of two amino sugars, *N*-acetylglucosamine and *N*-acetylmuramic acid dimer, with amino acid side chains which cross-link the chains together. They are digested at the glycosidic linkage, or at the peptide linkage, by glycosidases or peptidases. Murein contributes <10% of Gram-negative bacteria dry weight, but that of Gram-positive bacteria is 30–70% murein.

Proteins, lipids, nucleic acids and other cell components

These are the richest source of nutrient input into the soil and are most readily digested, because most species have enzymes to digest them. Therefore, their half-life in the soil is short, in the order of minutes to hours. However, a portion of peptides cross-react with other components of the soil organic matter to produce less accessible compounds. Furthermore, even some labile and soluble DNA can be stabilized from enzyme digestion by binding to clays (Stotzky, 2000). These enzymes are part of the metabolic pathways of most prokaryote and eukaryotes. Proteins are digested by endo- and exoproteases at the peptide bonds between amino acids, releasing more soluble and less hydrophobic peptides, or soluble amino acid monomers. Deamination reactions release NH_3^+ from amino acids or amino sugars. Lipases have low substrate specificity and digest lipids (sterols, phospholipids and other membrane lipids), releasing the fatty acids from the alcohol moiety. Nucleases digest nucleic acids into nucleotide monomers. These are a rich source of soluble phosphates that are easily absorbed by living cells. RNAs are naturally unstable and denature rapidly outside the cytoplasm. Intracellular decarboxylation through cell autolysis or bacterial digestion, especially from animal cadavers which are rich in proteins, releases a variety of nauseous amines (such as cadaverine from lysine, putrescine from ornithine and agmatine from arginine).

Hydrocarbons and phenolics

A variety of alkanes occur naturally as a by-product of metabolism, or from digestion of other molecules. Aromatic compounds accumulate from plant cell wall lignins. Other sources of alkanes and polyphenolic compounds are petroleum (naturally occurring or from pollution), and anthracene, asphalt, graphite or naphthalene. There are bacteria that can grow on these substances, usually through aerobic respiration. They tend to be enriched in chronically polluted soils or sediments. For example, a mixture of xylene, naphthalene and straight chain aliphatic hydrocarbons (C_{14}–C_{17}) was decomposed by natural soil microorganisms to low levels in 20 days (xylene), 12 days (naphthalene) and within 5 days (aliphatic hydrocarbons), under aerobic conditions (Eriksson *et al.*, 1999). The decomposition of substrates occurred above threshold temperatures. This may be due to inactivation of particular species or too slow metabolism of one or more participating species. For example, a study conducted on biodegradable plastics showed that both temperature and O_2 availability modified species that digested the plastics (Nishide *et al.*, 1999). Interestingly, only a small number of isolates could decompose the plastics in the laboratory. This emphasizes the cooperating role of multiple species co-occurring naturally, in providing a combination of enzymes to digest substrates. For example, in nitrogen-poor substrates, the role of nitrogen-fixing bacteria in the soil is important to provide an input of fixed nitrogen. Sometimes both functions can occur in the same species (Perez-Vargas *et al.*, 2000).

Overall, a useful working hypothesis is that one can always find a bacterium to digest almost any compound, under suitable conditions. There are even unpublished reports of bacterial growth in old glutaraldehyde, a rapid penetrating biological fixative used in light and electron microscopy. An area of bacterial nutrient acquisition in soil which remains understudied is the correlation of species that are active together on one substrate, and cooperate by **syntrophy** to digest litter. Activity of several species on one substrate provides a diversity of enzyme functions, that together more effectively digest complex molecules into soluble nutrients. This may be the prime mechanism of bacterial function in natural systems.

However, one must not ignore the role of other soil species in the decomposition of hydrocarbons and phenolic molecules. Most invertebrates and protists ingest the soil solution when feeding and are in contact with contaminated soil. Bacterivores and osmotrophs in particular can be severely affected as they ingest bacterial cells and the soil solution. These higher order consumers will accumulate the organic molecules to toxic levels if undigested. Many species of protozoa and invertebrates can be isolated that tolerate or digest hydrocarbons to various levels (Rogerson and Berger, 1981, 1982; Rogerson *et al.*, 1983).

Saprotrophic fungi

In terms of function (enzymatic activity on substrate), many of the enzymes found in primary saprotrophic fungi are similar to those described above for bacteria, and they are not repeated here. Several points are made below simply to distinguish between bacteria and fungi. Most notable is that as eukaryotes, fungi can synthesize larger proteins for secretion, in much greater quantities (see Chapter 1). The impact on the immediate microenvironment is far greater. Unless bacterial cells are protected from fungal proteases and other enzymes (through capsules or inhibitors), they will be partially digested. Unlike many bacteria which tend to focus on one substrate at a time, fungal cells can secrete enzymes for several substrates at the same time. This is advantageous because usually many compounds are mixed together. For example, plant cell walls contain cellulose, proteins, lignins, hemicelluloses and pectins mixed into a matrix (Table 4.7), and sometimes with additional components that can be defensive. To disentangle the cell wall and obtain soluble substrates, the whole structure needs to be dissolved. This can be achieved by bacteria if multiple species contribute their panoply of enzymes (syntrophy). Fewer species are required for simpler substrates or if a species can produce diverse enzymes.

Based on their ability to digest plant cell walls, these fungi are recognized as three groups. The **white rot** include about 2000 species of mostly Basidiomycetes, which can digest cellulose and lignin components of the plant cell wall. The **brown rot** include about 200 species of mostly Basidiomycetes which are unable to digest the lignin component, but only cellulose and hemicelluloses. The **soft rot** consist of those Ascomycetes and mitosporic species (deuteromycetes) which are efficient on cellulose and hemicelluloses, but digest lignin slowly or incompletely. Another important group of species is involved in the invasion and digestion of plant seeds (Watanabe, 1994). It is estimated that up to 80% of seeds are naturally decomposed by fungal digestion. These species need to break into the seed protective coats but, once penetrated, obtain a rich supply of starch, protein and oils that are common seed storage molecules.

Cellulose, xylan and pectin

Cellulase activity is from a complex of several enzymes which assemble into a cellulosome (Lemaire, 1996) similar to that in bacteria. It consists of several endoglucanases, exoglucanases and β-glucosidases (or cellubiases) which attack the polymer at multiple sites. The brown rot cellulase activity generally results in more complete digestion of the cellulose polymer than white rot fungi. The reason is believed to be the synthesis of H_2O_2 and ferrous ions which reach fibrils embedded in lignin and contribute to hydrolysis of cellulose (Moore, 1998). The accumulation of ferrous ions from ferric ions is through oxalate. This non-enzymatic process occurs through the Fenton–Haber–Weiss reaction and involves siderophores and lignin, which becomes chemically modified but not degraded. It helps loosen the cellulose fibrils, which are otherwise packed too tightly to allow the large cellulosome enzyme complex into the fibrils. Brown rot are common on coniferous wood, which is decomposed almost fully, leaving the lignin matrix on the forest floor. This chemically altered but undegraded lignin which remains is very recalcitrant to digestion. The reticulum of tunnels that remain functions as a sponge. It is an important sink for water and nutrients, and a refuge for microorganisms.

Xylanases are also complex enzymes formed from two endoxylanases and one β-xylosidase. Pectinases are particularly important in tissue invasion by parasitic fungi (and bacteria), as they involve loosening the pectin which holds adjacent plant cells together. They consist of polygalacturonases and pectin lyases, with arabanase and galactanase to hydrolyse pectin-associated sugars.

Lignin

The next most abundant plant cell wall component after cellulose and hemicellulose is lignin. It is particularly abundant in woody tissues, and becomes more abundant in older tissues, late season cells and other stressed conditions (Table 4.8). It provides reinforcement to the cellulose fibrils in the secondary cell wall deposition. It represents 18–30% of the dry weight of wood. Lignin is a complex compound derived from phenyl propanoid units, but polymerized variously through chemical reactions, not through precise enzymatic activity, so that its structure is variable (Fig. 4.7). The proportions of the lignin precursors coniferyl, sinapyl and

Table 4.8. Estimates of lignified material in roots and aerial parts of plants, from various ecosystems.

Biome	Roots (%)	Above-ground perennial parts (%)
Tundra, alpine	75	12
Desert, semi-desert scrub	57–87	2–40
Temperate grassland	83	0
Tropical savannah	28	60
Temperate deciduous forest	25	74
Northern coniferous forest	22	71
Tropical rainforest	18	74

Data from various sources (see Boddy and Watkinson, 1995).

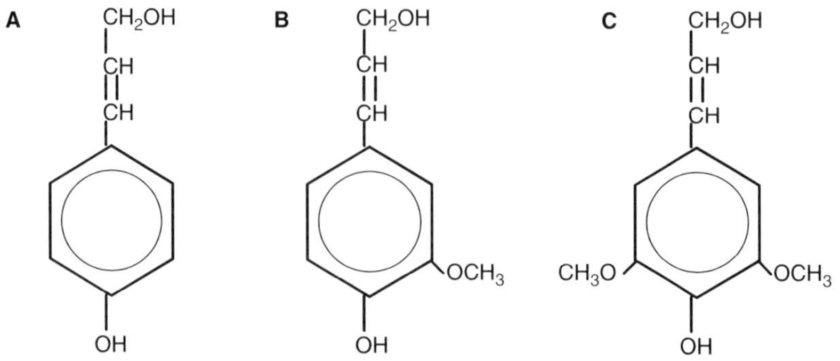

Fig. 4.7. Several lignin precursors. (A) Coumaryl alcohol. (B) Coniferyl alcohol. (C) Sinapyl alcohol.

coumaryl alcohols in lignin vary between angiosperms and gymnosperms (Moore, 1998). The reason for slow decomposition of lignified tissues is the relative scarcity of primary saprotrophs that are able to digest it. This implicates some Basidiomycetes, fewer Ascomycetes and some bacteria (Table 4.9). Most of the data on lignin digestion are based on two species, the Basidiomycetes (fungi) *Phanerochaete chrysoporium* which can completely digest lignin to CO_2 and H_2O, and the Actinobacterium *Streptomyces viridosporus*. The fungal enzyme is a complex of up to 15 lignin peroxidases (ligninase), Mn-dependent peroxidases and Cu-oxygenases (laccase) which oxidize *o*- and *p*-phenols. The enzyme activity has been reported to occur when available nitrogen sources are depleted, through a cAMP-mediated activation pathway. It is unclear if this is a generalization, as too few species have been investigated.

The white rot fungi which digest lignin are grouped in two functional groups (Blanchette, 1991, 1995). One group is non-selective and will digest both lignin and cellulose in varying proportions. Examples are *Trametes versicolor, Ganodermer applanatum, G. tsugea* and *Heterobasidion annosum*. The other group is more selective and will digest lignin more completely. Examples are *Phlebia tremellosa, Inonotus dryophyllus, Phenillus pini* and *P. nigrolimitatus*. There is variability between the proportions of cellulose, hemicellulose and lignin which are digested by different strains of one species, and even by the same strain through various regions of the substrate. The white rot also digest the middle lamella of pectins to loosen cells, and deposit MnO_2 and calcium oxalate at zones of digestion, and siderophores are involved. The white rot ligninases have a substrate preference for syringylpropyl against guaiacylpropyl. They predominate in angiosperm wood decomposition, but also occur in gymnosperms.

Table 4.9. Amount of remaining molecules in well-decomposed wood (% remaining) after decomposition by different types of fungi, compared with average initial plant cell wall content (g/kg).

Initial composition (g/kg)	Lignin	Cellulose	Hemicelluloses	Protein
Plant cell wall 1	41	386	463	110
Plant cell wall 2	117	352	477	64
In decomposed litter (%)	**Lignin**	**Glucose**	**Xylose**	**Mannose**
White rot (selective)	30	47	6	13
White rot (non-selective)	1	97	1	1
Brown rot	60	20	1	2
Soft rot	61	13	2	2
Bacteria	80	6	1	9

Data for lignin from Kleson analysis, and from HPLC for sugars: glucose from cellulose polymer; and xylose and mannose from xylan and mannan hemicelluloses. Data from Blanchette (1995) and Cadish and Giller (1997).

The soft rot fungi tend to predominate in excessively wet or dry regions where the brown and white rot species are not favoured. For example, they are more significant on the north American Pacific coast, in the temperate rainforests of the northwest USA and in British Columbia. These form cavities in conifer wood, or cell wall erosion in angiosperm wood (Blanchette, 1995). Implicating their role involves observations on the ultrastructure of the decomposed wood or identification of the fungal DNA, because there are few easily recognized morphological characters.

Starch, glycogen and chitin

These polymers also require several enzymes for complete digestion to monomers. Starch digestion occurs through enzymes which, in sequence, produce smaller molecules. These include (endo) α-amylases, (exo) β-amylases, glucoamylases (found in fungi only), pullulonase, and finally α-glucosidases which release glucose monomers. Glycogen (a fungal, protozoan and animal storage compound) is probably digested through the same enzymes. Chitinase activity is the result of glucan hydrolase on 1–4 glycosidic bonds, which release chitobiose disaccharides. These are hydrolysed by chitobiase to *N*-acetylglucosamine monomers.

Fungal proteases are mostly studied in two species which are used routinely in cell biology: *Neurosora crassa* and several *Aspergillus* species. Most other fungal enzymes were studied in parasitic species, to understand mechanisms of plant infection (de Lorenza *et al.*, 1997).

Osmotrophy

This is the only mechanism of obtaining substrates for metabolism and growth from the environment in primary saprotrophs. These include primarily species of hyphal fungi and yeasts, chytrids, prokaryotes and several taxa of chromista (such as Oomycetes, Labyrinthulids and Thraustochytrids). It involves passive diffusion, and primary and secondary active transport mechanisms through cell membranes (Fig. 4.8, see Chapter 1).

Small molecules and substrates of <600 Da molecular weight can pass through cell walls and cell membranes relatively easily. Smaller and more lipophilic molecules can pass through cell membranes most easily. The process occurs through **simple diffusion** along concentration gradients. It is not an important mechanism for nutrient acquisition, but more important in the uptake of water, ions, non-polar toxins, etc. The limiting regulator in allowing substances into cells is the cell membrane. The membrane contains substrate-specific proteins which function as enzymes to transport nutrient molecules into the cell by **facilitated diffusion** and **active transport**. Details of these mechanisms are explained in introduc-

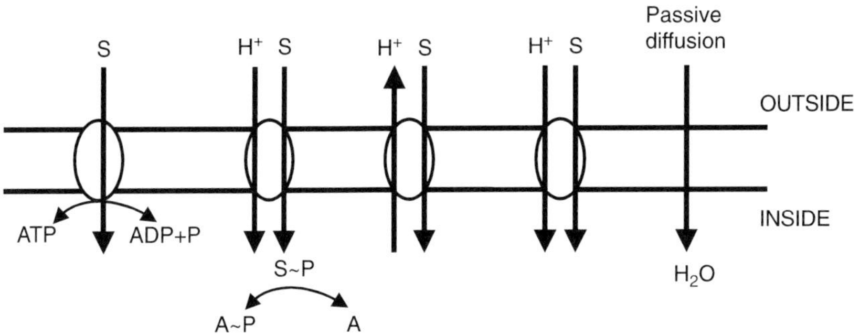

Fig. 4.8. Examples of mechanisms of substrate (S) entry into cells through the cell membrane. Carrier proteins transport substrate using an ion gradient (here H^+), substrate level phosphorylation or ATP-derived energy to accumulate transported molecules in the cytoplasm.

tory bacteriology or cell biology texts and will not be elaborated here (White, 1995; Lengeler *et al.*, 1999). Briefly, transport of a nutrient substrate through the cell membrane involves permeases and translocases, which are substrate specific. It is important to note that the exact mechanism of transport of a substrate into cells depends on the species, i.e. the same molecule can be transported by a different mechanism in a different species. The enzyme system that receives, transports and metabolizes each substrate, especially in bacteria, is usually inducible by the substrate, i.e. detection of the substrate initiates the synthesis of the proteins that will be necessary to obtain that nutrient. Bacterial species have a hierarchy of preferred substrates, so that the presence of a preferred substrate inhibits the synthesis of proteins that deal with less favoured substrates. Detection of the substrate and its transport through the membrane are linked to cell chemosensory detection, i.e. motile cells orient the direction of movement towards higher substrate concentration where more substrate is available. Where directional motility or growth occur towards a chemical, cells are said to respond by positive **chemotaxis**.

Facilitated diffusion involves binding of a substrate to a permease, which facilitates the transfer of nutrients through the cell membrane. Transport occurs along chemical concentration gradients and does not require cellular energy. Therefore, accumulation of the molecule in the cell above the external concentration is not possible. The in-flow halts unless the molecule is used rapidly in the cell. **Active transport** can be subdivided depending on the chemical processes that drive it. **Primary transport** involves the movement of ions across the membrane together with the substrate. This modifies the membrane electron potential. It is usually driven by respiration electron transfer reactions, ATP-driven ion pumps, through decarboxylation of metabolites (oxaloacetate, methylmalonyl-CoA and glutaconyl-CoA), or by photosynthesis. Because energy is used to transport the substrate into the cell (e.g. ATP hydrolysis), the molecules can be accumulated in the cell, against the chemical concentration gradient. **Secondary transport** is driven by the membrane electrochemical gradient. The electron potential is maintained by constantly pumping H^+ and Na^+ across the membrane (White, 1995). Modification of the substrate during or after active transport into the cytoplasm prevents the reverse transport out of the cell.

The substrate-specific permease and translocase enzymes involved in active transport are the principal mechanisms of nutrient acquisition. In bacteria and protists, it occurs at the cell membrane–environment boundary, but in multicellular organisms, such as the soil invertebrate species, it occurs along the digestive tract cells. The transport enzymes between species, for the same substrates, differ in their affinity for the substrate, as well as the rate of translocation, optimum condition for function and mechanism of substrate modification. Therefore, in the fluctuating soil environment, the presence of a diversity of species ensures that at any one time at least some species exploit the available nutrients, through a range of abiotic conditions. The diversity of mechanisms between prokaryote taxa that drive metabolic reactions is a fascinating subject that unveils the ecological success of these species.

Other means of obtaining nutrients from the soil solution exist. Several ciliates rely on dissolved organic matter for nutrition, while others supplement their ingestion of bacteria with soil solution. Many other small protozoa also rely on dissolved nutrients for part of their nutrition, by endocytosis. Some soil solution is ingested in the process of acquiring bacteria or small flagellates by phagocytosis. Similarly, nematodes that suck in soil solution to obtain bacteria, protists and other food benefit from the dissolved organic matter. In fact, if the solution is rich enough, such as with artificial media, some bacterivores (notably Rhabditida and several Diplogasterina) can be grown axenically (Yeates, 1998). Smaller mites and Collembola that feed on bacteria film, algae and other protists in the soil solution also benefit from the dissolved organic matter. These were (incorrectly) called filter-feeders (Walter and Kaplan, 1990) and include the *Histiostomatid* (Astigmata) mites. They have short generation times, and feed on wet decomposing litter and fruits. The chelicerae and pedipalps are brush-like and sweep solution towards a 5 μm mouth opening. Similarly, some Collembola are specialized to feed on bacterial films, such as *Archisotoma besselsi* and *Isotomurus palustris*.

Secondary Saprotrophs

In this functional group, we group species that are not involved in directly digesting soil litter. Primary saprotrophs were characterized by extracellular digestion of litter in the soil, and the absorption of dissolved nutrients through osmotrophy. **Secondary saprotrophs** are consumers of primary saprotrophs, or of litter partially digested by primary saprotrophs. Several functional groups can be distinguished, based on which type of organisms are ingested, and the mechanisms used for ingestion (Fig. 4.9).

Bacteriotrophy

Species that consume prokaryotes are referred to as **bacterivorous**. They include species from most taxa found in the soil. Ingestion of bacteria does not occur at random. Experiments show that, typically, protists and invertebrates exhibit clear preferences based on bacterial species, dimensions and other criteria. Therefore, even when bacteria are abundant, many bacterivores may not be active or may grow poorly, because preferred bacterial species are absent. Examples of bacteria choice are numerous in the literature, particularly for nematodes and protozoa.

Prokaryotes

Within the bacteria, many species of Myxobacteria release enzymes that lyse bacterial cell walls. They are the principal bacterivores amongst prokaryotes (Chapter 1). They grow as motile colonies that spread over bacterial colonies and digest them. Myxobacteria enzymes also digest other SOM molecules, which contribute to general decomposition and nutrient release. The role of bacteriophages should be mentioned because these viruses can infect and lyse susceptible bacterial strains and species that become infected. The lysis of bacteria reduces the popula-

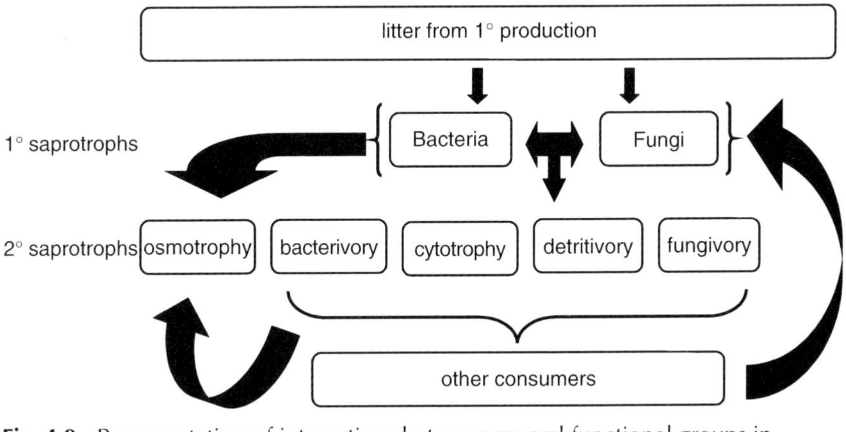

Fig. 4.9. Representation of interactions between several functional groups in decomposition. Arrows indicate direction but not quantity of flow.

tion and releases cytoplasm into the soil solution. The abundance and impact of bacteriophages in soil food webs are studied infrequently (Williams and Lanning, 1984; Lipson and Stozky, 1987; Angle, 1994), although they have been studied in aquatic food webs.

Protists

Protist consumers of bacteria are the most numerous bacterivores. These include the nanoflagellates (<12 μm long) from several phyla, which are abundant species of protozoa in the soil. Most protozoan species found in the soil are bacterivorous. However, not all consume bacteria exclusively. The smallest species complement growth by osmotrophy on the soil solution, or may be exclusively osmotrophic. Many small ciliates complement their nutrition by obtaining soluble nutrients from the soil solution. For example, *Tetrahymena* (a freshwater ciliate, that normally feeds on bacteria) can be grown in the laboratory on an axenic medium of sterile, rich solution of soluble nutrients. The small ciliate *Coleps* and several other genera of small ciliates aggregate about lysed cells and freshly dead or dying microinvertebrates to feed on the cytoplasm. These also ingest bacteria that grow on the lysate. Certain filamentous and saprotrophic fungi supplement their nutrition by bacterivory. This was observed in several ligninolytic species in the genera *Agaricus*, *Coprinus*, *Lepista* and *Pleurotus* (Barron, 1988). In these examples, saprotrophic enzymes are released from coralloid haustorial hyphae to digest the bacterial colony, and the nutrients are absorbed by osmotrophy.

The mechanism of ingestion of bacteria by protists is **phagocytosis**. Phagocytosis is stimulated through membrane receptors and mediated by the cytoskeleton, which invaginates the cell membrane to form a food vacuole (or phagosomes). Bacterivorous protists often have an area of the cell membrane which is specialized for capturing bacterial cells, called the oral region or **cytostome**. The cytostome can be a simple depression on the cell surface, or an elaborate structure with many specialized organelles associated with it. The cytostome can be an invaginated gullet, or funnel-shaped structure, as in many Ciliata, Cryptomonad or Euglenid. Capture of bacteria by the cytostome often involves movement of cilia (Fenchel, 1986; Balczon and Pratt, 1996; Boenigck *et al.*, 2001). In the simplest case, one cilium is used to direct bacteria on to the cytostome area, where phagocytosis is initiated. In other cases, one cilium can be used to hold on to soil particles, while the cytostome is held against a surface for phagocytosis of suitable bacteria. The cytostomes of ciliates are more invaginated in many cases and possess many specialized cilia, sometimes in bundles (Lee *et al.*, 2001). The oral cilia direct a current into the cytostome, and phagocytosis occurs inside, at the cytopharynx. Larger ciliates may have an elaborate cytostome, such as many stichotrich families which are common and ubiquitous in the soil.

Phagocytosis by amoeboid species is distinct because most of the cell membrane is sensitive to detection and capture of bacteria. Cell extensions (pseudopods) which explore the immediate environment capture and engulf bacteria. Many soil nanoflagellates can be partially amoeboid in film water, and some can extend **filopodia** to search their environment. The Sarcomonadea (Cercozoa or amoeboflagellates) in particular are abundant in soils, though understudied. Filopodia can be <1 μm in diameter and extended into narrow spaces to explore and obtain bacteria (Fig. 2.3). Therefore, even if the cell body cannot reach small spaces, the filopodia can. Pseudopods, including filopodia, are strong enough to exert force on soil particles. This is easily observed under the microscope, and causes some movement and repositioning of particles. Lastly, the Mycetozoa (protozoa: Amoebozoa), which includes the slime moulds (Dictyostelia), are abundant bacterivorous soil amoebae. Under suitable conditions, they excyst in large numbers.

Bacterivory and predation by protozoa in general were reviewed previously (Clark, 1969; Griffiths, 1994). Most experiments on feeding preferences of bacterivorous ciliates were carried out with aquatic species (Sherr *et al.*, 1992; Perez-Uz, 1996 and references therein). Bacterial selectivity by soil amoebae was addressed by Singh (1941, 1942). Selectivity of nanoflagellates and other soil protozoa has been addressed more recently. One study (Ronn *et al.*, 2001) compared the growth response of indigenous soil protozoa on two dissimilar species of bacteria, *Mycobacterium chlorophenolicum* and *Pseudomonas chlororaphis*. The former did not stimulate grazing in soil microcosms, whereas the *Pseudomonas* permitted growth of naked amoebae and heterotrophic flagellates. High-resolution video-microscopy has been useful in examining the mechanisms of prey ingestion, digestion and excretion of food vacuoles. These observations on various prey items are useful in determining feeding strategies, prey capture and retention, and optimal foraging (Boenigk and Arndt, 2000). Characteristics of bacterial processing can be used to differentiate between optimal feeding conditions and niche differentiation within microhabitats (Table 4.10). The process can be subdivided into time budgets for different prey items (contact time, processing time, ingestion time, refractory time and handling time) (Fig. 4.10). In some instances, a bacterium or particle that is partially ingested is ejected without food vacuole formation or digestion (Fig. 4.11). Ejection occurs if it is too large, too long or unpalatable. Important parameters that determine growth efficiency on a bacterium include differences in the time allocated to each portion of the time budget. They also include the nutritional and chemical composition of the bacteria (palatability). Actively growing species are often more palatable and selected for. This video-microscopy approach is suitable for modelling bacterivory rates and selectivity by a variety of nanoflagellates (Boenigk *et al.*, 2001) and other ciliates or amoebae.

Table 4.10. Bacterial capture in four nanoflagellates based on video-microscopy.

	Cafeteria roenbergensis	*Bodo saltans*	*Spumella* sp.	*Ochromonas* sp.
Cell volume (μm³)	18.5	25.8	75.4	163
Cilium beat (Hz)	39.5	47.9	61.1	68.7
Particle speed (μm/s)	105.6	91.6	50.3	112
Ingestion rate (particle/ind/h)	14.1	34.3	36.7	63
Non-capture rate (%)	82	87	88	63
Ingested volume (μm³/ind/h)	9.3	14.8	14.5	26.8

Data from Boenigk and Arndt (2000).

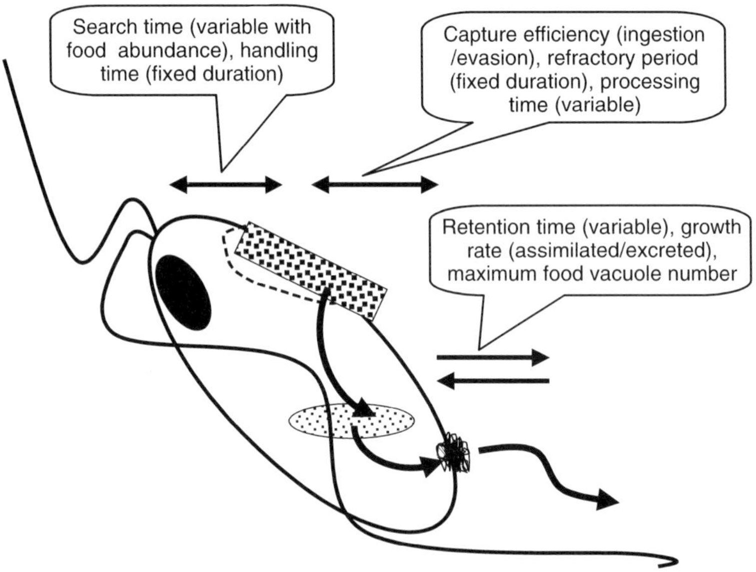

Fig. 4.10. Parameters that regulate the rate of food intake affecting optimal foraging, here represented with a biciliated cell ingesting a bacterium cell. The durations of fixed and variable periods both vary with temperature.

Invertebrates

Invertebrate consumers of bacteria are also common in the soil and surface litter (see Chapter 1). These include several families of rotifers (particularly the Bdelloides), gastrotrichs, tardigrades, nematodes and Collembola. Both the enchytraeids and earthworms are bacterivores indirectly, because they ingest whole soil or litter. Many invertebrate larvae in the soil will feed on bacteria directly or indirectly. The majority of bacterivorous invertebrate species lick or suck thin water films on soil particle surfaces. They feed by ingestion of bacteria and soil solution.

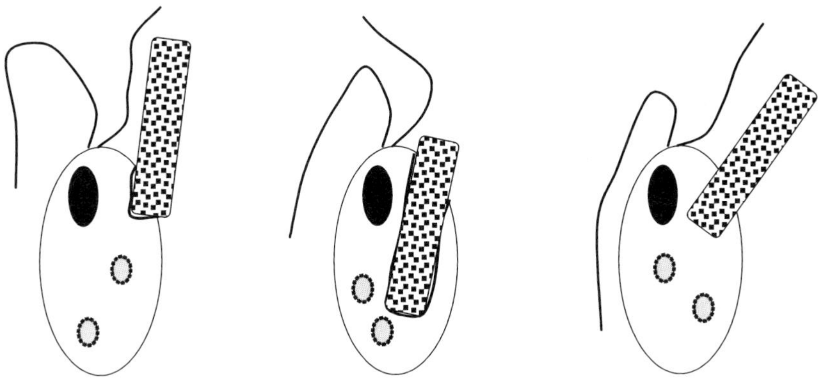

Fig. 4.11. Prey selection by protozoa. Contact and initiation of phagocytosis of a bacterial cell, followed by its rejection and evacuation, without phagosome formation or digestion.

The Bdelloides (Rotifera) and vorticellid ciliates depend on water currents created by the oral cilia and require at least capillary water in soil pores or litter surfaces for feeding. They encyst in less moist conditions. Although it is possible theoretically for soil mites to feed on bacterial films, there is scant evidence that it occurs or supports growth (Hubert and Lukesova, 2001). Feeding structures of many species, particularly in the smallest forms, could permit some bacterivory by ingesting soil films and slime layers. Most species probably ingest bacteria indirectly, by feeding on micro- and macrodetritus that would sustain bacterial growth. Small Collembola such as species of *Micranurida* feed on soil solution and films on particle surfaces through sucking mouth parts, ingesting bacteria, yeasts, protists and similarly sized particles.

Nematodes

Ingestion of bacteria is by sucking in soil solution (Fig. 4.12, Table 4.11). The fraction of cells lysed depends on the bacterial species. Some can even be excreted alive (undigested). A review of nematode selection of bacteria and growth efficiency on different bacteria was provided by Yeates (1998). These were elaborated further (Vanette and Ferris, 1998) by observing that some bacteria did not support growth of nematodes at all. Other bacteria could not support maturation of juveniles or egg-laying by females, although they could support growth of adults. In general, a mixed diet is probably more efficient. This study further points out that a minimum abundance of bacteria is required to sustain growth ($\sim 10^3$ bacteria/nematode/day). Optimal growth rates can be obtained with about 10^5–10^6 bacteria/nematode/day at 20°C. These values vary between species of nematodes and bacteria available. In particular, the nematode oral morphology, metabolic rate, reproduction rate, nutrient requirement, population size and abiotic conditions contribute to the

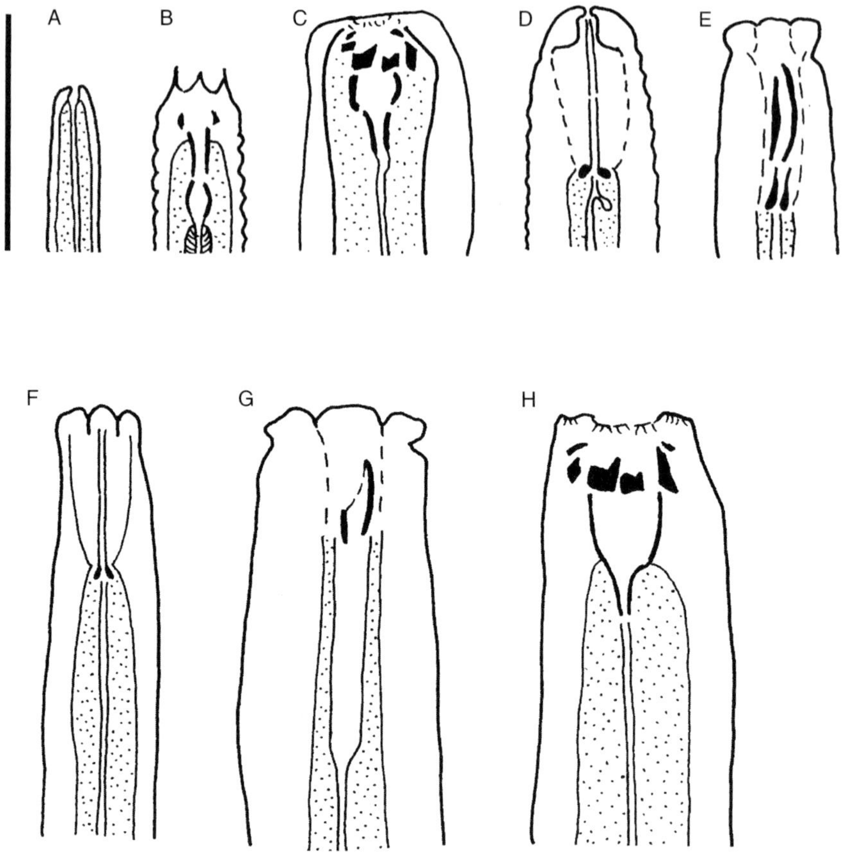

Fig. 4.12. Functional groups of Nematoda, based on the morphology of oral structures. (A) Bacterivorous without appendages. (B) Bacterivorous with some reinforcement of the mouth cavity. (C) Bacterivorous with denticles and reinforced cavity. (D–F) Tubular stylet for penetrating eukaryotic cells of protists, hyphae, invertebrates, roots or mosses. (G) Predacious with wide tubular stylet. (H) Predacious with denticles and reinforced mouth cavity. Scale bar 25 μm. Based on Yeates and Coleman (1982), Yeates *et al.* (1993) and Yeates (1998).

Table 4.11. Nematode functional morphology of stoma with feeding types and selected examples.

Morphology	Feeding types	Examples
Armoured pharynx	Bacterivory, cytotrophy,	Rhabditidae,
Dentate pharynx	Fungivory, roots, invertebrates	Diplogasteridae, Mononchoidea, Nygolaimidae
Lanceolate stylet	Roots, invertebrates	Longidoridae, Trichodoridae,
Tubular stylet	Hyphae, roots, invertebrates	Aphelenchina, Leptonchidae, Seinuridae, Tylenchina
No appendage	Bacterivory, cytotrophy	Araeolaimidae, Monhysteridae

efficiency of grazing on various bacteria. Nematodes vary in their sensitivity to bacteria prey variety and abundances. What is ingested by nematodes depends in part on the diameter of the stoma when stretched open (Yeates, 1998). Some nematodes with more complex or armoured pharynx, such as the dentate *Diplogaster*, can supplement their diet by feeding on bacteria in film water. When bacteria are scarce and bacterivores become more opportunist, other particles can be ingested, such as microdetritus, cysts, spores, diatoms, small testate amoebae or invertebrate eggs. However, ingestion does not mean that the particles can be digested, or that growth can be supported.

Cytotrophy

Species that feed on non-filamentous unicellular eukaryotes are referred to as **cytotrophic**. These include small invertebrates that suck in soil solution and capillary water, as well as many species of protozoa and several fungi. Some nematodes and tardigrades can prey on large and small testate amoebae by breaking into the test, or through the pseudostome opening. Lytic enzymes from primary saprotrophs, such as proteases and lipases, will damage or lyse cell membranes of protists. The lysate contributes to dissolved organic matter in the soil solution.

In general, most invertebrates that can feed on bacteria based on morphology probably also feed on small unicellular eukaryotes (Tables 4.2, 4.11 and 4.12). However, because protozoa are crushed and lysed

Table 4.12. Orders and suborders of Acarina found in soils, with generalized descriptions.

ORDER/Suborder	Description
ASTIGMATA	Some free-living species in the soil; though usually associated with other organisms, or found as pests in stored food
HOLOTHYRIDA	Adults have sclerotized cuticle with dense setae, 2–7 mm in length, found in the soil litter, on grasses, ferns, vegetation canopy. Considered predacious, feed on solution from decomposing or pierced organisms. About 30 poorly known species
IXODIDA	Ticks and ectoparasitic species; mostly 1.7–7 mm unfed, increasing to 2–3 cm in fed state; are vectors for disease, e.g. Sporozoa blood parasites
NOTOSTIGMATA	Resemble small Opiliones 1.5–2.3 mm in length, found under stones and in surface litter, in seasonally dry conditions. Hypertrophied chelicerae and serrated setae cut up food. Feed on pollen, fungi, microarthropods by ingestion of solids. Single family with nine genera and 20 species

Continued

Table 4.12. *Continued.*

ORDER/Suborder	Description
PROSTIGMATA	A heterogeneous group mostly 100–1600 μm in length, considered mostly predatory or detritivorous on microdetritus; the parasitic larvae of the Parasitengona become cytotrophic on pollen and spores, or predacious, or ectosymbionts on skin/scales/fur as adults
MESOSTIGMATA	Mostly 200–2500 μm in length and very diverse, 10,000 species. About half are free-living voracious predators, but ingesting fluids only; a membrane excludes entry of particulate matter
Sejina	Feed on pollen, fungal spores and similar sized particles, including invertebrate cuticle, with the less digestible parts visible in the digestive tract
Microgyniina	Found feeding on rotting wood and tree stumps (infested with hyphae)
Cercomegistina	Mostly soil species (?)
Antennophorina	Some predacious, others ectosymbionts (commensals) on insects and reptiles
Uropodina	Detritivores common in rich organic matter, such as forest floor organic horizons or animal excreta and composts; also fungivorous and predacious on nematodes
Diarthrophallina	Specialized parasites
Zercotina	Omnivorous predators in soil litter and organic horizons
Epicriina	Two genera found in forest litter and on decomposing wood
Arctacarina	Probably predacious in soil
Parasitina	Free-living (predacious) adults with sclerotized cuticle, cosmopolitan distribution in transient litter or detritus rich in organic matter, such as composts, manure
Dermanyssina	The most species-rich group of the order with free-living, parasitic and ectocommensals (on vertebrates, and arthropods such as Coleoptera and Diptera). Soil species are often predacious on nematodes and microarthropods; with families Rhodacaroidea and Ologamasidae dominating in temperate regions. Other species occur in marine/estuarine sediments
Heterozerconina	Associated with surfaces (ectosymbiont?) of millipedes and snakes in tropical/subtropical regions
ORIBATIDA	About 150 families. Adults with sclerotized cuticle, 200–1400 μm in length; primarily detritivorous, often supplemented by nematodes and microinvertebrates
Palaesomata	Mostly soil-dwelling species, with a majority favouring warm climates
Enarthronata	Found in forest and grassland soil, particularly in temperate climates
Parhyposomata	Known from soils and tree holes
Myxonomata	Soil species
Euptyctina	Burrow into and feed on twigs and conifer needles
Desmonomata	Found in forest soils, wet moorlands and bogs
Circumdehiscentiae	The largest suborder with cosmopolitan species, including the more significant portion of forest soil mites; also found in the tree canopy, freshwater vegetation and littoral zones

Data summarized from Evans (1992).

more easily than bacteria, it is impossible to identify them in the digestive tract or excreta without molecular probes. Most Gymnamoebae, Testacea and the nanoflagellates are slow moving relative to microinvertebrates, particularly nematodes. Ciliates and larger motile cells may be fast enough to escape if sufficient moisture is present. Prey preferences of invertebrates on protozoa have been studied with aquatic species, but no studies have been carried out in soil systems. Bacterivorous nematodes are known to feed on amoebae in microcosms (Elliott *et al.*, 1979; Griffiths, 1984). Species of soil mites are known to feed on prokaryotes (Cyanobacteria) and protists (Chlorophyta) if they are sufficiently abundant, as in mats and thick films (Worland and Lukesova, 2000; Hubert and Lukesova, 2001). Many are also known to ingest spores, cysts, pollen and similar sized particles. However, because of the protective cell walls, it is unclear whether these are digested to support growth. In feeding experiments, the latter are often excreted without damage, and a fraction can be shown to be viable after passage through the digestive tract (Behan and Hill, 1978; Pussard *et al.*, 1994; Hubert *et al.*, 1999, 2000 and references therein). However, some Collembola do feed on pollen, such as *Deuterosminthurus repandus*, *Folsomia fimetaria*, *Lepidocyrtus cyaneus* and *Sminthurides viridis*.

Certain larger species of amoebae, testate amoebae and ciliates can ingest protists, and some are specialized on a particular size range of prey particle size. However, it must be emphasized that a larger cell, or a larger cytostome does not indicate cytotrophy. Feeding is normally by size selection of prey, but motility and strength of the cell cortex are also important. Prey ingestion requires contact with pseudopodia or filopodia, capture by feeding currents, or contact with the cytostome. This is followed by phagocytosis and food vacuole formation. The dynamics of prey capture, ingestion, digestion and excretion can be measured in microcosms (Fenchel, 1986; Balczon and Pratt, 1996; Pfister and Arndt, 1998; Boenigck *et al.*, 2001).

The large stichotrich ciliate *Stylonychia* (an aquatic organisms) is similar to the ubiquitous soil species *Sterkiella* (also referred to as *Histriculus* or *Oxytricha*). These species are known to have prey preferences which vary with prey abundance, cell size and temperature (Adl and Berger, 1998). *Sterkiella* feeds preferentially on small cells up to 15 μm long. However, different *Sterkiella* species can ingest larger organisms including small ciliates, up to 30 μm long. These species also graze bacteria when they are abundant and when ciliated cells are at low abundance. The abundance of cytotrophic Ciliata increases dramatically in soils when prey nanoflagellates are abundant. This can be visualized in soil suspensions or dilutions fed with bacteria. First, bacterivorous colpodid and other small ciliates emerge from cysts and proliferate alongside the nanoflagellates. Next, cytotrophic ciliates, mostly Stichotricheae, emerge to prey on the nanoflagellates. Colpodids encyst when bacterial abundances are reduced, and the remaining ciliates reduce the abundance of nanoflagellates. Species such as *Sterkiella* continue grazing on the remaining nanoflagellates or bacteria and encyst later. Smaller cells can be predatory consumers on larger species. For example, *Spiromonas gonderi* attacks species of *Colopoda* (Ciliata) (Foissner and Foissner, 1984). The colpodid ciliate *Sorogena* feeds on other colpodids in the soil (Bardele *et al.*, 1991). This family is particularly interesting, because it forms a sorocarp similar to that of Myxobacteria (bacteria) and Dictyostelia (protozoa). Cells aggregate by the hundreds into a mass, which becomes wrapped in a protective secretion, on a stalk. Foissner (1987) estimates that about half of soil ciliate species belong to the class Colpodea, and most of the others are Stichotricheae (~37%) and Heterotrichiae. Of the non-colpodid species, about half are cytotrophic consumers on small ciliates, nanoflagellates, amoebae and testate amoebae. Ciliates in marine intertidal sands belong mostly to the class Karyorelictidea (Dragesco, 1960, 1962). Most of the common species feed on other protozoa and microinvertebrates, though some are bacterivorous or consumers of small cells and diatoms. Certain Thecamoebae and larger ciliates (*Frontonia*, *Loxodes* and *Sterkiella*) are known to ingest active testate amoebae indiscriminately (Chardez, 1985).

Testate amoebae are notoriously undersampled by soil biologists. One reason is that many are associated with the top soil and litter which are often avoided at sampling. Another reason is that they are not cultured on bacteria or by the most probable number (MPN) dilution methods. Growth of testate amoebae is greatly influenced by moisture conditions, temperature and prey species. They also have requirements for synthesis of the test material, which must be met. Many are not motile even when active, unless suitable prey species are available. The feeding habits of a few species were investigated (Bonnet, 1964; Schonborn, 1965; Laminger and Bucher, 1984; Chardez, 1985). Although bacterivory is probably a component of the diet of many species, it is insufficient to support growth of most species. Testate amoebae seem to be opportunist particle ingesters, which includes cysts, spores, yeasts and microdetritus. The Hyalosphenidae are reported to use pseudopods to penetrate other smaller species of testate amoebae and ingest the cytoplasm. For example, *Nebela* species can feed on *Assulina*, *Corythion*, *Tracheleuglypha* and *Trinema*. Smaller species such as *Hyalosphenia platystoma* attach to the test of larger species such as *Euglypha*, puncture a hole of a few microns and ingest the cytoplasm with pseudopodia. Some preferential grazing was observed with the testate amoeba *Trinema enchelys* (Laminger, 1978). This species feeds on bacteria and other small testate amoebae when there is sufficient moisture, but switches to grazing on microdetritus when soil is too dry.

Fungivory

Species that feed on filamentous fungi and yeasts are referred to as **fungivorous** or **mycophagous**. Feeding on fungi is distinguishable from other protists in that the fungal cell wall is difficult to penetrate and digest. Mycophagous species are specialized for penetrating or ingesting hyphae. In this section, we will distinguish between two types of fungivory. The first involves removing a length of hyphae, or hyphal bundle from mycorrhizae and fruiting bodies (**ingesters**). Ingesters remove hyphal lengths by ingesting the cell wall and chewing it. It is more common among microarthropods, such as a variety of mites and Collembola. Lengths of hyphae can be recognized along their digestive tract. The second involves piercing a hole in hyphae to ingest the cytoplasm (**piercers**). Piercers puncture a hole through the cell wall, which is otherwise intact. This avoids having to feed on the chitin wall which is difficult to digest. Instead, the cytoplasm is ingested and a length of hyphal wall empty of cytoplasm remains. This mechanism is more common in nematodes, amoebae and several colpodid ciliates. It is worth noting here that although many organisms can ingest fungal spores or whole yeast cells, these are not necessarily digested. Observations on recovered spores, cysts and yeast cells, from faecal pellets or excreted from amoebae and ciliates often report viability of a fraction of the material after egestion (e.g. Behan and Hill, 1978; Anderson, 1988).

However, in some cases, there is evidence for feeding on yeasts or spores. One study showed a reduction of spores placed on filter membranes set out in the field, which correlated with microarthropod faecal pellet accumulation on the membranes (Gochenaur, 1987). This resulted in an 80% decline of spores of one species in the autumn, reflecting some selective grazing. A second study showed that microarthropod grazing, primarily by Collembola on perithecia of pathogenic fungi, eliminated the infection the next year (Kessler, 1990). Two pathogens of black walnut (*Juglans nigra*), *Mycosphaerella juglandis* (causing leaf spots) and *Gnomonia leptostyla* (causing anthracnose) form perithecia in fallen leaves, which could be eaten by the microarthropod fauna. In areas with less surface litter, where soil microarthropods were not abundant, the perithecia survived to re-infect trees in the next year. Selective preferences for different yeast were reported for six mite species in beech forests (Maraun *et al.* (2000). With protozoa, Ekelund (1998) modified the MPN approach to enumerate those that would grow on yeasts and spores. That approach was not suitable to enumerate mycophagous amoebae.

Hyphae ingesters

A variety of organisms are known to graze on lengths of hyphae, or on bundles of hyphae. Collembola and mites are the better studied hyphae ingesters in the soil. Some can be found several metres deep along roots, grazing on mycorrhizae or dead roots. Other fungivores include some armoured nematodes (Fig. 4.12), enchytraeids, mesofauna invertebrates and small mammals (particularly on aerial fruiting bodies). There is some evidence that juveniles and smaller species of microarthropods are more dependent on bacteria and protozoa in the soil solution, whereas larger individuals can ingest hyphae (Bakonyi, 1989).

Collembola. Most Collembola in the soil are considered to be at least partially fungivorous, especially *Hypogastrura*, *Tomocerus* and *Sminthurinus*. Grazing interactions between Collembola and hyphae can affect both the vertical distribution of fungal species and the rate of decomposition. The Collembola *Onychiurus latus* feeds preferentially on hyphae of *Marasmius androsaceus*, rather than *Mycena galopus*, at a *Picea* plantation in England (Newell, 1984a). The two Basidiomycota which represent 99% of the sporocarps are restricted spatially by this grazing. The preferred prey *M. androsaceus* occurs in the H (or A_0) horizon, and the other in the A_1 horizon. When the abundance of the Collembola was increased experimentally at the site, the abundance of the preferred prey was reduced (Newell, 1984b). The less preferred *M. galopus* is a slower decomposer and, thus, there was accumulation of forest floor litter at the experimental site. Moore *et al.* (1985) showed that the Collembola species *Folsomia candida*, *Onychiurus encarpatus*, *O. folsomi* and *Proisotoma minuta* would feed on hyphae of the arbuscular mycorrhizal fungus *Gigaspora rosea*, none ate *G. mosseae*, and only *F. candida* ate *G. fasciculatum*. In a hyphae choice experiment, *F. candida* was offered a variety of arbuscular mycorrhizal (Zygomycetes and Glomales) and saprotrophic species (Klironomos *et al.*, 1999). The Collembola were followed over two generations, and the numbers of eggs laid and hatched were counted. The results showed that saprotrophic species offered were preferred to the mycorrhizal species of fungi. Furthermore, when fed on single species or less preferred hyphae, the reproductive success of the F_1 generation was reduced because the number of eggs laid was greatly reduced. One possibility for the bias could be hyphae diameter, so that finer mycelia are grazed preferentially by microarthropods (Klironomos and Kendrick, 1996). Another possibility is the secretion of glomalin by the Glomales, which may have a defensive role against grazers. In a separate study, *Onychiurus subtenuis* (Collembola) and *Opiella nova* (Oribatei) were found to prefer the larger diameter hyphae of the saprotrophic fungal species offered, when the fungi were cultured on agar (Kaneko *et al.*, 1995). This was not true when the fungi were grown on their natural substrate of pine needles. However, both species showed some food preferences even in field trials.

The overall effect of fungivory by hyphae ingesters was investigated in a microcosm grazing study, where the abundance of grazers was altered (McLean *et al.*, 1996). Despite the significant, and sometimes limiting effect of fungivory on hyphal growth, it is not clear whether grazing is intense enough to reduce species richness of fungi. This study concluded that even at twice natural abundances, their microcosm study did not change the number of fungal species, species frequency or species dominance, in pine needle fungi grown on agar. This example is contrary to the previous examples above (Gauchenaur, 1987; Kessler, 1990) where, in field studies, the Collembola grazer was shown to reduce species abundance. There are limits to interpretation of microcosm studies when not repeated *in situ*.

The interactions of fungi with their grazers is more complicated than one would imagine initially. For example, some Collembola are attracted to hyphae species by odour. However, a predacious species of mite is also attracted by the odour of hyphae, in order to find its Collembola prey (Hall and Hedlund, 1999). The effect of grazing on the hyphae can also modify the growth physiology of the fungus locally. After several days of being grazed by the Collembola *Onychiurus armata*, the Zygomycetes *Mortierella isabellina* switched to a faster aerial mode of growth, correlated with increased protease secretion (Hedlund *et al.*, 1991). It is unclear whether the protease was being secreted to favour growth or if it was defensive.

Hyphae piercers

Nematodes with narrow tubular stylet and mycophagous amoebae are the more abundant organisms in this category. Some colpodid ciliates are also known to specialize in piercing hyphae. The impact of fungivorous nematodes on hyphal growth was studied in a series of microcosm studies recently (Ruess *et al.*, 2000). The role of spore and hyphae-piercing amoebae was reviewed earlier (Old and Chakraborty, 1986). Some species of **Myxobacteria**, otherwise better known for their lysis of bacterial colonies, are also reported to feed on hyphae and conidia by digesting a perforation into the cell wall (Homma, 1984). Many of the Acarina with fine stylet-like chelicerae (piercing-sucking group) pierce hyphae to ingest the protoplasm (Walter and Proctor, 1999). Some mites feed primarily on fungi (such as Ameroseiidae) while others are facultative fungivores and predacious (such as the Mesostigamata *Proctolaeleps*, *Protogamasellus* and Uropodina). The Tydeidae are primarily fungivores, which are also important predators on nematodes.

Nematodes. In a series of feeding experiments between fungivorous nematodes and fungal hyphae, it became clear that different nematodes preferred to graze on different hyphae (Ruess and Dighton, 1996; Ruess *et al.*, 2000). Species of *Aphelenchoides*, *Aphelencus* and *Ditylenchus* were tested on a variety of hyphae on nutrient agar plates. Some fungal feeders would not feed on the hyphae offered; others grew well with preferences for certain hyphae. These types of studies usually stress the importance of mixed diets for long-term cultivation of fungivores. Preferential feeding on mycorrhizae would suggest that these fungivores should be more abundant in the rhizosphere of mycorrhizal roots than in non-rhizosphere soil. Overall, the nematodes tested grew better on mycorrhizal fungi than saprotrophic hyphae. Some saprotrophic fungi are known to have nematicidal substances (Cayrol, 1989). Interestingly, several plant feeders were observed to feed on the ectomycorrhizae, and one bacterivore could grow on the agar film (apparently axenically). It is from similar observations that nematode feeding groups based on stoma morphology alone provide a false sense of rigidity or specialization in food choice. Although many species are specialists and rely on one feeding mechanism, many other species are more flexible and supplement their diet with a variety of food sources, especially when preferred food items are less abundant.

Amoebae. A variety of amoebae genera are recognized to be fungivores at least some of the time (facultative), such as *Acanthamoeba*, *Arachnula*, *Gephyramoeba*, *Mayorella*, *Ripidomyxa*, *Saccamoeba*, *Thecamoeba*, *Theratromyxa* and *Vampyrella*, although some are obligate fungivores such as *Thecamoeba granifera minor* and *Cashia mycophaga* (Old and Chakraborty, 1986; Foissner, 1992). Most attach to the surface of a length of fungal mycelium or spore and dissolve a perforation through the chitin cell wall (Fig. 4.13). A pseudopodium is sent into the cell through the pore, usually 2–6 μm wide, although some as small as 0.2 μm are possible by species with filopodia. The perforation is probably caused by fusion of lysosomes with the cell membrane, digestion of the cell wall and penetration of the amoeba's pseudopodia into the cytoplasm of the mycelium. Ingestion of the cytoplasm is by phagocytosis and pinocytosis, and digestion is in food vacuoles. In some species, observations suggest that at first the amoeba attempts to ingest the spore or mycelium and, if it proves too large, it releases enzymes to digest part of the wall before invading the prey with pseudopodia. Old

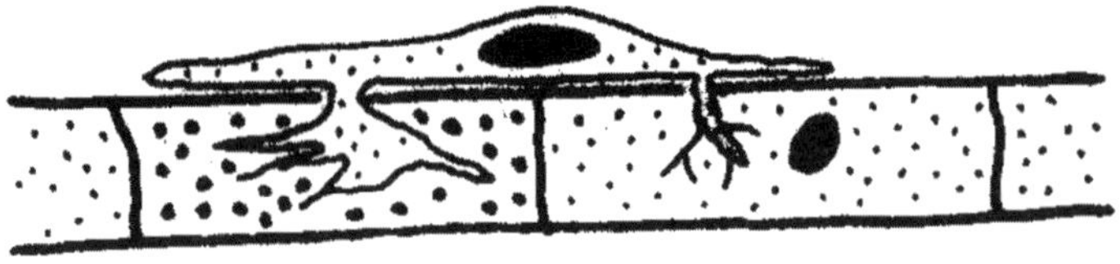

Fig. 4.13. Filose amoeba penetrating into fungus cells. The cell wall is digested to create a narrow opening and ingestion of the cytoplasm with invasive filopodia.

and Chakraborty (1986) reviewed several papers that observed ingestion of cysts and spores. However, not all studies confirmed continued growth on this diet, or checked the egested cysts and spores for viability. It is safer to assume that at least a portion of the cysts and spores are excreted viable at a distance (Wolff, 1909; Heal, 1963; Chakraborty and Old, 1982). None the less, many species, such as *Mayorella*, clearly are able to ingest and digest spores or cysts. There is some evidence that melanin pigmentation of fungal cell walls provides some resistance to lytic enzymes, but this is not a generalization, because the same fungi can be eaten by mycophagous amoebae (Old and Chakraborty, 1986).

Other amoebae that are not normally recognized as mycophagous, such as *Hartmanella* and *Schizopyranus*, also produce cellulase and chitinase (Tracey, 1955), to digest fungal cell walls and plant cell walls (in the microdetritus). Since other organisms also possess chitin, such as microinvertebrates, then it is not surprising that these are also preyed on by some mycophagous amoebae. For example, the well-known mycophagous genus *Arachnulla* (Cercozoa: Filosea: Vampyrellidae) and the giant amoeba *Leptomyxa* also attack and perforate nematodes and diatoms (Old and Darbyshire, 1978; Anderson and Patrick, 1980; Homma and Kegasawa, 1984). This mechanism of perforation feeding is also known to occur in amoebae feeding on filamentous algae (Lloyd, 1927). Many mycophagous species also feed on prokaryotes, although it may not be the preferred food item.

The impact of mycophagous amoebae was reviewed by Old and Chakraborty (1986). Most observations are based on laboratory incubations or pot experiments; much less is known of their impact *in situ*, where migration and predation also occur, and where conditions for growth are not always suitable. For example, amoeba grazing on ectomycorrhizae reduced the extent of fungal colonization of pine roots in pots (Chakraborty *et al.*, 1985). Although the experiments provided an idea of the impact it could have on seedlings and mycorrhizae, it could not establish whether the same would apply *in situ*. In fields that were infected with a plant pathogenic fungus, the mycophagous amoebae could reduce but not eliminate the abundance of spores in the soil (Chakraborty, 1985; Chakraborty and Warcup, 1985). Therefore, in agricultural soils that would support amoebae, i.e. with adequate organic matter to support microorganisms, there is the potential to reduce infectivity greatly in successive years.

Some families of Prostigmata and Endostigmata (both Acarina) have stylet-like chelicerae that puncture hyphae and other prey, such as microinvertebrates and protists (Table 4.12), if they can be immobilized (Walter, 1988; Kethley, 1990; Curry, 1994). The drained solution and cytoplasm are externally partially digested and sucked in by the oral structures.

Detritivory

We refer to **detritivory** as the ingestion of microdetritus and macrodetritus, i.e. fragmented litter and excreted organic matter derived from litter. As the nutritional quality of this detritus is variable, its value to the organism is unpredictable. Therefore, detritivory is often opportunistic, and ingestion depends on size selection by the feeding structures, although some chemotaxis cannot be eliminated, and some species have preferences for particular leaf litter or wood species. Detritivory and particle size selection are accompanied by ingestion of similar sized particles, such as cysts, spores and invertebrate eggs. Many invertebrates engage in detritivory, particularly millipedes, several onychophorans and the enchytraeids (Tables 4.2 and 4.12). Some nematode species can be seen to engage in some detritivory on agar plates depleted of bacteria or prey species. The more omnivorous species, such as many Dorylamidae, probably complement their diet with microdetritus (Yeates, 1998). Detritivory by nematodes probably is more common than reported in the literature. Ingestion of microdetritus (including organisms in it) is followed by crushing–breaking of a portion of the cell walls by denticles and armature when present, and the pharyngeal bulb musculature.

Testate amoebae can be seen to engage in a substantial amount of detritivory (Heal, 1964; Schonborn, 1965; Foissner, 1987). Many species are found in the litter layer, some even buried inside leaf fragments. These can be maintained in culture on microdetritus or litter fragments, in otherwise sterile soil (Schonborn, 1965). It is likely that pseudopodia of Difflugiid and Centropyxid species participate in the excretion of cellulases to break into cell walls of plant or algal cells (Chardez, 1985).

Walter and Proctor (1999) recognize comminution grazers (detritivores) among the Oribatei and Astigmata. They note that although the smaller species may feed on spores, cysts and hyphae, larger species or individuals can feed on animal and plant litter colonized with primary saprotrophs. Some shredding of this litter is possible in species with large chelicerae. They also note that these organisms are generally opportunists that can also feed on protists and small invertebrates. The oribatid endophagous species that burrow into litter include the Carabodoidea, Cepheidae, Hermaniellidae, Liacaroidae, Ptyctima (box mites), Xenillidae and *Rhysotritia minima*. Some Collembola are also found feeding in invertebrate corpses and detritus.

Other Consumers

In the sections above, we described the grazing of secondary saprotrophs on primary saprotrophs, and on each other. In addition to those, there are other trophic interactions between the interstitial saprotrophs.

These trophic interactions include predation on secondary saprotrophs. In this section, we group trophic interactions that are mostly between microinvertebrate predators. Several trophic interactions have been studied in particular, whereas many others are poorly described. The roles of predatory rotifers (Monogomontes), the tardigrades and onychophorans are poorly quantified and neglected (see Chapter 1). One needs to remember that one species does not necessarily belong to any one trophic level, but can also be found at other levels. For example, some of the ligninolytic fungal species complement their nitrogen-poor diet by capturing nematodes.

Nematotrophy

In this section, we consider primarily feeding on nematodes. The mechanism of capturing a nematode prey is the most studied; however, the same mechanisms operate on rotifers, tardigrades and similar sized organisms that can be captured or penetrated. It is unclear what the extent of **nematotrophy**, or feeding on microinvertebrates in general, is in the soil. The phenomenon is common but has been quantified insufficiently. A variety of microarthropods are known to feed on nematodes (or microinvertebrates) or at least to supplement their diet with nematodes (Walter, 1987) (Fig. 4.12). These include pseudo-scorpions and several families of mites and Collembola. Many soil Collembola probably supplement their diet with nematodes (Hopkins, 1997). Among the Acarina, these include the groups listed below as predatory. However, the nematodes are often the preferred prey, especially for Mesostigmata, Prostigmata (Bdellidae and Cunaxidae) and some Endostigamata (*Alycus* and *Alicorhagia*). Certain predacious testate amoebae, such as *Nebella*, are known to hold nematodes with pseudopodia, immobilize the prey and engulf the whole organism. Sometimes, only the tail is held and the rest of the nematode breaks free (Yeates and Foissner, 1995). In all the above cases, access to nematodes in soil pores is not restricted by the size of the predator as much as by the diameter of the amoeba pseudopodia, the chelicerae of mites or other appendages. Several interactions involving hyphal fungi (Basidiomycetes and mitosporic species), Chytridiomycetes, Zygomycetes and the Oomycetes (chromista and pseudo-fungi) are considered below (for further details and references, see Barron, 1977).

One method of attacking and digesting nematodes involves monociliated dispersal cells which swim towards prey. The dispersal cells do not hold many reserve nutrients and have a limited time of activity to find a new prey. The cells probably track a solute from the prey's path by chemotaxis. Penetration into the prey can be through an orifice or through the cuticle. Once attached to the prey cuticle, the cilium is lost and the cell encysts. From the cyst, cytoplasmic extensions grow into the host and form a mesh. The prey is invaded by the hyphae which secrete digestive enzymes, absorb the solutes and empty the cuticle. When nutrients run out or the hyphae become crowded, spores form which grow an exit tube and swim off. In dry conditions or through tight exit tubes, cells can be amoeboid. This mechanism is common in Chytrids (fungi) such as *Catenasia*. The chytrids also attacks similar sized organisms such as rotifers, tardigrades, invertebrate eggs or macrodetritus consisting of animal corpses or parts thereof.

A modification of this mechanism occurs in the Oomycetes (chromista) *Myzocitium*, where the dispersal ciliated cells encyst in the soil and become sticky. Attachment on to the cuticle of a passing prey is the stimulus to extend hyphae into the prey. In *Haptoglossa*, the cyst discharges a coiled tube into the passing prey. The coil is 5–8 μm long and 0.5 μm wide, and discharged in about 0.1 s. No infection occurs if the coil is discharged into a cuticle which is too thick to traverse. In Zygomycetes, such as *Meristacrum*, at the end of the growth phase in the prey, some mycelia extend out of the cuticle and form adhesive conidia spores, which are then picked up by passing prey. Some of the spores remain inside the old cuticle. The Basidiomycetes *Nematoctonus* also form adhesive conidia spores which then attach to the cuticle of a new prey. The protoplasm migrates out of the hyphae inside the prey and accumulates in conidia bearing branches outside the prey. Several mitosporic species form adhesive dispersal spores, such as *Cephalosporium*, *Mesia* and *Verticillium*. Ingestion of infective spores can also stimulate growth. In the case of *Harporium* species, the shape of the spore prevents passage through the nematode pharynx into the middle intestine. Sporulation occurs in the oesophageal muscle, and invasive hyphae grow through the organism. One difference between nematotrophy by soil hyphal fungi (such as the Basidiomycetes genera mentioned here) and others such as Chytrids and Oomycetes is that the latter two tend to grow into the prey, with very few vegetative hyphae in the soil. In contrast, Basidiomycetes grow extensive saprotrophic hyphae in the soil and only supplement their nutrition with nematodes, or other microinvertebrates. This is particularly important in some lignin-decomposing fungi (such as the oyster mushroom, *Pleurotus*), which are otherwise nitrogen limited. The *Pleurotus* and *Hohenbuchelia* genera (family Pleorotaceae, order Agaricales, Basidiomycetes) are nematotrophic white rot fungi (Thorn *et al.*, 2000). They secrete a non-adhesive nematotoxic substance (*trans*-decenedioic acid) to immobilize nematodes, which are then invaded with hyphae through the orifices. The *Hohenbuchelia* also produce adhesive knobs, as for the anamorph *Nematoctonus*.

Hyphal fungi in the soil and litter rely on adhesive mycelium or rings for capturing nematodes and similar sized prey (Fig. 4.14). Barron (1977) distinguishes between six methods as follows.

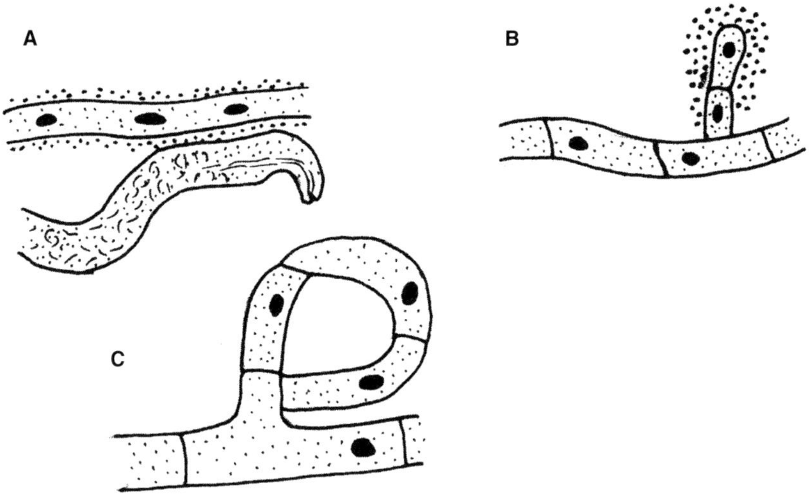

Fig. 4.14. Examples of protist and invertebrate traps used by filamentous fungi. (A) Zygomycete hyphae coated with a sticky substance, showing a glued nematode. (B) Sticky knob with a coat of sticky substance. (C) Loop of collar cells which constrict around a passing organism.

1. Adhesive hyphae of the Zygomycetes, such as *Cystopage* and *Stylopage*, secrete a sticky substance. Trapped invertebrates will fight it to exhaustion. Branching mycelia grow into the prey. The protoplasm moves out of the hyphae in the prey and grows outside the prey. These vegetative hyphae in the soil near the prey form conidia.
2. Short adhesive branches or rings occur in only a few mitosporic species. For example, *Dactylella tylopage* captures amoebae on its sticky surfaces, although other species in the genus capture nematodes, which become stuck at several points. Only a few seconds of contact are sufficient, and mycelial growth begins rapidly. Nematodes can withdraw quickly enough to avoid being trapped, often succeeding. Sometime they try to proceed only to be caught again.
3. Adhesive nets are formed by ubiquitous and abundant fungi, such as *Arthrobotrys*. These form more extensive adhesive hyphae branches and rings. Some species secrete toxins which help to immobilize or kill the prey while mycelia grow into it. As in other groups, the protoplasm retracts from the hyphae in the prey and grows outside the cuticle which is left almost empty. Conidia form outside the cuticle from these later emerging branches.
4. Adhesive knobs are formed by several Basidiomycetes, such as *Nematoctonus* and *Hyphoderma*, and mitosporic fungi such as *Dactylaria*. These consist of single adhesive cells at the end of short branches, which can be broken off by a struggling prey. Each adhesive cell, once attached to a nematode or cuticle, will grow invasive hyphae. This mechanism

combines prey invasion with dispersal. The adhesive substance of *Nematoctonus* is very strong and the knob will not break off. A struggling nematode can escape at the cost of losing some epidermal cells and cuticle (causing a wound).

5. Many mitosporic species that form adhesive knobs also form rings that consist of three cells. These break off, so that the nematode carries a collar of three cells which grow mycelia through its cuticle to digest the nematode.

6. Constricting rings also consist of three cells, with a fourth supporting cell and a fifth which branches from the main mycelium. These also occur in the genera *Arthrobotrys* and *Dactylaria*. The collar is about 20 μm in diameter and constricts inward within 0.1 s by rapid uptake of water. The collar cells constrict around the nematode, then grow into the caught prey.

Predatory microinvertebrates

Numerous genera of Acarina are implicated in predation, or facultative predation (Table 4.12). These would include many of the species of piercing-sucking mites and the nematophagous mites mentioned previously (Walter and Proctor, 1999). For example, the Cunaxid mite *Coleoscirus simplex* (piercing–sucking) can move 20 cm/min in the laboratory at 25°C to attack a prey. The dominant predatory mites are the Mesostigmata (suborders Dermanyssina and Parasitina) and several families of Prostigmata. The Mesostigmata and the Rhagidiidae (Prostigmata) species are fast and constantly searching for prey, so that time spent foraging is important in understanding their behaviour. Some of the Prostigmata families include the Bdellidae, Cunaxidae (Bdelloidea); the slow moving Labidostommatidae; seven of the eight Anystina families; both Erythracoidea; the adults of the Trombidiioidea (red velvet mites); one Chyletinae subfamily; and nine families of Raphignathoidea, particularly in drier soils. Among the Mesostigmata, the Laelopidae include particularly aggressive predators (not all are edaphic species) that are used in greenhouses for pest control, such as *Hypoaspis*, *Geolaelaps* and *Stratilaelaps*.

Among the Collembola, *Frisea* species are known to prey on tardigrades, rotifers and invertebrate eggs. The Antarctica species *Cephalotoma gradiceps* preys on other Collembola, as well as *Isotoma macnamarai* (Denis, 1949; Hopkins, 1997).

The main predators of microarthropods and of other predatory microinvertebrates are larger litter invertebrates which are part of the meso- and macrofauna (Coleman and Crossley, 1996). The predominant ones are Opiliones, hunting spiders and those with webs in the litter, pseudo-scorpions and ants. Some beetles are specialists on particular prey. For example, setal and antennal traps of carabid beetles or *Stennus comma*

(Staphylinid) are effective predators on Collembola. Predation of the carabid beetle *Notiophilus biguttatus* on the Collembola *Orchesella cincta* showed that there was positive density-dependent mortality of the Collembola (de Ruiter *et al.*, 1988). This suggested a Holling (1959) type III sigmoid functional response. It also suggests that the beetle can, in theory, control the abundance of the prey species. The beetle also exhibits some prey preference and optimal foraging behaviour (Ernsting and van der Werf, 1988).

Earthworms

Earthworms have a diurnal rhythm and tend to be more nocturnal, probably to avoid ultraviolet light, solar radiation and desiccation. They are active when there is sufficient moisture and become inactive as the soil dries, by dehydrating and entering dormancy. The earthworms are generally placed into three functional groups based on the observations of Bouché (1977) on Lumbricina. **Epigeic** species tend to tunnel through the surface litter and organic horizons near the surface. They have a preference for partially decomposed litter and ingest rich organic matter and surface litter. Some species can also be found in animal dung (some species of *Lumbricus* and *Aporrectodea*), composts or under tree bark. **Anecic** species reside in deeper permanent burrows and emerge at night to drag selected leaf litter into their tunnel. The litter is eaten in the tunnel and in some species the excreted material, the earthworm **cast**, is deposited on the soil surface. **Endogeic** species tend to remain in the mineral soil horizon, where there is less organic matter. They can feed on rootlets and ingest whole soil to digest the organic component, including the living organisms. Truly endogeic species avoid composts and soils rich in organic matter. Therefore, some earthworms can be found in tunnels or burrows several metres below the surface, though most occupy the top 30–60 cm of the profile. Many species do not fit neatly into these categories and are best described as intermediates. Others may be predators, such as the *Chaetogaster* (Naididae) which are reported to hunt and prey on microinvertebrates (Avel, 1949, p. 387). The integument of earthworms is permeable to the soil solution through the coelomic pores, and bacteria or protist cells potentially can enter this space. Richard and Arme (1982) demonstrated that the soil solution could enter the coelom, but it was insufficient to be a source of food. Mark recapture has been possible using a variety of labels, such as dyes, fluorescent stains and radioactive tracer, and it contributes to understanding the behaviour and function of individuals.

The main earthworm predators vary with the ecosystem and regionally. When they are out of the soil, particularly when the soil is wet, they are preyed on by a variety of birds (during the day), and surface predatory arthropods, such as ants, centipedes and Carabid beetles.

Otherwise, common predators are mammals such as rodents and some carnivores such as foxes and badgers, species of terrestrial molluscs, flat worms and moles (MacDonald, 1983; Blackshaw, 1997). The latter have been known to wound earthworms which they accumulate in their burrows for a later feeding. Some species of earthworms could be predatory on other earthworms, such as in *Fridericia*, *Agastrodilus* and *Bipalium kewensis* (Lavelle, 1983a,b).

In a laboratory feeding experiment, Lavelle *et al.* (1980) showed a preference for soil of different depth by three species from one area. Each species has a preferred location in the profile, and an optimum growth rate and ingestion rate depending on the soil depth. The litter-dwelling *Millsonia anomala* occupies the top 10 cm and prefers to feed on organic matter from the top 5–10 cm. *Millsonia ghanensis* was found in the 10–40 cm profile, and showed optimal growth on soil from 10 to 25 cm depth. *Dichogaster terrae-nigra* occupied the top 30 cm of soil and also grew best on soil from 10 to 25 cm depth. The estimated calorific value for the soil depth sections were 17,000 kJ/g for litter, 376 kJ/g for 0–10 cm, 210 kJ/g for 10–30 cm and 167 kJ/g for 10–40 cm soil. These values do not reflect the difference in nutrient composition of SOM with depth. The quality of the organic matter is also of importance, and species can be choosy about what is ingested. In feeding experiments (Mangold, 1953; Edwards and Heath, 1963; Satchell, 1967, Cooke, 1983), it was shown that loss of phenolic molecules in leaching or with primary saprotroph activity renders some leaf litter more palatable and acceptable to epigeic and anecic species. With hunger, individuals become less discriminative.

Since soil litter and whole soil contain many individuals from many species, ingestion of a mouthful of litter or soil by an earthworm inevitably includes other organisms, such as nematodes, protists and bacteria. This contributes to the mineralization of nutrients immobilized in cells. It is an important source of nutritious molecules for endogeic species, where the nutritional quality of the SOM associated with mineral particles is lower. Food preferences of earthworms are therefore affected by the type of microdetritus and macrodetritus, as well as by the biotic composition of the soil. In feeding experiments, it was shown that fungal hyphae are also eaten selectively by earthworms (Bonkowski *et al.*, 2000b). The authors conclude that earthworms use early successional fungal species as cues for fresh organic detritus. Several papers reported the impact of earthworm grazing on reducing the abundance of protozoa and nematodes in microcosms (Piearce and Phillips, 1980; Dash *et al.* 1980; Yeates, 1981; Rouelle, 1983). Similarly, soils that have an abundant earthworm population have reduced general protozoa abundance (Bachelier, 1963; Davis, 1981; Block, 1985; Meisterfeld, 1986). Earthworms tend to inhibit or reduce the growth of bacteria and fungi, when cultures are placed together in the laboratory (Kobatake,

1954; Van der Brael, 1964; Ghabbour, 1966). This may be through secretion and accumulation of antibiotic substances with the mucus or from the coelomic fluid. The purpose may be to protect the coelomic cavity and inside wall of burrows from infection. The excreta contain some undigested bacteria, protozoan cysts and fungal spores. In general, cysts and spores which are desiccated are not digested. These can become activated in the excreta, but for many it is a mechanism of dispersal (Rouelle, 1983).

However, in microcosms or from field samples, both microbial activity and decomposition rate of litter are stimulated by earthworm activity. The increased activity is in the cast or excreted material, compared with the surrounding soil. The excreted material contains partially digested plant debris from the earthworm gut, released unabsorbed but somewhat homogenized and partially digested. This provides a hydrated substrate that can be digested further by soil saprotrophs. The growth of primary saprotrophs in turn provides cells for the secondary saprotrophs and their consumers to prey on. For example, this was demonstrated with the litter-dwelling *Eisenia foetida* fed on leaf litter in microcosms (Anstett, 1951). This species also requires its gut protozoa for efficient growth (Miles, 1963). The role of the gut bacteria and protozoa needs to be reassessed, taking into consideration: (i) whether the cells are from the anterior undigested portion, or from the posterior portion; and (ii) whether the symbiotic ciliates and other organisms are contributing to the digestion, or are simply obtaining nutrients. This needs to be done without causing lysis of the anaerobic protozoa (see Chapter 3).

Omnivory

When organisms feed on more than one functional group, or at different trophic levels, they are considered to be less specialized and are called **omnivorous**. Another term sometimes used is polyphagous. The occurrence of omnivory allows more opportunist feeding on a variety of food items, as the availability of specific food items fluctuates over time. This does not mean that omnivorous species do not have food preferences, but it does provide for a broader diet and it is an adaptation to the fluctuating environment.

When trying to understand the role of species in a sample, several facts need to be known about them. These include what they are feeding on; environmental cues which activate, maintain activity and inactivate them; and which species prey on them. These details can be used to reconstruct diagrams of trophic interactions (Pimm, 1982). Most diagrammatic representations isolate a small number of interactions from the rest of the community or ecosystem. These result in short linear diagrams that are easily understood. This representation produces inconsistencies. For example, larger or better known species are often represented by a single box, whereas smaller or less known species are also grouped into a single box, as if they were one functional species (Coleman *et al.*, 1983).

More realistic food webs try to represent species with similar function as single trophic groups irrespective of taxonomy. However, there is a serious technical limitation because we do not know enough about the role of most soil species. Numerical estimates of the amount of energy transferred from nutrients between trophic groups is then limited by the amount of knowledge we have on the interacting species. It is limited further by our knowledge of the interacting links between the boxes in diagrams that represent trophic interactions. Mathematical models that try to capture the dynamics of soil biology need to balance the amount of knowledge necessary to reconstruct the interacting links with the amount of knowledge that is available. In some cases, there are sufficient data to predict trends and outcomes (Smith *et al.*, 1998; Moore and deRuiter, 2000). However, models that have some predictive validity at one scale may not be adequate at different scales of time or space.

In reality, soil trophic interactions are not easily isolated into short chains (Walter and Proctor, 1999, p. 98–103). The more serious problem in soil food webs arises from the apparent rarity of specialist species, and the abundance of generalist and omnivorous species. Most trophic studies use a mean chain length of 2.71–2.89 interacting boxes (Polis, 1991). They further assume that omnivory is rare and can be ignored in most situations. In soil food webs, with opportunist feeding and species switching food preferences, links between boxes representing trophic groups are more variable. In other words, the nature of the interactions varies as abiotic conditions change (temperature, moisture, litter quality and abundance), as well as more direct biotic changes (abundance of food, prey or predators, and competitors). Most species have to be placed in several trophic groups, although they may not be active in all, all the time. The mathematics and the dynamics of such interactions are more complex. It is more difficult to construct correct diagrams, and mathematically difficult to predict end-points and trends. In these interactions, specialization of species may be more to a set of abiotic conditions and a habitat, and only secondarily to available food items.

Symbionts

Symbiosis with animals

Among the Collembola, *Lepidosinella armata* (Entomobryinae) and Cyphoderidae are specialized symbionts in the nests of ants and termites, where they feed on detritus (Hopkins, 1997). *Calobatinus rhadinopus* are associated with termites (*Bellicositermes netalensis* and *B. bellicosus*) in the sawdust imbibed with saliva or on the termite mouth parts (Denis, 1949). Numerous species of mites are ectosymbionts on a variety of animals (Evans, 1992) (Table 4.12). Some form close associations with soil and litter organisms. For example, the majority of species in the suborder Antennophorina are found living alongside insects, but primarily ants. *Antennophorus* species will stroke the mouth parts of the ant *Lasius* which disgorges liquefied food for the mite (Donisthorpe, 1927). Soil mites themselves harbour internal and external symbionts. In a study of five species of Oribatida and Acaridida, the culturable bacterial population between the surface and the digestive tract of the mites was found to be different. Moreover, the surface bacteria differed between mite species that reside in the same soil (Smrz and Trelova, 1995). This study also identified bacteria in the somatic tissues, resembling the mycetocyte of insects. Mycetocytes are also known from the somatic tissue of lower termites, such as *Mastotermes darwiniensis*, which harbours symbiotic bacteria in mycetocytes within the fat body. The role of these bacteria in the somatic tissue is not clear. The bacteria associated with the cuticle of soil invertebrates could be picked up randomly from the soil, alongside spores, cysts and pollen. These would benefit from transport and dispersal by invertebrates (Klironomos and Moutoglis, 1999). However, the differences observed between bacteria on invertebrate species suggest that there is some selective transport, or a closer ectosymbiotic association.

The symbiosis between internal bacteria and protozoa populations with their invertebrate or protozoan host has been studied the most in termites. Termites are responsible for the ingestion, digestion and mineralization of a large portion of woody litter in temperate and tropical regions. The digestion of cellulose and lignin in the lower termites occurs only through the cooperation of the termite with its indispensable protozoa and bacteria symbionts. Most of the cellulase activity occurs in food vacuoles of specialized protozoa that occur only in termite hindgut. The Trichomonadida, Oxymonadida and Hypermastigida ingest the woody microdetritus from the hindgut and digest it in food vacuoles (Breznac and Brune, 1994; Rother *et al.*, 1999). The remains are excreted into the hindgut, mostly as glucose, acetate, CO_2 and H_2 for the bacteria to use. The process involves detritivory by the termite and the protozoa, and osmotrophy by the bacteria. The role of the protozoa, which contribute one-third of the termite weight, was confirmed in axenic cultures (see Breznac and Brune, 1994). In contrast, cellulolytic bacteria comprise <100 cells/gut only. The main role of the digestive tract bacterial symbionts is in the utilization of the acetate, hydrogen and carbon dioxide, to produce methane and carbon dioxide ($<5\%$ of global methane emissions is from termite guts). There are also other metabolic pathways represented, including sulphate reducers. Detailed metabolic studies of termite digestive tracts showed that it was a highly structured environment, with steep gradients of oxygen and metabolites, with specific and structured communities of bacteria and protozoa (Tholen *et al.*, 1997). The bacterial community consists of aero-tolerant species, facultative aerobes, aerobic and anaerobic species. The protozoa are amitochondriate anaerobes. The presence of oxygen occurs through diffusion from outside, and it is required for digestion of lignin and mineralization of aromatic molecules. The concentration of oxygen decreases to zero away from the gut wall, due to a high consumption rate. The position of the bacteria and protozoa is thus dependent on the oxygen and substrate levels. Some of the spirochetes and rod-shaped bacteria occur as symbionts on the surface of the Hypermastigida and other protozoa, but also inside the cytoplasm (Rother *et al.*, 1999; Patricola *et al.*, 2001). This type of symbiosis between bacteria, protozoa and an invertebrate also occurs in wood-eating cockroaches (Cleveland and Grimstone, 1964) and in Oligochaeta (including earthworms). The latter are known to possess specific symbiotic ciliates (Astomatidae and other ciliates) in the digestive tract, which are not found anywhere else (Small and Lynn, 1988; Ngassam *et al.*, 1994).

The process is different in higher termites (Termitidae) which usually have symbiotic bacteria only. Species of the subfamily Macrotermitinae actively culture the fungus *Termitomyces* (Basidiomycota). The fungus is cultured on faeces of the termite, where it completes the digestion, for re-feeding by the termite (Breznak and Brune, 1994).

Symbiosis with plant roots

Many species of Basidiomycetes and Ascomycetes, as well as most Glomomycetes form tight symbiotic associations with roots of vascular plants and rhizoids of non-vascular plants. This association between the fungal hyphal network and the plant tissues is called a **mycorrhizal association**. The roots and rhizoids of most terrestrial plants, and some aquatic species, form varying degrees of mycorrhizal symbiosis. There are several types of mycorrhizal associations based on the overall morphology and cellular arrangement between the fungal–plant cells. The role of this association is to mediate competition for nutrients between plant species and fungal species. Under the continuously fluctuating soil environment, at times, the fungus will support plant nutrition and, at other times, the plant will support the fungus. The symbiotic advantage is generally clearer when one or more nutrients, or water, become limiting. Moreover, the weight of recent evidence suggests that in many cases, the fungal hyphae mediate between individual plants. The hyphae can connect between plants of different species and of differing root morphology. The advantage is most evident in times of water or nutrient stress, when a more competitive plant species will become a source for another individual, through the mycorrhizal hyphae. We will consider below several types of mycorrhizal associations. The reader is directed to Smith and Read (1997) for a detailed survey of the subject.

Types of mycorrhizae. Woody perennial species, especially in northern temperate forests, tend to form **ectomycorrhizae** with Basidiomycetes (and a smaller number of Ascomycete species). These include beech, birch, larch, oak and pine genera in symbiosis with a variety of Agaricales and Boletales species. Many of the fungal species are not only saprotrophs but also mycorrhizal species with particular tree species. Thus, reproductive structures of a fungus are often found near trees it associates with. Of the 6000 or more fungal species that form ectomycorrhizae, about 4500 are epigeous and the remainder are hypogeous. It is estimated that about 3% of seed plants form ectomycorrhizae. However, they are species of economic importance, such as the Pinacea and Fagacea in northern temperate regions, the Myrtaceae in the southern temperate and subtropical regions, and the Dipterocarpaceae in the tropics of South-East Asia. The ectomycorrhizal association involves penetration of hyphae between plant cells, without intracellular invasion. This network of hyphae within the plant tissues is called the **Hartig net**. At the periphery of the root, hyphae grow transverse to the root axis in a dense mycelium called the **mantle**. The exact morphology of the mantle and Hartig net is characteristic of a fungal genus on a particular host (Agerer, 1987–1993). There is variation between the relative extent of the Hartig net and mantle mycelium. In mycorrhizae with reduced mantle and/or more extensive Hartig net, the association is called an **ectendomycorrhizae**. Some intracellular penetration of the root cells occurs, especially in older or senescent roots. There are variations between the morphology of mycorrhizae formed by one fungal species with different plant species. For example, species that typically form ectomycorrhizae in some species will form ectendomycorrhizae with the same plant under different conditions. These could be caused by nutrient stress or senescence. Alternatively, the fungus may form ectomycorrhizae in conifers, but arbutoid mycorrhizae in other trees. Sometimes, whether an arbuscular or ectomycorrhizal association forms with the roots, as in *Salix* and *Populus* (Saliceae), depends on the fungus species which colonize the root.

By far the most common type of mycorrhizae are **endomycorrhizae**. These fall into several categories as classified by Smith and Read (1997): arbuscular, arbutoid, ericoid, monotropoid and orchid. The **arbuscular** mycorrhizae occur in most vascular and non-vascular plant families. Indeed most plants are capable of forming arbuscular mycorrhizae. The fungus species responsible are obligate symbionts that cannot be grown without the plant host. They occur in the Archemycota taxa

Acaulosporacea (*Acaulospora* and *Entrophosphora*), Gigasporacea (*Gigaspora* and *Scutellospra*) and Glomacea (*Glomus* and *Sclerocystis*). These types of mycorrhizae typically consist of sparse and loose hyphae on the root or rhizoid surface, with most of the mycelium inside the plant tissue or in the soil. The hyphae extend in the SOM some distance from the plant. The hyphae inside the plant grow between the cell walls, and plant cells are also penetrated. The hyphal cell membrane does not break through the plant cell membrane. Both membranes remain in close apposition even when highly invaginated or ramified. The hyphal extensions into plant cells branch profusely and repeatedly, and resemble a small tree – thus the name arbuscule. In some cases, the intracellular hyphae grow into a tight coil. In about 80% of cases, the symbiosis is accompanied by formation of intracellular 'vesicles' inside some plant cells. The vesicle is a terminal hypha which is an intracellular propagule, like an intracellular spore. It contains a reinforced cell wall and storage material (lipid droplets, glycogen and proteins), as do the chlamydospores formed in the soil matrix.

Three other types of mycorrhizae are recognized. In the Ericales, which dominate many heathlands in the northern hemisphere, or the Epacrideceae in southern ecosystems, a characteristic **ericoid** morphology is recognized. The roots normally have few cell layers that are delicate. The cortical cells are colonized by dense intracellular hyphae. Epidermal cells that are colonized do not form root hairs. Dispersal of some fungal species is by arthroconidia, whereby the hyphae break into nucleated segments by septation. A variant form is observed in some Ericales, notably in *Arbutus*, called arbutoid mycorrhizae. These are formed by Basidiomycetes species which often form ectomycorrhizae with other plant species. The last two forms of mycorrhizae involve plants that are achlorophyllous for at least a part of their life history, namely in the Monotropaceae and Orchidaceae. Both of these rely to a large extent on their mycorrhizae for survival, as they are not capable of obtaining their own carbon. The fungus mediates between SOM or a second plant host, transferring organic molecules into the parasitic host. Often, the fungal species are Basidiomycetes forming ectomycorrhizae with other plants, or saprotrophic and parasitic fungal species.

Mycorrhization. When considering these mycorrhizal associations, one tends to observe a general lack of specific interaction between the fungus and plant species. However, even though in most cases there is a broad choice of possible plant–fungus species interactions, there tends to be some preferences for particular genera or families of fungi by the plant, especially when Basidiomycetes are involved. In considering the plant–fungus compatibility, and for comparative purposes between experiments, it is important that the genotypes and alleles of both the plant and the fungus are considered. The recognition and compatibility occur at the interface of the plant and fungus cell walls, through surface proteins and possibly also through chemical signals.

The regulation and acceptance of the mutual association between the plant and fungus occur at several levels. In the soil, plant-derived molecules, such as terpenoids, flavonoids or sterols, can have an effect on the germination of spores, hyphal branching frequency and direction of growth. Once the fungal invasion of root tissues begins, the amount of colonization and morphology of the appressorium formation against the plant cell can be regulated. In return, many species of ectomycorrhizae and Glomales can produce at least some of the classic auxin, cytokinin and gibberellic phytohormones (Beyrle, 1995). These are recognized to be implicated in initiating, maintaining and regulating the plant–fungus interaction (Barker and Tagu, 2000). However, elucidation of the details of concentration gradient changes, and of the hormonal message exchange is still in its infancy. Usually only one hormone is monitored and the manipulations of concentration are crude. None the less, this is an exciting area with many explorative research opportunities to be exploited this decade.

The significance of the early communication between the two species should not be underestimated. The symbiosis develops from cooperation between two species, before there is any advantage to the plant from the fungus. The plant must accept this invasion as friendly, even though the fungus could release its panoply of digestive enzymes. This is an adaptive co-regulated process, in which both species make an investment against future suboptimal abiotic and biotic conditions. In fact, plant cells adjacent to the hyphal appressorium are perfectly capable of activating their molecular defences against pathogenic intrusions. Therefore, accepting the presence of the fungus within the root tissues is not due to a general systemic failure to respond to infection. Instead, it is a local interaction based on a mutual acceptance of a very close association, leading to living cell to cell, sharing the same space and exchanging nutrients. In this close association, there is some flexibility so that with changing environmental factors and nutrient limitations, the relationship can vary between mutualism and mild parasitism of one species on the other.

The mycorrhizal growth is affected by certain bacteria found commonly in the soil. This phenomenon was recognized from the differing performance of mycorrhizal growth from soils that differed only in their edaphic community (Garbaye and Bowen, 1987). It led to the recognition of 'mycorrhization helper bacteria' (MHB) which are mostly fluorescent *Pseudomonads* (80% of isolates in one study). The effect of isolated bacterial strains on mycorrhizal formation was clearly demonstrated in poor media (Duponnois, 1992). The correlation coefficient of 0.9 was highly significant ($P = 0.01$), with some isolates clearly enhancing mycorrhizae, while others were neutral or inhibitory. The interpretation of these experiments led the authors to conclude that metabolites from certain bacteria were responsible for the hyphal growth enhancement. Different bacterial biochemical pathways would provide specific contributions to the digestion of SOM which benefit the hyphae by correlation. In fact, replacing the bacteria with filtrates of the bacterial cultures or with solutions of the enzymes had the same effect. However, much of the detail of the interaction between the positive effect of some bacteria on mycorrhization remains speculative and hypothetical (Garbaye, 1994).

Nutrient translocation. In light of the trophic association between the soil mycelium of fungal species and root cells, it is necessary to consider briefly the nature of this nutrient exchange. The mycelia feed on SOM and participate in its digestion, as well as forming mycorrhizal associations with plant root cells. Hyphae within a mycelium are often connected to several plants. Hyphae in one region are able to translocate nutrients through the mycelium to different hyphae. The exchange of elements at the mycorrhizal cell–cell contact is generally agreed to consist of a net flow of C into hyphae, and a net flow of P, N, S, Ca, Zn and other minerals into the root cells (Smith and Read, 1997, Chapter 14). The flow of various elements is described from isotope tracer studies that have shown hyphal uptake from the soil solution by osmotrophy of ^{32}P, ^{15}N, ^{35}S, ^{45}Ca and ^{65}Zn. The minerals are obtained in their ionic form by hyphae through active transport. They are translocated through the mycelium and distributed to areas of new growth of both hyphae and new roots. The minerals are transferred into root cells at the areas of cell–cell contact between both species. The transfer rate is maintained by a concentration gradient in favour of the plant. The ionic form of P obtained by fungi is assumed to be $H_2PO_4^{2-}$ as in other protists. It can be stored in the cytoplasm in the form of short polyphosphate chains. The ionic form of N obtained is more variable and depends on the fungal species. Moreover, according to Smith and Read (1997), it also depends on the specific plant–fungus symbiosis. The rate and preference for each form of nitrogen vary with the fungal taxa. In general, amino acids, amines, amides, ammonia and nitrate are known to be absorbed through membrane transport mechanisms into fungal cells. They are incorporated into glutamine and glutamate for subsequent anabolic metabolism and translocation. The carbon received from root cells is mostly hexose sugars, which are incorporated into fungal trehalose, glycogen and triacylglycerol storage molecules. Carbohydrates and lipids are also translocated to the hyphae in the soil matrix, through the mycelium.

The role of mycorrhizae in decomposition is twofold. One function is the removal of nutrients from the soil solution, for transfer into both hyphae and roots. It is a final sink which removes soluble forms of N, P, S and other minerals from the decomposition food webs, by translocation into plant cells for photosynthesis-driven primary production. Another function is the transfer of organic carbon from photosynthates into soil mycelia. This provides a steady source of usable organic molecules for fungal growth, which is available to fungivory and the decomposition food web. The amount of transferred carbon is not trivial. It represents about 4–20% of fixed carbon from photosynthesis, or about 5 billion Mg C/year (see Bago *et al.*, 2000). The extent of translocation of nutrients from one plant to another, through the interconnecting mycelia of mycorrhizae, has been debated. In an important study, seedlings pulse labelled with $^{13}CO_2$ and $^{14}CO_2$ were shown to translocate the fixed labelled carbon bidirectionally (Simard *et al.*, 1997). The seedlings were grown in the forest in groups of three (consisting of the two ectomycorrhizal species *Betula papyrifera* and *Pseudotsuga menziesii*, and a vesicular–arbuscular species, *Thuja plicata*), each group planted 0.5 m apart. The ectomycorrhizal plant species were interconnected by the same species of ectomycorrhizae. Each seedling of *B. papyrifera* and *P. menziesii* was pulse labelled with $^{13}CO_2$ or $^{14}CO_2$. The authors observed a net bidirectional transfer in the third year seedlings (not in second year seedlings). Furthermore, the authors demonstrated that the direction of translocation was greatly affected in proportion to the amount of shading a seedling received. In other words, the amount of C translocated by the mycelia was greater in the direction of a seedling grown in shade, with less photosynthesis-fixed CO_2. However, it is unclear whether the organic molecules translocated remain in the hyphae or actually cross into the root cells (see Bago *et al.*, 2000). It is worth remembering here that some ectomycorrhizal fungi are capable of substrate decomposition, so that they are not entirely reliant on the root C supply, but use it as a supplement. Orchids obtain their nutrients from the soil hyphae which decompose SOM. In orchids, nutrients are transferred in a controlled way from the fungal hyphae to the plant cells, through intact cell membranes (Alexander and Hadley, 1984, 1985; see Smith and Read, 1997, Chapter 13). (However, orchid mycorrhizae are more complicated by the fact that in some cases, the plant or the fungus may digest the other partially, for nutrients.) Also, in other non-photosynthetic plants with mycorrhizae, the transfer of organic nutrients is from photosynthetic plants linked together by the same mycelium (Smith and Read, 1997, Chapter 14). Therefore, in some cases, there clearly is transfer of organic molecules into root cells from fungal supplies. It remains to be seen whether this ability occurs exceptionally in some mycorrhizae or if it is restricted to certain fungal taxa, or if it is more generalized. This level of resolution will require autoradiography to determine the path and location of labelled nutrients.

The benefit of the mycorrhizal association in experiments is usually demonstrated under growth conditions where one or more nutrient, or water, is limiting to the plant or to the fungi involved. These experiments are relatively elaborate, and many parameters must be considered, namely the affinity of the fungal hyphae cell membranes for the nutrient (enzymatic K_m value), concentrations in the mycelium, translocation rates towards the roots, rate of transfer or flux into the root cells and relative ratio of the nutrient in root cells at the interface with the fungal hyphae. These parameters fluctuate with varying conditions in the soil, and through seasons with changing substrate availability and changes in the plant nutrient requirement patterns. For example, some of these issues were demonstrated with phosphate uptake and translocation into root cells (Van Tichelen and Colpaert, 2000; Jakobsen *et al.*, 2001). The absorption of P from soil through decomposition, osmotrophy or from clay mineral solubilization releases a pool of P which would not otherwise be available to the plant roots. The hyphae explore a larger volume of soil pores and grow more quickly into new areas than root growth. It is believed that hyphae compete with the soil biota for P more effectively than plant roots. This is certainly true with respect to plant roots, which have a lower affinity (higher enzymatic K_m value) for P absorption. These kinds of studies with mycorrhizal roots in intact soil cores are promising, especially because they lend themselves to more detailed analysis of nutrient translocation, e.g. by autoradiography (Lee *et al.*, 1999). They clearly show the symbiotic advantage and that responses vary with the plant–fungus interaction. Moreover, in natural soils, each root system can be colonized by several species of mycorrhizal fungi, so that the overall benefit to the plant can be more extensive. More specifically, as nutrient and water availability fluctuate, one fungus–plant association may become more beneficial than another, in the same root network.

Lastly, the response of mycorrhizal associations in natural soil depends to a large extent on the edaphic biota (Fitter and Garbaye, 1994; Hodge, 2000). The hyphae in the interstitial pores, and those associated with roots, are in competition for resources with other fungi, protists and bacteria. The hyphae are also grazed by a variety of invertebrates (Collembola, other microarthropods, enchytraeids and earthworms), protists and several Actinobacteria. As we saw earlier, there is some preference for different species when grazing. Moreover, the interconnecting mycelium between different plants redistributes resource sinks within the mycelium network, according to plant species nutrient requirements. One should account for unexpected interactions that occur with the rest of the soil community. For example, in one study, the microcosm set-up revealed that Collembola in the soil were being captured and digested by the ectomycorrhizal fungus *Laccaria bicolor* (Klironomos and Hart, 2001). This provided a rich source of N to the fungus which becomes available to the roots for plant growth. Similarly, anaerobic free-living nitrogen-fixing bacteria are assumed to be unimportant, because their contribution to the soil nitrogen cycle is very small. However, it is clear that under nitrogen-limiting conditions, their activity makes the difference between fungal activity and inactivity on N-poor substrates, such as woody debris and crop litter in agricultural fields (Hill and Patriquin, 1992). This typifies the paradox that faces soil ecologists. On the one hand, one strives to simplify the system to understand the complexity. On the other hand, simplified experimental systems may fail to detect the complexity of interactions that sustain soil food webs. The latter then leads to erroneous or naïve conclusions.

Opportunistic Parasites and Parasitism

Parasitic species require a host in order to complete their reproductive cycle, or life cycle. The host is a source of nutrients, and the parasitism may or may not be lethal, or even pathogenic, to the host. However, some strains or certain parasitic species are virulent and cause great harm to the host. Many human parasitic species have part of their life cycle in the soil, as spores, cysts, eggs, larvae or in soil organisms. These are deposited in the soil through cadavers (a previous host), faeces or as part of the life cycle through other soil-dwelling species. Transmission to humans is by contact, ingestion or inhalation. In some cases, the pathogen is transmitted through arthropod bites, such as with tick-borne diseases (Acarina Ixodidae). Some better known human parasites transmitted through soil are diseases caused by nematodes and flat worms (Platyhelminthes). The phyla Microsporida (fungi) and Sporozoa (protozoa) consist solely of parasitic species. There are also numerous bacterial and fungal species that are parasitic to humans or soil organisms. Common parasites of soil invertebrates are fungi, nematodes and sporozoa.

Parasites contribute indirectly to decomposition, in that epidemics increase the rate of return of living biomass to the decomposition process. Parasites have a competitive advantage over saprotrophs, from which they probably evolved. These species are able to feed on (or 'decompose') living organisms before they are dead, i.e. before decomposers obtain access to the nutrients. However, successful parasites must first penetrate the host and avoid its antibiosis against xenogenic intrusion. A large number of species are opportunist pathogens, in that they do not require a host for reproduction, and are not specialized in host invasion. These species are part of the normal soil biota and normally contribute to decomposition. They are opportunist in that if accidental access to a suitable living organisms is provided, they will feed on it.

Opportunist parasites represent an evolutionary transition stage between primary saprotrophs and parasites. Selected taxa are described below as examples. For more detailed explanation of parasites and relevant references, the reader is directed to standard texts such as *Basic Medical Microbiology* (Boyd, 1995).

Soil organisms are regularly in contact with fungal conidia, spores, including chytrid spores, and bacteria. Some of these represent parasitic species which are activated in contact with the host. Spores and bacteria can be carried on the surface of moving organisms, deposited elsewhere or contaminate new hosts by coming into contact with other individuals (Zimmerman and Bode, 1983; Klironomos and Moutoglis, 1999). Some entomophagic fungal species that can infect and kill microarthropods are species of *Acremonium*, *Beauveria bassian*, *Conidiobolus coronatus*, *Metarrhizium anisopidae* and *Verticillium lecanii* (see Keller and Zimmerman, 1989). We do not know to what extent parasites of soil organisms affect population abundances and dynamics. Rare studies on Collembola suggest that infection rates are 2% by Microsporida (fungi) infections, and a further <1% each for other fungal and bacterial parasites (see Lussenhop, 1992). The rate of infection was higher in areas exposed to SO_2 deposition from pollution. One must bear in mind that every species of protist, invertebrate and bacteria is also susceptible to infection by their viruses. We do not know what the impact of viral infections in the soil habitat is on the saprotrophic community.

Human pathogens occur in the soil as opportunist parasites and as transient residents. Several species of the Gram-positive endosporic rods, *Bacillus* and *Clostridium* that are natural to soils, have opportunist species. Those that are potential pathogens are of the most virulent bacterial infections, such as *Bacillus anthracis*, a facultative anaerobe that causes anthrax. The species normally decomposes animal hides. The *Clostridia* species are obligate anaerobes with motile flagella. They are the agent of botulism in spoiled food, tetanus and gas gangrene, through the secretion of toxins. Several species of *Mycobacterium* in the soil can cause pulmonary or cutaneous infections. The better known parasitic Mycobacteria species cause leprosy or tuberculosis. The Actinobacterium *Nocardia* genus contains several soil saprotrophic species that can infect immunocompromised hosts, causing nocardiosis. The spirochete genus *Leptospira* are saprotrophs but include several strictly parasitic species.

Among the protists, we have mentioned the Microsporida and Sporozoa (Gregarina and Coccidia) which release infectious spores in the soil through previous hosts. Numerous amoebae can colonize the digestive tract, and some will cause mild to severe amebiasis. For example, *Entamoeba histolytica* causes clinical symptoms in 10% of human cases, of which 2–20% will be invasive. *Naegleria fowlerei* is common in soils and aquatic sediments. If it accidentally reaches the nasal passage, it will find its way into the brain and central nervous system. It will proliferate rapidly by feeding on the nervous tissue and cause death. An abundant soil amoeba genus *Acanthamoeba* is also known for causing damage to nervous tissue if accidentally infected. *A. castellanii* is known for infections through contact lenses. Growth on the eye can cause blindness in a few days. Infection originates from air-borne cysts.

Fungal infections from natural soil saprotrophs also occur, with transmission through spores. Systemic infections occur through inhalation of spores. Most spore inhalations are asymptomatic; however, some species can cause infections, such as *Aspergillus* sp., *Blastomyces dermatitidis*, *Coccidioides neoformans*, *Cryptococcus immitis* and *Histoplasma capsulatum*. Cutaneous infections are often caused by primary saprotrophs of skin such as *Microsporum* and *Trichophyton* genera, although several species are obligate parasites. Subcutaneous infections occur through skin punctures. Treatment of this type of infection often requires removal of hyphae-infested tissue by surgery. Agricultural labourers are particularly exposed to these through injury and soil contact. Typical cases involve *Sporothix schenkii*, *Wangiella dermatitidis*, *Cladosporium carrionii*, *Fonsecaea* sp., *Loboa loboi* and *Phialaphora verrucose*.

Summary

Saprotrophs are species that participate in decomposition food webs. They are responsible for the digestion of fresh litter and partially decomposed organic matter. These organisms feed on and excrete the organic matter, as well as feeding on each other by predation. The digested and excreted material in turn becomes a substrate for further digestion and excretion by primary and secondary saprotrophs. The trophic interactions release a steady trickle of soluble nutrients in the soil solution, and release carbon dioxide through metabolic respiration. Both the soil solution and carbon dioxide become available again for plant growth, which drives the primary production food webs. Saprotrophs are categorized into functional groups based on what they feed on. The initial stages of litter decomposition are called primary decomposition. It occurs in the surface horizons and can be studied from litter bags. Secondary decomposition involves the further decomposition of microdetritus and the amorphous humus. Primary saprotrophs are responsible for the external digestion of litter and SOM. Secondary saprotrophs graze on primary saprotrophs, on microdetritus and more decayed organic matter, as well as on each other. Other consumers are predacious on secondary saprotrophs and other predatory species. Earthworms tend to feed on whole soil or litter, with a preference for particular stages of decomposition. Most saprotrophs have a range of preferences for particular types of litter or prey species. Some species are more specialized on a particular diet, but many are opportunists that feed on a variety of prey species or substrates. Omnivory adds to the complexity of soil food webs. Many species participate in several functional groups by opportunism or at different times through their life cycle. Syntrophy has been little studied, but is probably very important in bacterial function within peds on microdetritus and humus. Symbiosis of fungal species with plant roots (mycorrhizae) and of protozoan and bacterial species with invertebrates is a very common adaptation in the decomposition subsystem. In the future, we need to move away from studies based on a taxonomic grouping (such as nematodes or microarthropods), in order to focus on functional groups. The main task before soil ecologists in the years ahead is to integrate these interactions between coexisting species.

Suggested Further Reading

Cadish, G. and Giller, K.E. (1997) *Driven by Nature: Plant Litter Quality and Decomposition*. CAB International, Wallingford, UK.

Clark, F.E. (1969) Ecological associations among soil micro-organisms. *Soil Biology, UNESCO – Natural Resources Research IX*, 129–131.

Coleman, D.C. and Crossley, D.A. (1996) *Fundamentals of Soil Ecology*. Academic Press, New York.

Darbyshire, J.F. (1994) *Soil Protozoa*. CAB International, Wallingford, UK.

Dix, N.J. and Webster, J. (1995) *Fungal Ecology*. Chapman and Hall, New York.

Paul, E.A. and Clark, F.E. (1996) *Soil Microbiology and Biochemistry*, 2nd edn. Academic Press, New York.

Smith, S.E. and Read, D.J. (1997) *Mycorrhizal Symbiosis*, 2nd edn. Academic Press, New York.

Swift, M.J., Heal, O.W. and Anderson, J.M. (1979) *Decomposition in Terrestrial Ecosystems*. University of California Press, Berkeley, California.

Spatial and Temporal Patterns 5

The physical structure of the soil provides a reticulate porous habitat for saprotrophs. As we have seen in Chapter 2, this matrix is structured by the organic component derived from surface and interstitial sources of litter, and from the mineral component. The soil profile is stratified into a more or less continuous series of layers which affect the physical and chemical properties of the soil with depth. Biological organisms that inhabit this matrix are affected both by its organic composition through the profile and by the physical and chemical properties that change with depth. The main changes through the profile include an increasingly decomposed litter, an increase in compaction and reduced mean pore diameter. With depth, the macrodetritus and microdetritus are replaced increasingly by amorphous humus and mineral soil. These changes influence the composition of soil air, of soil solution and water retention, which are the immediate habitat of saprotrophs. With depth, as the mean pore diameter decreases, the amount of habitable pore space decreases, aeration decreases and there is an accumulation of CO_2 and decrease of O_2, compared with surface air. Also, smaller pore sizes retain capillary water longer and therefore remain water-filled longer. Water retention is also affected by the organic matter abundance and composition, as well as by the mineral texture of the soil (Figs 2.4 and 2.5). Organic matter increases water retention, and loamy or clay soils delay drainage (Fig. 2.6) These parameters which affect the soil structure provide a stratified habitat that determines species composition through the profile. With depth, the soil air, solution, physical matrix and food resources vary. Species that occur through the profile are adapted for a range of variations in these parameters. This restricts their range of activity in the profile. Moreover, species are limited in activity by diurnal and seasonal changes in temperature and moisture. Some migration through the profile is possible in order to escape changes in abiotic parameters, but biological activity depends on the presence of adequate (gas, liquid and solid) resources. Therefore, moving out of the range of adequate resources is only a temporary refuge from fluctuating abiotic conditions. Species activity depends on adequate resources and abiotic conditions, which must occur together. Species are limited to that zone in the profile where these conditions are met. Furthermore, reproduction of invertebrates may have additional requirements to provide a substrate for egg deposition, survival through quiescent periods, or food for immature stages. In this chapter, we will consider trends in the distribution of species through the soil profile, across the heterogeneous landscape and through time.

Regulation of Growth

The simplest description of growth of a single species on a single substrate is through the **logistic growth equation**. In this formulation, *growth rate* varies with the size of the population according to the relationship $r = r_0(1 - N/K)$, where r is the rate of growth, r_0 is the initial intrinsic rate of growth when the population size N is very small, and K is the carrying capacity of the culture or the ecosystem. The growth rate r decreases as the population size increases and as resources are depleted. Using the equation for unlimited growth $\mathrm{d}N/\mathrm{d}t = rN$, and substituting r with the equation for growth rate above, we obtain the logistic equation $\mathrm{d}N/\mathrm{d}t = r_0N(1 - N/K)$. Therefore, the ratio of population size N to the carrying capacity K determines whether a population is increasing or decreasing in numbers, and at what rate (see Krebs, 2001). The typical graph of the logistic equation is S-shaped, with an accelerating rate of increase followed by a decelerating rate, until an equilibrium number is reached at the carrying capacity (Fig. 5.1). This representation of growth has been very useful in understanding the fundamental factors that regulate population growth. It has been applied to growth of a single predator on single prey, to a grazer on its resource, to axenic growth on culture media, and many other examples. Its limitations in natural studies are that normally there are many interacting species with alternative resources living in an open system, i.e. outside the simplified confines of a flask or microcosm. The natural oscillations of population abundances through time are more complicated, and tend to mimic logistic growth over short time intervals only (see Krebs, 2001).

Prokaryotes

The growth of primary saprotrophs and other osmotrophs depends on a steady source of nutrients obtained from the soil solution. The nutrients are absorbed efficiently at low concentrations because osmotrophs

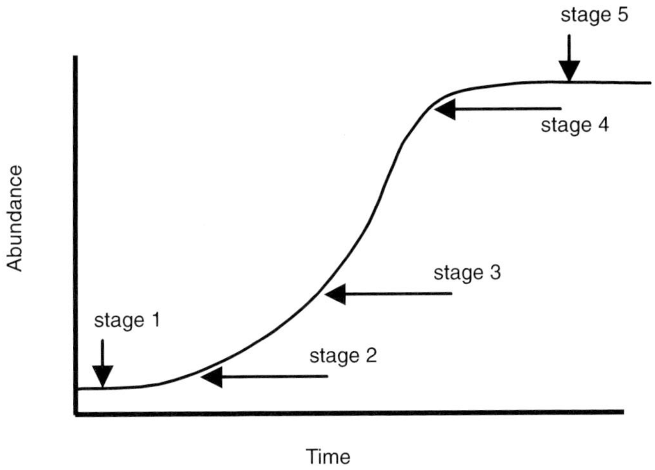

Fig. 5.1. Representation of the logistic growth curve, with stages of bacterial growth on a single substrate.

are also in competition with plant roots. For example, uptake of amino acids by soil bacteria is saturated at millimolar concentrations that occur naturally (Vinolas *et al.*, 2001). Nutrient uptake is constrained by the diffusion rate of nutrients from microsites. The rate of diffusion varies with soil compaction, moisture, ped particle size and micro-organism locomotion through the porous reticulum (Focht, 1992). Bacteria respond physiologically within minutes to changes in growth conditions or solution composition. Molecular biologists are very familiar with their sensitivity to slight changes in conditions, from culturing transformation-competent cells. When preparing cultures for transformation with DNA, slight variations (minutes) in the time to cool growing cells, or to sediment cells, or abundance of cells, greatly affect their competence. Cells that are not cooled in the active state quickly enough are already inactivated before they are prepared for transformation. Handling these cooled cultures too roughly also contributes to inactivating cells. Roughly a 5–10 min delay can reduce active (competent) cells by up to 100 times. If the same manipulations are carried out at room temperature, most cells become inactive within minutes of handling. Therefore, when soil is sampled, handled, mixed, diluted or partially desiccated at room temperature, inactivation of the bacteria is unavoidable. This is particularly significant if the cells are transferred from the soil solution to a diluted solution or water. It is difficult to cool intact soil rapidly enough to maintain active cells in the active state. Mostly, species that are described as active from handled samples are those that are active under the new conditions, and not necessarily all those that were active when the soil was sampled.

The growth of a bacterial culture on single substrate is useful to understand some of the underlying physiology that regulates growth (Fig. 5.1). Under suitable conditions, a stimulus such as the addition of nutrient resources will initiate cell activity. Cell activation from the inactive (non-growing) state requires a lag period for the molecules to be activated or synthesized (stage 1). The duration of the lag period depends on how much growth is required before cells can enter cell division. An increase in cell numbers is observed when cells have grown sufficiently to begin cycles of cell divisions (stage 2). Cells grow and divide until nutrients are limiting. At this point, a decrease in food resources and the accumulation of metabolic by-products decrease the rate of growth (stage 3). Very few species are able to accumulate storage material, especially under normal soil conditions. There are exceptions; for example, *Arthrobacter crystallopoietes* can accumulate a glycogen-like compound. Eventually, nutrients are insufficient to maintain the cell cycle and divisions end, although cells continue to remain active (stage 4). After some time of inadequate growth, cells become inactive (stage 5, late stationary phase). At this stage, some species overproduce or lose capsule material into the environment. This is easily observed in stationary phase cultures, and contributes to soil aggregate formation and biofilms. When soil conditions are inadequate to support metabolic activity, the cytoplasm becomes inactive and bacteria are said to be quiescent. Some species are able to form spores to survive unfavourable conditions. Once bacteria are stressed by lack of nutrients or unfavourable physical conditions, the cytoplasm becomes quiescent. To be reactivated and grow again, they need to be stimulated by suitable conditions. The conditions required to reactivate the cytoplasm are not necessarily the same as those favouring optimal growth. This may require a different nutrient solutions or changes in the physical environment (such as temperature). The longer bacteria are quiescent or encysted, the greater the loss of viability.

From a molecular perspective, growth of bacterial cells during the lag phase (stage 1) involves the activation of the genes necessary to transcribe the metabolic proteins and cell membrane transport proteins. These are regulated at the level of the operon (co-regulated genes), and at the level of the regulon (co-regulated operons). Entry into the growth and division cycle requires the accumulation of sufficient ribosomes, proteins and nucleic acid precursors (stage 1). There is a minimal size requirement to maintain cells in the cell cycle. It is generally observed as an increase in cell size from stationary phase cells (stage 5). During the period of rapid growth (stages 2–3), cells accumulate further cytoplasmic molecules and ribosomes, and reach an optimal size. The duration of the period of DNA replication and cell separation (or filamentation) is independent of growth rate, and does not vary under fixed conditions. The period of growth to minimal size varies with the rate of nutrient intake and temperature. Therefore, not only the size of individuals but also the duration of the cell cycle (growth phase to division) varies as conditions change.

When concentrations of adequate nutrient substrates are too low, cells may continue to take in molecules, but at a rate too low to maintain growth and division. Thus, even though cells are not increasing in number, they can be metabolically active. The threshold growth rate and nutrient abundances vary for each species and its substrates. Depending on the soil physical environment and nutrient availability, bacterial growth rates can vary from 20 min under optimal laboratory conditions, to >3000 h (see references in Kjelleberg, 1993). Under conditions where the soil solution offers poor growth, or under nutrient stress, large cells or microcells develop over time. This involves a change in both volume and chemical composition of the cell. The aim is to maintain an optimal amount of active cytoplasm until the conditions for growth are better. Variations in size and morphology caused by starvation are studied mostly in marine and freshwater species, particularly in marine *Vibrio* (Kjelleberg, 1993; Lengeler *et al.*, 1999). Many Gram-negative species reduce their energy needs by decreasing cell size. There are similarities in this process to spore formation in Gram-positive species. At first, there is a decrease in the copy number of plasmids, followed by changes in the protein composition and membrane lipid composition. Many enzyme pathways are inactivated and other enzymes synthesized. In the initial phase (30 min), proteolysis begins and *starvation-induced proteins* are synthesized. The accumulation of starvation-induced proteins imparts more resistance to heat and ultraviolet light. It is accompanied by switching to new nutrient substrates. In the following hours, membrane lipid composition changes, the flagellum is lost, and high-affinity substrate transport proteins are synthesized, as well as new starvation-induced proteins. In the third phase, metabolic activity decreases and cell size is gradually reduced. There is degradation of ribosomes and proteins, and cell wall material become a source of nutrients to maintain the cytoplasm. These become a source of phosphate and nitrogen for metabolism. This period continues for several days, and it is accompanied by synthesis of further high-affinity transporters and extracellular hydrolytic enzymes. The microcells can be about 30 times smaller than well-fed cells and can pass through 0.2 μm membrane pores. The morphology of the cells with starvation varies, according to what the nutrient limitations are, and some nutrient limitations lead to larger cells (Fig. 5.2). If only certain nutrients are deficient, metabolic activity can continue but without cell division.

Attempts at studying single species of bacteria on single substrates, or mixed species on single substrates, are not very informative about the mechanisms of natural growth on complex substrates. Natural samples usually contain a variety of species which contribute to decomposition,

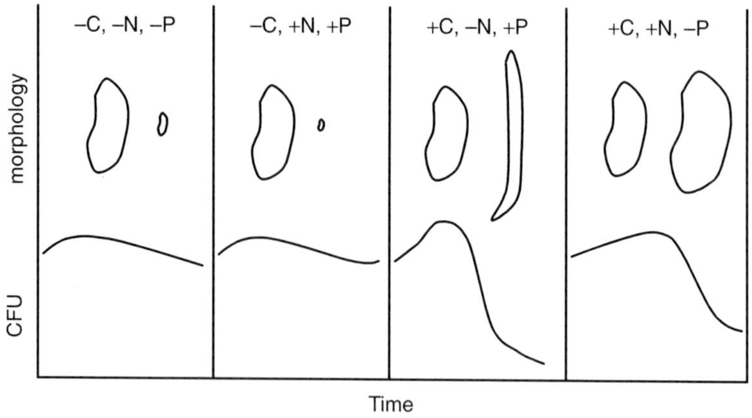

Fig. 5.2. Size variation in *Vibrio* with different nutrient limitations over 100 h, and effect on viability (CFU: colony-forming units) (data from Kjelleberg, 1993).

and cells grow on complex substrates that have a variety of molecular components. In the presence of multiple substrates, cells will switch from growing on a preferred substrate to another, as each is exhausted sequentially. Moreover, cultures are materially closed systems without exchange of nutrients with the environment. They are also shielded from the positive (syntrophy with other decomposers, symbiosis) and negative (predation, defensive molecules) effects of interactions with other co-occurring soil organisms.

Protists

Fungal growth from dispersal spores or from hyphae is guided by chemotaxis. Growth occurs in the direction of adequate resources. Species can respond to a variety of tropisms (light, temperature and pCO_2). Hyphae extend in length with few branch points when searching or escaping. With adequate resources, the extending hyphae branch profusely and remain in the same bulk of material. This forms a dense mycelium. The frequency of branching (new growing tip initiation) and distance between branches reflect the nutritional value of the substrate. On adequate substrate, the number of branches (growing tips) increases exponentially and in proportion to the overall mycelium biomass. The **hyphal growth unit** measures the ratio of hyphal length (or volume) to the number of branches. It approaches a constant value in balanced growth conditions. The mycelium-**specific growth rate** measures the biomass change through time (µg/h). As resources are exhausted locally, the mycelium spreads outward. Old hyphae in the centre of the mycelium become empty as cytoplasm moves to growing areas, where cytoplasmic growth continues. In Archemycota, the movement of cytoplasm towards growing tips is more complete and can be rapid (Giovannetti *et al.*, 2000). In Basidiomycota, the outwards growth of the mycelium away from the centre of the colony can be seen as 'fairy rings'. The aerial reproductive structures emerge seasonally in the region of mycelium growth. The annual increase of diameter in these mycelia is about 10–15 cm, with about 15–20 μm/min under optimal conditions. Therefore, large 'fairy rings' that are several metres in diameter are decades old. Some mycelia can be centuries old. For example, one individual of *Armillaria bulbosa* was described to be 1500 years old, to weigh 10 t and cover an area of 15 ha (Harold, 1999). Not all species continue to grow in successive seasons. Many species are r-selected and grow for a short duration on existing resources. Once resources are exhausted, these species reproduce to release spores, and the mycelium dies.

Initiation of sporulation or encystment depends on abiotic parameters to activate the quiescent cytoplasm. To sustain growth, the cytoplasm also requires adequate resources. Species have different preferences for resources and abiotic parameters. Therefore, the activation of spores and cysts is often seasonal, when the combination of suitable food resources and abiotic parameters occurs. Different species become activated at different periods through the year, and at different depths through the soil profile. Extension of the hyphal tip requires deposition of a chitin cell wall behind the growing tip. Vesicles with inactive chitin synthase enzymes (chitosomes) are released at the cell membrane behind the growing tip. Activation of the enzymes polymerizes chitin monomers into the cell wall. The extensively cross-linked polymer (with hydrogen bonds) provides a rigid and resistant cell wall against the soil matrix and fluctuations in osmotic potential. Loosening of the cell wall at branch points is promoted by localized chitinase activity. The dominant period of the replication cycle in filamentous species is the growth period. The period of DNA replication occurs during the growth phase and lasts about 20 min at room temperature. Extension of the hyphae is continuous during growth. Mitosis occurs in <10 min and is not necessarily accompanied by septation in filamentous growth. On average, one or two nuclei occur in the same compartment in Basidiomycota, four in Ascomycota, and up to 50 in the apical growing compartment. In the Archemycota, septation is infrequent and irregular, except in thallus forms (such as Chytridiomycota) that have one nucleus and form septa only in reproductive phases. Septa also occur in older hyphae, to seal the growing hyphae from the older abandoned mycelium.

The active portion of the mycelium extends through a large volume of soil, especially in older individuals. Some parts of the mycelium will experience conditions very different from other parts further away. There will be variation in moisture, nutrient availability, temperature, predation and competition. However, since the mycelium is a continuous meshwork of cytoplasm, cytoplasmic streaming occurs from high to low moisture areas. This contributes to redistributing soil moisture, as water is absorbed from one region and translocated to another region in the soil. Nutrients are redistributed throughout the hyphae from high to low resource regions of the mycelium by cytoplasm transport mechanisms. Hyphae that are in suboptimal soil are abandoned, and hyphae in more suitable regions stimulated. The cytoplasm, organelles and storage compounds are salvaged from old hyphae before being abandoned. This salvage of components is by translocation to the growing mycelium, especially in Archemycota. In other fungi, autolysis of older compartments releases nutrients which are then used by growing hyphae. It is important to distinguish between the activity of these autolytic fungal enzymes, which are active in nutrient-poor soil or older hyphae, from extracellular saprotrophic enzyme activity from the growing mycelium.

Growth of non-filamentous unicellular species also obeys several rules that are fundamentally similar. Most research on the molecular regulation of eukaryotic cell growth has focused on two yeasts (*Saccharomyces cerevisiae* and *Schizosaccharomyces pombe*). However, a large amount of non-molecular work exists with species in fungi and in other phyla. Several rules can be described based on experiments with a variety of ciliates, green algae, yeasts and amoebae (*Amoeba proteus*, *Dictyostelium* and *Physarum*) (Adl and Berger, 1996; Adl, 1998). Species have an optimum size range that supports growth and division cycles. When cells are too small, they need to grow and increase biomass until a critical minimal size threshold (first size threshold) is reached. Organelle replication during the growth phase also requires a minimal size threshold (second size threshold), often expressed as a cell mass to cell nuclear DNA ratio. Lastly, a third control operates at commitment to cell division. Whereas the first two size controls monitor that nutrients are accumulating to support organelle replication, the third determines whether the organelles were replicated. The rate of cell growth is limited by the most limiting nutrient. Cells continue to grow in size, for as long as necessary, until the most limiting nutrient is acquired sufficiently. These cells are large relative to more abundant nutrients. Therefore, duration of growth and cell size at division are reduced in more balanced solutions. Once the second size threshold is satisfied, organelle replication begins. The order varies somewhat between species, but includes replication of chromosomes, basal bodies, centrioles (or cytoskeleton-organizing centre), contractile vacuoles, cytostome and other organelles. When organelle replication is completed successfully, a third control point commits the cell irreversibly to separation of organelles, ending with cytokinesis in most species. This final segment of the cell growth and division cycle varies very little in duration (at constant temperature).

Earlier segments of growth vary in duration, and continuously adjust the optimal cell size and cell cycle duration as growth conditions change. When cells grow in nutrient-poor medium, cell size is adjusted down over successive cell divisions, and less growth (accumulated new biomass) is needed prior to organelle replication. However, duration of growth is extended until the size thresholds are reached. When nutrients are less limiting, the size thresholds are adjusted upwards. Variations in optimal size and adjustments in duration of the growth period are predictable, but are sometimes counter-intuitive.

Invertebrates

Similarly, growth of invertebrate species is affected by cell division rates and cell growth rates in the immature or larval stages. The rate of growth (cell divisions) depends on the rate of feeding on suitable resources to maintain a balanced cytoplasm. As in protists, the optimal size of adults depends on the size of the component cells. Cells are smaller when juveniles grow on poor resources, but the final number of cells (of cell cycles completed) is less variable for developmental reasons. This affects the fitness of the adults, and the number of gametes they produce. If the growth rate is too slow and cell size thresholds are not met, development of juveniles is aborted and the organism dies. The adults are no longer dependent on cell divisions and they are sexually mature for reproduction. Feeding in the adult is to sustain reproductively fit individuals, through to fertilization of gametes and to completion of egg-lay. In some species, such as the Onychophorans, the adult is also responsible for caring for the young after hatching. The dynamics of proliferation of invertebrate species depend on many variables and are observed over longer time spans than for unicellular species. Adults that grew in nutrient-poor conditions produce fewer gametes (eggs and spermatozoa). The gametes also have fewer storage reserves for embryogenesis. The development is more likely to be arrested or to yield less fit immature stages. Therefore, the quality of growth of the parent (F_0 generation) is reflected in the fitness of the immature stages of the next generation (F_1 generation). Furthermore, if the parent developed in the presence of heavy metal toxicity or other toxins, a grandparent effect can be predicted. That is because the adult (F_0) egg (F_1) contains the cytoplasm with poison from which the next germ cells form (F_2). This effect is particularly pronounced in species with early germline determination. For these reasons, growth conditions of one generation have repercussions into successive generations. This affects the number and fertility of individuals in subsequent generations. It will also affect the abundance of predators, prey and nutrient resources in the immediate food web.

A more detailed look at the underlying parameters that regulate the abundance of particular invertebrate species depends on understanding the biology of the species. For example, a review of variations in life history strategies of microarthropods (Siepel, 1994) provides an indication of the many factors that must be accounted for when considering invertebrates. The review identifies four categories that are considered separately: (i) type of reproduction mechanism; (ii) factors that control development; (iii) synchronization with the environment; and (iv) effectiveness of the dispersal mode. These are summarized briefly below from Siepel (1994). The types of reproduction are diverse, especially in microarthropods. Parthenogenesis is common and varied in mechanism (arrhenotoky, thalytoky and amphitoky). Arrhenotoky are cases where the unfertilized haploid eggs become male, while the fertilized diploid eggs become the females in the population. This mechanism does not occur in Oribatida and Collembola, but is known from species in Mesostigmata, several families of Tarsonemida, Tetranychinae (Actinedida), Eryophioidea and several other families. Thalytoky are cases where eggs develop into female adults only. The male does not exist. Eggs are fertilized by selfing of haploid nuclei (automictic) or without meiosis (chromosome number reduction) or nuclear fusion (apomyctic). The latter is known from Oribatida in primitive lineages, and some species in Mesostigmata, Tullbergiinae, Actinedida, Tarsonemida, and in some Collembola (*Isotoma notabilis* and *Isotomiella minor*). Amphitoky refers to species where a combination of the above mechanisms operates, as in certain Tarsonemids. Sexual reproduction involving a male and female gamete occurs, but even here there are variations. For example, some species reproduce sexually in winter, but by parthenogenesis in summer months. This shift also occurs in Rotifers, which may be sexual in winter, shift to thalytoky in summer, then to arrhenotoky in autumn. Semelparity refers to the special case where there is only one short period of egg deposition, during the optimal season. That is the basic type in Oribatida. Iteroparity refers to cases where egg deposition occurs over a broader time frame. It is a safeguard against fluctuations in the environment, so that the probability of egg and immature survival is increased. Iteroparity is more common in general, and permits variation in the frequency of ovideposition and number of eggs laid. In Collembola, a population becomes sexually active in synchrony, and individuals aggregate to reproduce once during the period of activity of that species.

Development is modulated by the environment. Changes in temperature and moisture affect the rate of cell divisions during embryogenesis, and the rate of development of immature stages. The effect of these changes on development is a trait that varies at the family level. Typically, egg fertilization leads to embryogenesis of a larva, protonymph, deuteronymph, tritonymph and adult. Not all stages occur in all families. The duration of development varies between species, and with the temperature and moisture conditions. Juvenile feeding stages also require adequate food resources (abundance and quality). Development into the adult may take 3 days in some species, or as long as 3 years. Therefore, the period to first ovideposition of the adult may be as short as a few days, or several years. Typically, it is several months, as most invertebrate species require several weeks at ambient soil temperatures to complete development, reach sexual maturity and reproduce. Microarthropods in soil food webs lay 1–13 eggs/day, with some variation in the duration of the period of ovideposition. Besides semelparous species (Oribatida, Collembola and Actinedida), edaphic species continue to reproduce throughout the seasons. Therefore, the number of generations produced each year varies with the growth conditions that year. The fitness and fertility of populations each year also depend on past growth conditions (see earlier).

Most edaphic species do not have a developmentally determined **dormancy** period during the year. Dormancy is a mechanism of survival through long periods unfavourable to growth. When it occurs, it tends to be limited to an immature stage. However, periods of **quiescence** occur (metabolic inactivity) whenever conditions are temporarily inadequate. This may be caused by lack of oxygen, food resources, or moisture and temperature extremes. Migration of individuals from unfavourable conditions is spatially limited by the soil pore reticulum. Individuals are known to migrate vertically to accommodate diurnal shifts in conditions. Dispersal of edaphic species by water runoff and drainage has not been documented. Dispersal by wind may occur over very short distances, but it is more likely to cause desiccation. Transport by more mobile animals is possible, but that remains hypothetical. Attempts by several authors to describe r-, k- and s-adapted edaphic species have so far been only partially satisfactory. There are clearly species that fit in one or other category. However, most do not, and for many other species we do not know sufficient biology. This description will require some further thought as to the correct time scale and probable dispersal range of species before it can be applied to interstitial saprotrophs. The time scale and dispersal range to apply to different families, classes and phyla are necessarily different, as the rate of proliferation, survival between seasons and mechanisms of dispersal vary.

Periods of Activity

Besides food and temperature requirements of species in an adequate habitat, soil moisture is the most direct regulator of species activity. The composition of the soil solution and soil air varies with soil moisture content. Fluctuations in temperature, surface illumination by sunlight, and dew or rain deposition change soil moisture content. These changes are rapid, and cells must respond (physiologically) immediately if they are to survive these variations. Soil moisture fluctuates through the day, as conditions in the soil and above-ground change through the day. It is a matter of survival that individuals resist excessive desiccation before it occurs. Soil species are adapted to withstand desiccation by encystment, migration and physiological adaptations. The most immediate response is a physiological modification of solute composition in the cytoplasm. This mechanism is called **osmoregulation** and determines whether a species will remain active or inactive through drier periods. Different species have different tolerance to desiccation and different thresholds for inactivation of metabolism and quiescence.

Osmoregulation

The water potential of pure water without any solutes is zero by definition ($\pi = 0$). If solutes are added, the potential becomes negative relative to pure water. Therefore, cytoplasm and soil solution have a negative water potential relative to pure water. The negative sign indicates that pure water will diffuse into the solution, until an equilibrium solute concentration is reached. When comparing two solutions, the one with more solutes is more negative. Again, water will diffuse into the more negative solution (the more concentrated solution), until an equilibrium solute concentration is reached. The concentration of solutes in the cytoplasm tends to remain the same for physiological reasons. However, that of the soil solution fluctuates continuously. Since cell membranes are semi-permeable membranes (see Chapter 1), the direction of water diffusion will be to the more negative solution. If the soil solution becomes more concentrated than the cytoplasm, water will be drawn out, and that would desiccate the cell. This occurs as soils dry. If the solution is less concentrated than cytoplasm, water will accumulate inside the cell. This occurs after rainfall. Therefore, cells must adjust their solute concentration to adapt continuously to these fluctuations.

The soil solution is usually hypotonic relative to the cytoplasm. The inflow of water in eukaryotes is countered by water expulsion through the contractile vacuole. In species with cell walls (bacteria and fungi) which lack contractile vacuoles, the pressure is countered by resistance from the cell walls. Bacterial cell wall can withstand an increase in external water pressure up to +10 MPa. Failure to excrete excess water would cause cell lysis. In response to changes in the availability of water from the soil solution, as water drains, evaporates and is removed by roots, the concentration of ions and dissolved solutes increases. The osmotic potential of the soil solution becomes more negative. At some point, water is no longer available to living cells because solutes are more concentrated in the soil than in the cytoplasm. At this point, water is lost from the cytoplasm and flows into the soil solution by diffusion. The cell membrane is permeable to water and there is relatively free exchange of water between the cytoplasm and the environment. However, the solutes do not cross the membrane and become more concentrated in the cytoplasm. If all the water is removed from the cells as the soil dries, soil organisms would die. The cytoplasm must therefore be protected from excessive dehydration. Once hypertonic, protective mechanisms are activated. These include membrane permeability changes, transporters and carriers, as well as *de novo* synthesis of molecules. In bacteria, the periplasmic space outside the cell membrane occupies 20–40% of the overall cell volume (e.g. *Escherichia coli*). This space does not change volume because the cell wall is rigid. However, the cytoplasmic (living) volume can shrink significantly in hypertonic soil conditions. This difference between the real and apparent cell volume can account for large biovolume errors under the light microscope, especially if cells are re-suspended in water.

To measure and calculate osmotic potential, both the volume of solution and the amount of dissolved solutes must be considered. **Osmolarity** is calculated as the sum of the concentration of all solutes in solution in mol/l. The **osmolality** of a solution cannot be calculated. It can only be measured, and it is the osmotic pressure at a given temperature, expressed in mol solutes/kg solvent (osmol/kg). The osmotic potential π is calculated as follows, $\pi = (RT/V_w) \ln M_w$, where R is the gas constant, T is the temperature, V_w is the *partial molal volume of water*, and M_w is the *mole fraction of water* (in other words, the fraction of water in the solution). The osmotic potential can be approximated as $\pi = (RT/V_w)\ln M_w$, where M is the molar concentration of solute.

There are several common mechanisms of osmoregulation, which depend on the accumulation of osmoprotective solutes (Csonka, 1989; Csonka and Hanson, 1991). The solutes that are accumulated need to be non-toxic in the cytoplasm, even at high concentration, and must not interfere with metabolism and cellular activity. The osmoprotectants tend to be metabolic end-points which are not metabolized further. There are several physiologically inert molecules which are found over and over again in diverse phyla. They have been termed *osmotic balancers*. The accumulation of these molecules is accompanied by changes in the concentration of ions, such as K^+ and Na^+ that are co-transported into the cell. These osmoprotectants are absorbed from the soil solution (with great efficiency) or synthesized *de novo* if necessary. They are trehalose, glycine-betaine, proline, ectoine, glutamine and putrescine^{2-} exchanged for $2K^+$ ions. The internal concentration of K^+ is important to the functioning of the cell membrane, and for maintaining the electrochemical potential difference across the membrane. Therefore, the ion concentration must be restored and maintained during the physiological adaptation to desiccation. The removal of excess ions occurs with water outflow. The most effective and potent is usually glycine-betaine, because it is a non-ionic zwitterion. Trehalose, which is also abundant in fungi, helps to stabilize the cell membrane against desiccation. The accumulation of glutamate is necessary in some species. Where it occurs, it can account for up to 90% of the free amino acids in the cytoplasm. Species often have a hierarchy of responses to desiccation. For example, the accumulation of glycine-betaine can inhibit the accumulation of proline and ectoine, which are a back-up system in certain species. Several others also occur in prokaryotes, such as sucrose, D-mannitol, D-glucitol, L-taurine and small peptides.

The response of bacteria to desiccation occurs in three stages (Wood, 1999). First, during the first 2 min, there is a drop in cell turgor pressure, accompanied by loss of water to the environment. In 20–60 min, rehydration begins as the cell accumulates osmoprotectants. The physiological response includes excretion of putrescine (from amino acid metabolism), and nucleic acid-binding ions (such as Mg^{2+}) are replaced. Cell growth begins in stage three, once the osmoresponsive genes are activated. When water returns, the cell must adjust back by expelling the osmoprotectant molecules, or metabolize them into anabolic pathways. The adaptation of *Corynebacter* to hypersomotic shock has been described recently and provides an example of the elaborate molecular response (Peter *et al.*, 1998). Once a bacterium is quiescent after stress (for nutritional, temperature or osmoregulation reasons), it may require a different medium composition or temperature stimulation before it will reactivate. These cells may show months of poor growth subsequently, in the laboratory. For these reasons, it is not easy to distinguish between quiescent and dead bacterial clones. The latter lose their ultrastructure and eventually appear devoid of cytoplasm within a cell wall.

In eukaryotes, similar osmotic balancers occur. Polyols (such as glycerol) are common in fungi, algae, microarthropods and insects (Csonka, 1989; Nevoigt and Stahl, 1997; Zischka *et al.*, 1999). Proline and glycine-betaine accumulate in plant cells during osmotic stress. Many eukaryotes also accumulate betaine, taurine and myo-inositol. These osmoprotectants do not seem to occur in *Dictyostelium* (Amoebozoa) (Zischka *et al.*, 1999). Instead, cells respond by decreasing cell volume up to 50% or by encystment. In the diplomonad (Protozoa) *Hexamita inflata* during osmotic stress, glucose is metabolized to alanine, ethanol, lactate and acetate, which accumulate in the cytoplasm (Biagini *et al.*, 2000). Normally, the cell contains 214 fl of H_2O for a mean cell volume of 262 fl (80% water). If the soil solution shifts from 300 mosmol/kg (an average value) to 350–400 mosmol/kg, the cell shrinks within 1–2 min and remains small for at least 30 min, until the cytoplasm adjusts its osmotic pressure through the production of osmotic balancers and excretion of ions and soluble molecules. If the solution shifts from 300 to 150 mosmol/kg, the cell swells over 3–4 min but returns to normal by 12 min mostly through excretion of excess water by the contractile vacuole. The authors calculate that 70 fl of water was lost by co-excretion of amino acids (10.5 fmol in 7 min), 40 fl was lost through excretion of 6 fmol K^+ (~30% of cell K^+), and a further 40 fl water was lost through the excretion of anions. In many protozoa, the common osmoregulator is alanine, and to a lesser extent other amino acids, such as in *Giardia*, *Acanthamoeba*, *Leishmania* and *Paramecium*.

Differences in the ability to withstand desiccation determine which species remain active as soils dry. Fungi remain active in drier soil than bacteria (Griffin, 1969). Bacteria require more moisture to function, reflecting their limited osmoregulation compared with terrestrial eukaryotes. Their locomotion is also restricted as soil dries (Wong and Griffin, 1976a,b). This affects the outcome of competition between fungi and bacteria (see Wardle and Yeates, 1993). The competitive ability of hyphae for water comes in part from the large surface area that is colonized by a mycelium. It is enhanced further by the ability of the connected cytoplasm to distribute water throughout the mycelium. In turn, this property of mycelia promotes root water uptake and plant performance in mycorrhizal systems (see review by Augé, 2001). Amoebae are reported to function between –1 and –100 kPa matric water potential (Chakraborty, 1984). At –100 kPa, pores with openings larger than 3 μm diameter are drained of capillary water. Some species or functional groups may be limited to a more narrow range. For example, the mycophagous *Arachnula impatiens* could only perforate fungal cell walls between –1 and –20 kPa (Homma and Cook, 1985). In dry soils, soil amoebae are dominated by very small cells <10 μm in diameter. It reflects in part the necessity to remain immersed in a thin film of water, and that some amoebae become small osmo-tolerant cells as an adaptation to water stress. They have been called Platyamoebae (Amoebozoa) based on morphology, but it may be a physiological response of larger species to desiccation.

Growth response dynamics

When active, individuals of a species are actively searching for prey, or adequate substrate for growth. The exceptions are immotile species, mostly prokaryotes, that are active only when resources are available in the immediate vicinity. Even fungal hyphae search the soil for adequate substrates and respond directionally by chemotaxis. Feeding on primary saprotrophs is best described as **grazing**, and resembles **herbivory** by terrestrial consumers. Grazing is different from predation by consumers because it does not usually cause death or depletion of the grazed individual. Fungivores only graze a portion of the mycelium or thallus, or ingest cytoplasm from a small number of hyphae. As for herbivory, the fungus mycelium is immobile relative to the mobile grazer. It is possible under intense grazing pressure by consumers for the eaten individual to die. These are extreme events that occur when consumer populations are exceptionally abundant, or the grazed species exceptionally rare. In **predation**, consumption of an individual usually leads to death because the whole individual is killed and consumed. An unsuccessful predation attempt (prey capture) may cause wounding or escape of the prey. Predation occurs as cytotrophy on motile protists, or as consumption of nematodes, microarthropods or other invertebrates. In bacterivory and fungivory on yeast cells, it is almost impossible for the consumers to ingest all the clonal descendants. The soil matrix provides too many refuges from consumers (see below). Whether consumption of a prokaryote or eukaryote cell is considered predation or grazing depends on the scale of observation. If ingestion of a single cell is considered, predation is correct. If consumption of bacteria or protozoa in general is considered, grazing is probably the correct term.

The relationship between a predator and its prey is described theoretically by the **functional response** (Holling, 1959). Three types are hypothetically possible (Fig. 5.3). In the simplest case, the rate of prey ingestion increases linearly with the abundance of prey individuals, without an apparent upper or lower limit (type I). Realistically, most consumers have **minimum prey abundance** requirements (to sustain growth and reproduction), as well as **maximum ingestion rates**. This relationship is described in part in the type II response, where prey capture rate increases rapidly at first but levels off. As more prey are available for capture, **search time** decreases and **capture frequency** increases. However, **handling time** and ingestion rate become rate limiting, and the predator may become satiated. These cause the levelling off of the curve to a maximum rate. The type II response does not account for a minimum prey abundance requirement. Its shape is similar to that of enzyme kinetics as described by the Michaelis–Menton equation, and several derivations have been proposed through the years (Williams, 1980). At very low prey abundances, search time is so long between captures that predator growth may not be sustainable. In the natural habitat, prey abundance below a minimum level for an extended period causes a behavioural shift in the predator. The predator can switch to an alternative prey, it can become quiescent (or encyst) or it can enter a dispersal phase. The type III response is often the best descriptor of predator–prey interactions. Its S-shape is the same as that of the logistic growth equation. The type III functional response is more common than assumed, even with microscopic species. It is often an indication of **optimal foraging** behaviour by the predator or grazer (Krebs, 2001).

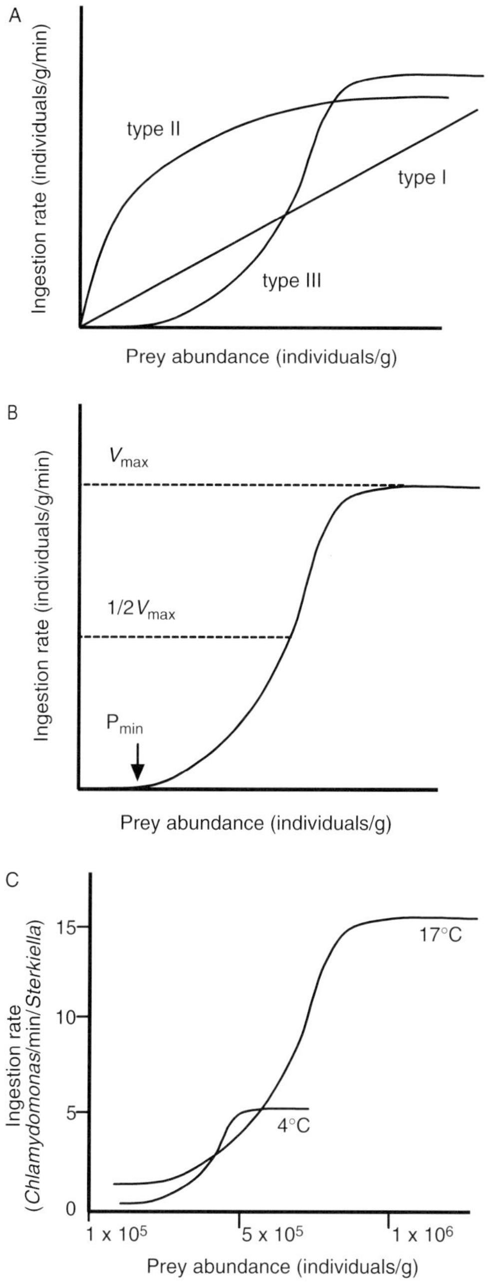

Fig. 5.3. Functional response curves. (A) Theoretical functional response curves (types I, II and III) observed in biological systems. (B) Type III functional response, with similarities to enzyme kinetics and the logistic growth equation. (C) Growth of *Sterkiella histriomuscorum* on the soil alga *Chlamydomonas reinhardtii* at two different temperatures.

The response of the common soil ciliate *Sterkiella histriomuscorum* to variations in conditions can illustrate predator–prey dynamics. This species can ingest bacteria and protists of a variety of size ranges up to about 25 μm. When grown on single species, cell cycle duration and cell size are affected. Most efficient growth results in large cells with short cell cycles (Fig. 5.3). Less efficient growth results in smaller cells with longer cell cycles. When grown at different temperatures, the preferred prey varies. Furthermore, the minimum required prey abundance, maximum ingestion rates and growth efficiency vary with changes in temperature. This is because the handling time and encounter rate also vary with temperature for different prey species. When grown in the presence of several prey species, *Sterkiella* show a clear preference for one prey over others. This preference depends on the temperature and on the cell size at the beginning of the experiments, as well as the abundance of each prey. These functional response curves are similar to enzyme kinetic curves in their response to temperature changes, prey abundance changes and initial predator:prey ratio. One simply substitutes the prey for the substrate and the predator for the enzyme. Prey preference at different temperatures is then analogous to activation energy barriers in kinetic reactions: reactions with high activation energy are less probable (slower rate) as the temperature decreases.

In the soil habitat, the dynamics of population interactions are complicated further by the heterogeneity of the habitat. Dispersal distance through the soil reticulum is slow, and patchiness (heterogeneity) is observed at all scales from submillimetre distances and larger. At the microscopic scale, each piece of microdetritus is a microhabitat. There are microclimatic effects which vary with soil depth, shade from a tree, presence of a rock or pebble, aspect of a hill slope, and so on. The accumulation of a particular type of litter becomes a resource patch. A twig, pine needle, dead rootlet, shed insect cuticle or empty test of a testate amoeba become the habitat for other species. One can quickly imagine a multitude of microhabitats and microclimates in the soil, over short distances and depth. These have repercussions on the natural dynamics of interstitial populations. It is unclear what dispersal distances and dispersal rates are inside the soil matrix. Clearly, dispersal occurs as decomposition of litter proceeds every year, and succession of local species activity through seasons is observable. However, patch dynamics and local immigrations of saprotrophs in decomposition food webs have received very little attention.

Why so many species?

Samples of soil are notoriously rich in biodiversity. A handful of soil will contain numerous species from many phyla (Chapter 1). We have seen examples of this diversity and abundance in previous chapters, and more examples are outlined here. The abundance of individuals in a population (of any one species) depends on the duration that adequate growth conditions persist. It also depends on the interactions with other species, through predation, symbiosis or competition. However, the diversity of species depends on **niche diversity**. A niche describes the combination of abiotic parameters (such as pH, ion concentrations, temperature and moisture) as well as the *range of tolerance* for each parameter, that affect the activity and survival of a species. The niche is also delineated by specific nutrient requirements and biotic parameters. For example, syntrophy with other species requires that one or more other species be active at the same time or in close succession. Ecological theory predicts that species number reflects niche diversity (for a full discussion, see Krebs, 2001). Also, when food resources and space are not limiting, the carrying capacity increases. Both the available space and resources are allocated to more species. For example, a desert soil with limited resources will contain fewer species than a pasture with more abundant resources. However, the number of species can be grouped into a smaller number of functional groups (Chapter 4). Both the desert soil and the pasture would have the same functional groups, although with fewer species each in the desert soil. Functional groups cross taxonomic boundaries, unlike **guilds**, which represent species within the same functional group and the same taxonomic group that occur in the same habitat. Therefore, an ecosystem provides a range of abiotic conditions, resources and a spatial habitat for species. The space and other resources are allocated to species, which occupy a niche. Species can be grouped into a smaller number of functional groups which fulfil the same ecological function. Through the year, with seasonal cycles, the abiotic conditions, food resources and space vary in quality and quantity. Therefore, the niche occupied by a species becomes more or less abundant, or restricted with time. These changing conditions cause species to become quiescent at regular intervals. Their activity is replaced by other species which become active on the opportunities vacated. The shift in active species through time displays a succession of adaptations to changing abiotic parameters and resources. Species that become quiescent are not replaced by species that fill exactly the same niche. In part, it is because the resources change in quality and abundance as a result of decomposition by previous species. There is a stochastic element in which species are activated and become dominant.

The soil habitat is also notoriously heterogeneous in structure. With depth, the composition of the physical environment changes (Chapter 2). Horizontally, there is variation in litter composition at the landscape scale down to microscopic distances at the submillimetre scale. With depth, it is not just the litter decomposition state that varies. The depth profile also reflects the history of the origin of litter deposition. At a precise sampling point, this year's litter is followed by litter from previous years. However, the composition of litter at that location may be different between years (Fig. 5.4). This creates heterogeneity of resources at the microscale with depth and reflects the horizontal litter accumulation differences. Microscale heterogeneity creates microhabitats which provide conditions for different species activity, and different species succession.

The coexistence of many nematode species was considered in an important paper, in relation to soil spatial heterogeneity (Ettema, 1998). Comparing nematode species composition between adjacent samples usually reveals little species overlap. Species composition similarity is low between samples short distances apart. This suggests a patchy distribution over several centimetres. Each soil core contains a small number of dominant species, with many rare species. The dominant species are different between soil cores. However, the same species occur across the site, at a larger scale (forest, grassland and agricultural field), so that 10–100 species represent that site. These observations reinforce the idea of many overlapping patches (microhabitats) in close proximity. Each microhabitat then supports different guilds of nematodes that reflect species composition. Some will be competing for the same resources, and with species outside of the guild in similar functional groups. This competition further reflects individual abundance between species. Those species that compete best and survive predation best are more numerous, given a set of abiotic and resource parameters. As conditions change with time in the microhabitat, the succession of species in each is also different. The succession pattern is then determined by activation of quiescent individuals in the vicinity, immigration of other species and predators, and changes in the abiotic parameters (that include weather changes). Therefore, there is an element of stochasticity in activity patterns, succession and abundance, even at the microscale level.

The following example attempts to demonstrate how one can reduce the apparent large number of species down to a variety of niche allocations. Let us postulate that we observe that a taxonomic group is

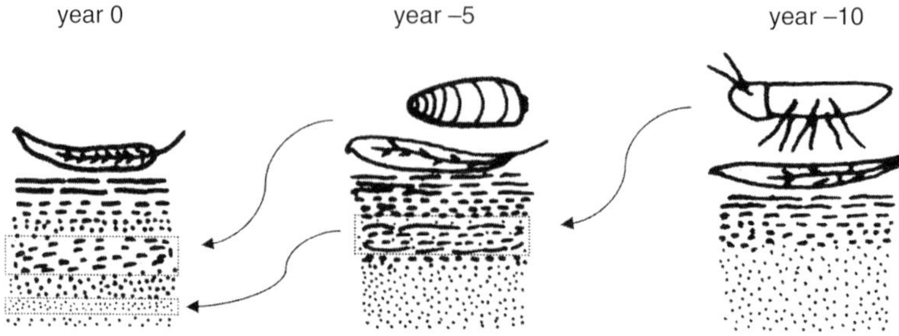

Fig. 5.4. Microhabitat development with depth through successive years.

represented by 100 species at a field site (total number of species at that site). Let us then observe that 30 species are active in summer, 40 in autumn, ten in winter and 20 in spring. Let us observe further that in each season, some species are active only on moist and warmer days, others on moist and cooler days, others on drier days. This series of observations already reduces the number of active species to a smaller number on each given day. If we then consider that some species are active in the top horizon, and others in successive horizons, we obtain a further spatial reduction in the number of coexisting species. Next, we have to note that species belong to different functional groups. The remaining coexisting species can be delineated further into a niche by considering their functional group and food preferences. By considering these parameters, along temporal and spatial stratification, accounting for succession in microhabitats is possible, without being burdened by too many competing species. The remaining species in the same functional group that appear to compete can be segregated further developmentally. For example, the development rate to maturity and resource requirements can be out of synchrony between competing species (Vegter, 1987).

In summary, within core samples, observed coexisting species can be differentiated into a variety of niches. Niche delineation requires consideration of microhabitat preference, functional group, development rate, dispersal strategy and other pertinent parameters (Krebs, 2001). Arguments to this effect were proposed for nematodes (Bongers and Bongers, 1998; Ettema, 1998), and species were proposed to fit into five basic categories. These could be differentiated further by considering spatial and temporal segregation in the habitat, e.g. activity periods in relation to weather and season, pore space restrictions, and food preferences within functional groups. Field sampling can reflect niche allocation of species only if sample analysis is stratified sufficiently, at the correct spatial and temporal scales (see Chapter 3). The interpretation of data and understanding niche differentiation between species are limited by how much biology is known about the interacting organisms.

Patterns in Time and Space

Any soil sample contains active individuals from many species, representing a diversity of phyla and functional groups (Fig. 5.5). It will also include numerous inactive species that do not grow under the habitat conditions at the time of sampling. In the sections below, we will consider examples of the distribution pattern in the soil and how that changes through time. Species occurrence in the soil is spatially restricted by abiotic and biotic parameters. Some species are adapted

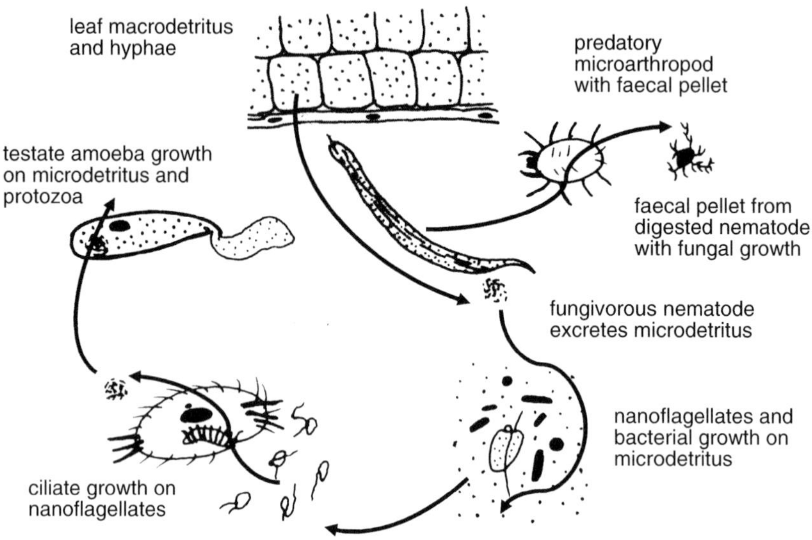

Fig. 5.5. Stratification of individuals and functional groups in soil on different food resources and microhabitats.

for low-oxygen microhabitats, others prefer high moisture content, or others may be limited by size to large pore spaces. Species within functional groups (such as bacterivores) also have different food preference, developmental and dispersal strategies. On short time scales (minutes to days), individuals respond to changes in aeration, moisture, temperature, and resource abundance and quality. On longer time scales (days to months), predation, parasitism, and seasonal changes in abiotic and food resources affect which species are active or quiescent. On even longer time scales (years to centuries), changes in soil physical structure and above-ground succession affect the belowground decomposition food web. Above-ground species succession changes root and litter input, and chemistry of the litter. Therefore, both species composition and number of active individuals are continuously changing and adjusting to new conditions. Within a guild, samples represents species that are decreasing and increasing in abundance with time. That is because there are species that were adjusted to past conditions (decreasing in number of individuals) and those that are adjusted to current conditions (increasing in number of individuals). As a result, samples contain a small number of species that are abundant, and others that are rare. The latter consist of species that are adjusted to past conditions, as well as those newly active in the microhabitat. This activity pattern can be observed at different time scales, from diurnal responses to weather changes, to successional trends over decades.

Primary Saprotrophs

Spatial organization in bacteria

Growth of bacterial cells in colonies on adequate substrate, under adequate abiotic conditions, occurs on the surface of soil particles and in the soil solution. Bacterial growth on a substrate forms colonies that are more or less compact depending on the genera (see Chapter 1). Growth of colonies is arrested by substrate limitation, spatial congestion, excessive metabolic by-products or growth inhibition by other organisms. Grazing by bacterivores reduces the colony population and often disaggregates the colony. It contributes to dispersal as well as to reducing the population growing on the substrate at a microsite or in a pore. Elimination of all individuals of a colony is unlikely, especially if the colony is disaggregated by the disturbance. The soil porous reticulum provides ample refuge for at least some of the bacterial cells to escape grazing (Vargas and Hattori, 1986; and see Hattori, 1994). These remaining cells continue to grow and divide on the substrate as long as the conditions are suitable. Dispersed cells settle on new substrate that may or may not be suitable for growth. Regular reduction of colony size reduces substrate depletion rate at first, and reduces secondary metabolite accumulation locally. As a result, the period of activity on the substrate is extended, as the colony continues with a growth phase. Displaced cells on adequate substrate can grow into new colonies, which contribute to the increase in biomineralization observed in the presence of grazers. These are mimicked *in vitro* by mixing soil microcosms.

The soil particles provide a variety of substrates and microclimates over very short distances, at the submillimetre scale. In fact, the microscale heterogeneity of the soil is such that in grasslands, which are less heterogeneous than forest soils, only 1 g of soil can represent the more abundant bacteria at the site (Felske and Akkermans, 1998). Since 1 g of soil will tend to have 10^7–10^{10} bacterial cells, it suggests that most colonies are probably very small. The distribution of bacterial colonies is best described as consisting of many small colonies on various substrate surfaces, representing an array of species and functional groups. A statistical study from soil thin sections predicted that 80% of the microcolonies existed in pores of <1 μm, and >90% were in pores of <5 μm in diameter (Kilbertus, 1980). Different species that are active at any one time are growing on the various substrates available. Within a ped, aeration is limited and the centre can be anaerobic. Therefore, over short distances (10–100 μm), there are gradients of microclimates, with aerobic and anaerobic activity in close proximity (Focht, 1992). Many species represented in each sample are inactive because the abiotic conditions are not suitable, or their substrate is absent. Many individuals are inactive simply because they are not on a suitable substrate, even though other individuals of the species could be growing in the vicinity. A study of bacterial diversity at the submillimetre scale revealed that at the microhabitat scale (<50 μm), isolates were very different between samples (Grundmann and Normand, 2000). The authors identified the chemolithotrophic *Nitrobacter* species using specific primers. They identified 20 isolates from 5 g of soil, and 17 isolates from several microsamples. The genetic distance between isolates, compared with reference strains, showed there was as much diversity between isolates in microhabitats from the same field as we know for the whole genus.

There are attempts at investigating the role of the bacterial assemblage in the soil at this scale of resolution. One approach is to use tagged cells which can be identified from samples. For example, in an elegant, study Jaeger *et al.* (1999) designed *Erwinia herbicola* with a reporter gene to determine the presence, activity and location of this strain in experiments. The reporter gene was the ice nucleation gene *inaZ* from *Pseudomonas syringae*. The assay for the presence and abundance of transformed cells is by a simple droplet-freezing protocol, followed by counting the number of ice nucleation sites/mm according to established procedures. The study showed that this approach was adequate to map the abundance and location of the cells and their substrate along root tips. In this study, both tryptophan and sucrose abundance in root exudates could be mapped along the root by monitoring the active tagged cells.

Other innovative techniques are being exploited to visualize soil microheterogeneity. These methods include computed tomography and magnetic resonance imaging techniques (Gregory and Hinsiger, 1999). These techniques can already provide resolution of soil physical structure at 100 μm. Minirhizotron video-microscopy techniques are also improving in magnification and are increasingly useful in monitoring decomposition. Infrared spectroscopy has also been applied to identify strains and species in very small growing colonies, without disturbing the colony. Fourier transform infrared absorption spectra (FT-IR) can discriminate between bacterial species and strains on agar plates within hours of culture establishment (Choo-Smith *et al.*, 2001). This assay relies on comparison of the spectra against reference spectra, and requires a small number of cells.

Another method is to use microcosms which are sampled by obtaining thin slices of soil, with relatively low disruption to the matrix structure. This approach was exploited using labelled litter that could be traced into the microcosm as it decomposed (Gaillard *et al.*, 1999). The authors note that since regular mixing or disruption of soil–bacteria suspensions or of soil microcosms increases the decomposition rate (organic matter mineralization), this may be caused by redistribution of cells on to suitable substrate regularly. In the natural soil, this mixing and redistribution of peds is carried out by the invertebrates and particularly by Oligochaeta species. Fractionation studies of microdetritus and macrodetritus into various aggregate size categories showed that soluble substrates as well as particulate microaggregates (or peds) of 2–50 μm were the site of this metabolic activity (Chotte *et al.*, 1998). These studies cannot determine the zone of influence of a microdetritus point source across peds, since the soil matrix is disturbed. In order to do this, Gaillard *et al.* (1999) devised replicated compact soil core microcosms. Each soil core contained labelled (^{13}C and ^{15}N) wheat straw (1 cm clippings) as the source of litter, and was incubated at 15°C for up to 100 days. The soil cores were fractionated at regular intervals to remove the straw, the adhering soil, and the top and bottom half of each core. The bottom half was sliced into 800 μm sections. The data from these soil cores were compared with controls without added straw. The results of the time course incubation indicate that almost all the activity due to straw occurs within 4 mm of the straw (Fig. 5.6). The straw ^{13}C accumulation in the soil was higher at the end of the incubation. In contrast, the ^{15}N accumulation was higher at the beginning of the incubation. Lastly, the straw and adhering soil were observed under low-temperature scanning electron microscopy (SEM) for visual observations of the spatial arrangement of bacterial cells and hyphae. There was a clear gradient of decreasing abundance and size of colonies away from the straw, with fungal hyphae dominating adjacent to the straw. The 3–4 mm zone of influence, or detritusphere, identified in this experiment probably varies with soil bulk density and porosity, as well as the composition of the litter from which the microdetritus or macrodetritus is derived. The detritusphere is probably greater in fresh leaf litter which leaches more nutrients.

In the environment, more complete decomposition occurs in the presence of a complex community of primary saprotrophs that contribute a variety of enzymes (or various functional groups) (Filip *et al.*, 1998). This decomposition is sensitive to the composition of the soil solution and soil organic matter (SOM). For example, humic substances from the organic matter added to nutrient broth stimulated growth of mixed cultures up to six times the control levels, without humic substances (Filip *et al.*, 1999). It is more difficult to obtain bacterial growth from soil isolates in single species cultures on natural substrates. In part, this is due to limitation of the enzymatic panoply that would otherwise be available to decompose complex substrates. For example, when bacteria are diluted then cultured, much less growth occurs below a threshold number of species, whereupon it is assumed the enzymatic or metabolic panoply has been reduced too much (see Salonius, 1981; Beare *et al.*, 1995; Martin *et al.*, 1999). Another limitation is whether corresponding bacterial physiotypes are active. This point is illustrated in a study of methanogenesis in anoxic rice fields (Chidthaisong and Conrad, 2000). The study considers the metabolic pathways of the bacteria that would be active under the anaerobic conditions of submerged rice fields. Bacteria that use ferric ions, nitrate,

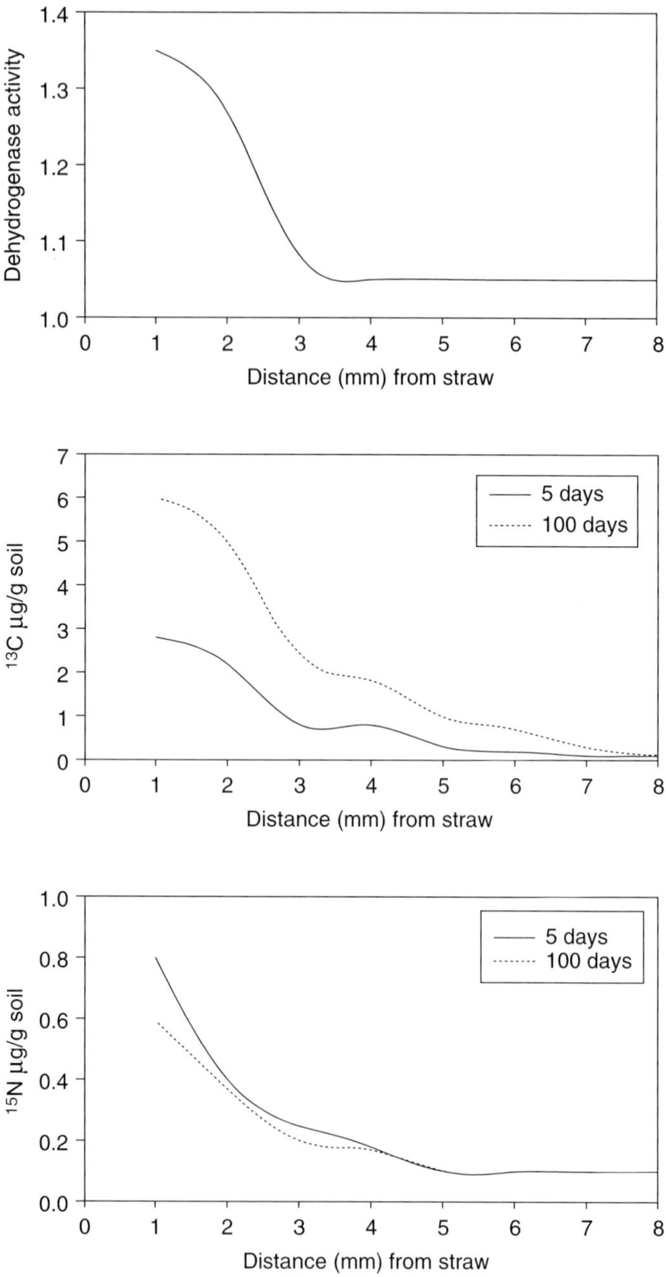

Fig. 5.6. Primary saprotroph activity in the detritusphere of straw buried in soil. The eukaryote cellular activity is reflected in the mitochondrial dehydrogenase enzyme activity; the amount and distance of labelled carbon and nitrogen from straw extend <5 mm from the straw, over the incubation period (data simplified from Gaillard *et al.*, 1999).

sulphate or CO_2 as electron acceptors utilize nutrients from organic matter decomposition. The order of electron acceptors used first is determined by the redox potential of reactions, the highest redox potential being used first. This predicts the order nitrate, ferric ions, sulphate and lastly CO_2. Under these anaerobic conditions, H_2, alcohols, acetate and fatty acids are produced during decomposition. Acetate and H_2 are in turn converted to methane or to CO_2, depending on the conditions. The latter is produced only in the presence of sufficient ferric iron, nitrate or sulphate. Competition for the available H_2 to use in synthetic reactions is high, and generally the nitrate reducers out-compete the others. Therefore, H_2-dependent methanogenesis is inhibited when the other reducers are active. When these reducers are active enough, they out-compete methanogens for acetate, producing CO_2 and not CH_4. The authors further point out that the outcome of these reactions are also dependent on which species are active. For example, the two acetate-utilizing methanogens in the experimental soil were *Methanosarcina* and *Methanosaeta*. Each has a different preferred temperature, and the relative abundance of each changes during incubation experiments. The kinetics of acetate uptake also vary, with a lower K_m value of *Methanosaeta* for acetate. In conclusion, this study indicates, as do other similar studies, that an understanding of the function and dynamics of bacterial communities requires consideration of syntrophy of the bacteria on natural substrates, under simulated natural conditions, or *in situ*. A promising method for *in situ* studies relies on intact soil core microcosms. These intact soil cores can be modified in the laboratory and followed both in the laboratory under controlled conditions and in the field (Gagliardi *et al.*, 2001). The results are reproducible, and variability can be investigated further under controlled conditions.

At a different scale, heterogeneity of the landscape, plant species distribution, riparian areas and other factors affecting the microclimate influence bacterial species composition. Species of bacteria and fungi and their growth physiology are affected by the plant roots and plant species composition. One study, conducted near Pullman in Washington State and Durham in North Carolina, compared the microbial community of three naturally occurring plant pairs at each site (Westover *et al.*, 1997). The isolates were tested for 13 physiological attributes, and the growth behaviour of isolates from each plant pair was compared statistically. The results show that each plant pair rhizosphere microbial community was significantly different from the others. When plant pairs were reproduced in greenhouse trials, the rhizosphere soil community of each plant pair changed under the influence of the plant pair rhizosphere. Initially, the soil microbial composition was similar between pots, but differentiated over time in response to being grown with different plant pairs. When plant pairs were planted in the field, again the initial microbial physiological profile was modified after 70 weeks. The authors suggest that plants may interact with the soil microbial community in species-specific ways that may affect plant fitness or competition by interacting with the soil community. This influence on the soil community is affected by the overall plant species composition. For example, the spatial map of ground cover is influenced by the microclimate effect of the tree cover (Saetre, 1999). The tree cover modifies the ground cover composition, which together affect the carrying capacity of the soil community, as measured by total microbial biomass and soil respiration. The nature of the cellular interactions between the roots and the bacteria is unclear, but it is most probably through root litter chemistry and root exudates. Both of these are likely to modify the decomposer food web, and the relative abundance of consumer species.

Variations in the microclimate and vegetation over several metres is therefore sufficient to cause detectable change in the bacterial species composition at the community level. In sampling for a particular bacterial enzymatic function or species over hundreds of metres, variation can be detected along transects as the vegetation or abiotic conditions change. Superimposed on this variation, there is the variation in activity caused by microscale soil heterogeneity. The latter is responsible for the large variations detected between samples. In part, the microscale activity difference reflects variations in the syntrophic activity of the microbial community, in what has been termed local 'hot-spots' of particular enzymatic or functional activity. However, the soil microbial community activity also varies with seasons, as the quantity and quality of the litter input change, and the abiotic parameters also change. These fluctuations were illustrated in a study that followed three grassland locations in Britain (Grayston *et al.*, 2001). At each location, three adjacent sites were selected with different management intensity and soil fertility. Soil samples were obtained through 1 year and measured for variations in pH, moisture, total microbial biomass, total respiration, bacterial and fungal colony-forming units, phospholipid fatty acid (PLFA) profile and BIOLOG carbon utilization profiles. The data identify specific management effects on soil characteristics between sites, that fluctuate through the seasons. The primary saprotrophic colony-forming units reached a maximum during August at all sites. The carrying capacity (total biomass, g C/m^2) and activity (respiration, CO^2 l/h/m^2) varied much less between sites across seasons, except one very high result at one site in April. Particularly interesting was the change in C utilization profiles through seasons and between management type at each location. The patterns changed significantly and there were detectable shifts between the three locations and each management type with time of sampling. A similar study, conducted over five seasons in a New Jersey forest, describes similar patterns (Rogers and Tate, 2001). There were significant differences between soil types in the forest site, as well as between seasons, in the metabolic activity of the bacteria. However, these changes did not affect the carrying capacity of the soil microorganisms, as measured by microbial biomass. The dehydrogenase activity, which more closely reflects cellular activity, varied with seasons and depth, but did not correlate with total organic matter. Studies conducted at this level of resolution are more difficult to interpret. Dramatic changes in bacterial species activity or species abundances are not reflected in total microbial biomass estimates, which also include the other soil organisms. Furthermore, it is difficult to compare biomass estimates from chloroform fumigation between sites, unless the K_c transformation constant is known at each site (Dictor *et al.*, 1998). However, PLFA, BIOLOG plate assays and similar crude analysis of the community profile do provide an indication of changes in bacterial community function, if the changes are large. In a review of the literature, Wardle (1998) compared 58 published data sets from forest, grassland and agricultural fields from various latitudes. The data from these studies were analysed for trends between sites over set time periods. Beyond the expected trends with climate over microbial biomass, several trends involving pH, soil organic C, N and latitude were detected. The author notes that biomass turnover increases with latitude, as seasonal variability increases. However, with decreasing latitude, increased mean annual temperatures also increase the decomposition rate and biomass turnover. These two components work in antagonism. Temporal variability in the microbial biomass C correlated best with soil N content in forests, where there is ample organic matter input. In grasslands, with less litter input, soil C, pH and latitude correlated best with microbial biomass temporal variability. In agro-ecosystems, only pH and latitude correlated well. The study did not identify differences in microbial biomass between agroecosystems, forest or grassland habitats.

In order to detect changes in the functional structure of primary saprotrophs, it is best to assay for the enzymatic activity of species on natural substrates, over a time period. One such study correlated the activity of bacteria, fungi and several microinvertebrate by placing each into functional groups (Dilly and Irmler, 1998). The study used black alder litter placed in 5 mm mesh bags at two forest sites in the Bornhöved Lake area, near Kiel in Germany. The two sites were 25 m apart on different soils with different moisture contents. During the 1 year study, mass loss was from 10 to 6.5 g at the drier site, and to 2 g at the wetter site. The succession of bacterial and fungal activity on the litter was determined by growing extracted species from the litter on selective nutrient agar media. Bacteria were extracted from the litter by washing with sodium hexametaphosphate (0.2% (w/v), pH 7). The functional composition of the bacterial species was determined by culturing on nutrient broth (general culturable bacteria activity), starch (amylolytic), gelatine (proteolytic), Tween-20 (lipolytic), filter paper (cellulolytic), natrium polypectate (polygalacturonolytic), ammonium sulphate (ammonia oxidizers) and sodium nitrite (nitrite oxidizers). The fungi were obtained by culturing litter fragments on cold soil extract agar and malt extract agar. The cultures were assayed for their enzymatic potential for digesting various substrates (amylase, protease, lipase, xylanase, polygalacturonase, pectinlyase, cellulase and laccase). Dry-extracted microarthropods were enumerated and their gut content described. These taxa were placed into functional groups based on gut content and some guesswork about their feeding habits. Wet-extracted taxa (rotifers, nematodes, enchytraeids and protozoa) were not described in this study. Over the 1 year time period that the litter was assayed, there was a clear seasonality and succession of functional groups that digest the black alder leaf litter through primary decomposition. An example of this succession is shown for bacterial functional activity (Fig. 5.7). It is interesting to note that when bacterial activity on a substrate decreases, it may persist or begin in fungal species. This maintains the functional activity through several seasons, when conditions that are inappropriate for one type of primary saprotroph remain adequate for another. For example, decomposition of substrates other than cellulose showed no overall seasonal effect, when all taxa were considered. The results were developed further by comparing the correlation between particular fungal and bacterial enzymatic activity (Dilly *et al.*, 2001). Several other positive correlations were identified, such as the simultaneous presence of ammonium oxidizers with cellulolytic species and with certain macrofauna species. Also, lipolytic bacteria

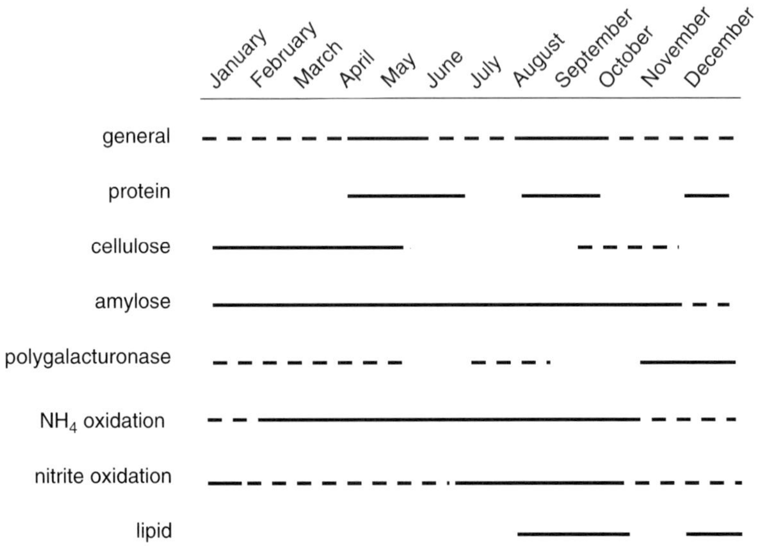

Fig. 5.7. Periods of highest (solid line) and reduced (dotted line) bacterial enzyme activity on various substrates, through 1 year on leaf litter (data from Dilly and Irmler, 1998).

occurred together with diplopod, isopod and microdetritivorous microarthropods. There was an increased frequency of predatory microarthropods when microarthropod abundance was higher. The results of this study provide a map of functional group co-occurrence and succession on this leaf substrate. The significance of these correlations should be explored further in subsequent experiments.

Organization of fungal species

The hyphal biomass of forest soils, especially in the organic L–F–H layers, predominates in the living saprotroph biomass. The hyphae proliferate through the litter, and the mycelia are responsible for clumping together the macrodetritus and microdetritus with the mineral particles and humus. A variable portion of the hyphae is connected to roots and rhizoids in mycorrhizal association. Fungi are less prominent but certainly very abundant in other ecosystems, in proportion to the amount of organic matter accumulated. Soils that are regularly disturbed, such as agricultural soils, tend to be much lower in fungal biomass. This is due in part to the reduced amount of SOM, but also to the regular shredding and disruption of the mycelia. The fungal species diversity at each site can be very large, especially in forest soils which provide a diversity of substrates and microhabitats. Different species are found in different substrates, in different habitats, and these vary with seasons. For example, if one follows a linear gradient from a forest floor, to a grassy clearing, to a riparian edge, one would find a sequence of species characteristic of the litter in each habitat. Some fungal species would occur only at one habitat, but others would be less restricted in habitat or litter preference. If this walk was repeated later in the year, one would expect to find changes in the species abundances and occurrence at each habitat. In general, fungal species occurrence is limited by certain biotic factors and abiotic parameters.

The biotic factors include the presence of adequate substrate and substrate chemistry. Some species are specific in substrate choice and will not sporulate or grow into alternative litter. For example, there are species that occur only after a fire event, on the resulting charred litter. The post-fire litter is higher in mineral ion content and has higher pH. The agaric *Pholiota carbonaria* and the discomycete *Geopyxis carbonaria* are typical examples. Another example is *Rhizina undulata* which only sporulates between 35 and 45°C and invades pine roots. Other species may be specific to particular species of leaf or wood, or other substrates. For example, *Retiarius* species are specialized pollen saprotrophs. Others such as *Clavaria*, *Collybia*, *Marasmius* and *Mycena* are stimulated by flavonoids from pine needles, which provides them with a competitive edge in the presence of this substrate.

The abiotic factors include the local pH, pO_2, pCO_2, temperature, moisture and microclimate. Species have a tolerance range with an optimum value for each of these parameters. Through weather fluctuations and seasonal changes, these parameters change and affect the ability of species to grow. Moreover, superimposed on changes in litter quality and abundance through the year, one can see how the activity period of fungal species is restricted through the year. Through the profile, with depth, the abiotic parameters change and affect the presence of species. Litter chemistry and nutrient availability also vary with depth. These gradients impose a species stratification through the profile. The horizontal species variation depends more on the heterogeneity of litter substrate and local microclimate.

On each substrate, several parameters affect the co-occurrence and succession of fungal species. One mechanism of excluding competing fungi is the secretion of secondary metabolites or fungicides against sensitive species. This mechanism is probably common in soil and occurs in *Trichoderma viride* and several Ascomycota and mitosporic species. However, there are not many real examples of the benefit from these metabolites in a natural setting. Another mechanism is the contribution of enzymes from coexisting or preceding species. Different enzymes modify the chemistry and physical properties of the substrate. A co-occurring species may facilitate the digestion of the substrate by providing enzymes missing in another species. With time, the substrate become softer and more porous, changing water retention by capillary forces. This reduces the dissolved oxygen content, and the substrate matrix can become water-logged, which modifies the abiotic and biotic parameters. The effect of microarthropod and other invertebrate species that co-occur further modifies the structure of the matrix. For example, wood in early stages of decomposition that contains 26–33% moisture gradually becomes water-logged as ligninolytic and cellulolytic species decompose the substrate. At first, water can be limiting and the hyphae rely on water molecules released from the hydration reactions of polymer decomposition. With time, as the microfauna accumulate and the matrix becomes more porous, capillary water also accumulates, decreasing pO_2 and other changes in abiotic parameters, and excessive moisture content may inhibit the initial fungal activity.

A study conducted on pine needles in central Japan illustrates some of the parameters that affect fungal species composition. The experiment was conducted in a forest of *Pinus densiflora* on a moder soil, near Sugadaira (Tokumasu, 1996, 1998). There were changes in species composition through seasons, with depth in the profile, and between interior/exterior pine needle colonizers. Unlike sites at the same latitude in Europe and Canada, where the L and H horizons remain distinct through the year, here the L layer disappears over the summer and the H layer is indistinct. The author demonstrates by surface sterilization and washing that the hyphae inside the needles are different from those growing on the exterior of the pine needles. With depth in the litter, the composition of the saprotrophic fungi in the needles varied in composition (Table 5.1). The main interior colonizers varied depending on the season during which the needles fell. Most species were inactive through the winter. Over the summer, dead needles still on the tree were internally colonized by *Aureobasidium pullulans*, while those on the ground were colonized by *Chaetopsina fulva* (which coexists with *Verticicladium trifidum*). In late autumn and spring, the needles were internally colonized by *Selenosporella curvispora*. In the summer months, *Thysanophora penicillioides* colonized the exterior of the needles. Other soil fungi decomposed the exterior during other seasons.

In a similar study, also in Japan but in a forest of *Pinus densiflora* with a warm temperate monsoon climate, pine cone decomposition was described (Kasai *et al*., 1996). The pine cones were surface sterilized or water washed, then placed on malt extract agar to identify cultured species. The field cones were followed for 65 weeks, at which time <30% dry mass loss was observed. The authors describe various species that colonized the cones, noting a preference for different periods after senescence. Some species were early colonizing saprotrophs that preferred coniferous substrates or pine cones. Others were opportunistic parasites on wounds of sick trees that also occurred on the cones at senescence. Interestingly, the authors discuss the effect of a colonizing species on determining subsequent species composition.

The effect of other fungal species on succession can be categorized simply into three types (Fig. 5.8). One type is the *mutual tolerance* and coexistence on a substrate of two or more species. This is observed particularly, but not exclusively, when nutrients are not limiting. Another type of interaction is *facilitation* of colonization by a former species for a later species. In this situation, one fungal species may remove a physical or chemical barrier to the growth of another species which then proliferates. Lastly, there can be *inhibition* between fungal species, through chemical inhibitors or complete removal of an essential nutrient by a former species. In this scenario, an early colonizer uses resources in such a way as to affect subsequent colonists. Of course, on natural substrates in the environment, all these interactions occur at the same time between different species.

Table 5.1. Frequency of fungal species occurrence inside pine needles at different depths in the litter.

Horizon	L	O_L	F_H
Chaetopsina fulva	31.0%	16.0%	5.5%
Trichoderma koningii	15.5%	54.0%	81.8%
Mortierella rammanians	1.8%	11.0%	23.6%

Data from Tokumasu (1996).

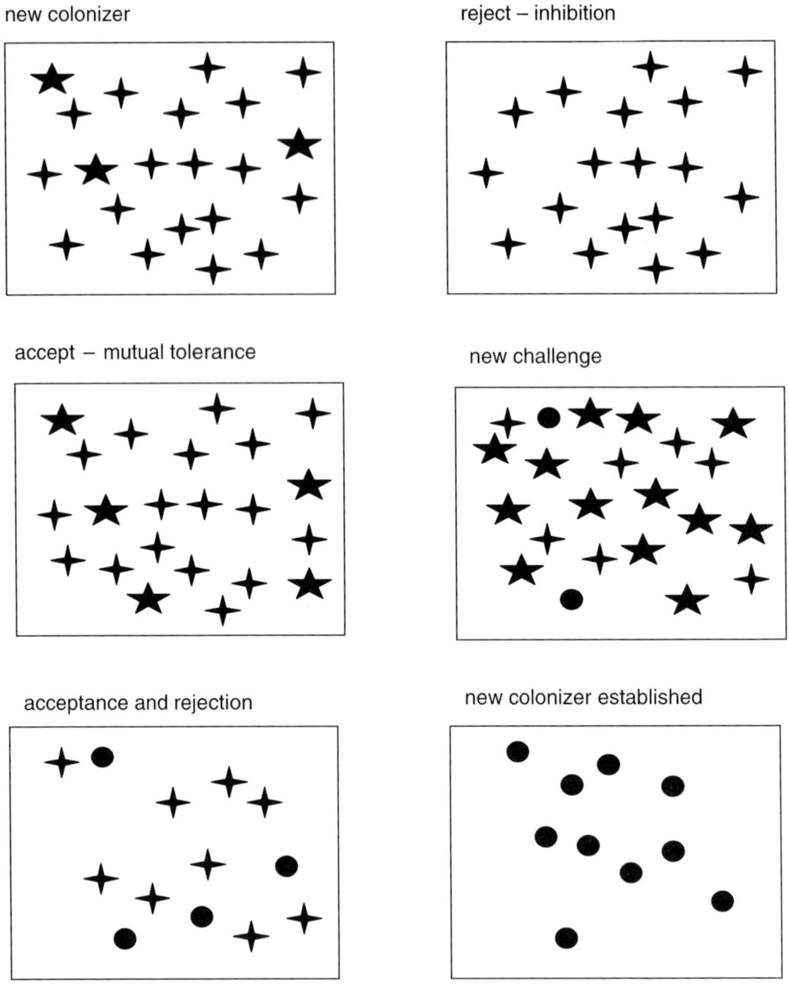

Fig. 5.8. Example of interactions between fungal colonies during succession.

Several authors have tried to describe fungal succession in primary decomposition, beyond simply naming species in sequence. One study in northern Germany followed the activity of fungi over 1 year in black alder forests (Rosenbrock *et al.*, 1995). In this study, the function of active hyphae was described after washing off spores and other dispersal and inactive propagules. Two media were used in the laboratory: a cold water soil extract agar for slow growing species, and a malt extract agar for faster growing species. The role of fungal species was determined by enzymatic assay *in vitro* on various substrates. Although this assay determines whether species possess enzymes to digest a substrate, it does not determine whether that enzyme was being used when it was sampled. However, it provides an idea of the functional sequence of species succession on the leaf litter. The study was able to correlate species succession with changes in the enzymatic potential. The functional succession was similar to results obtained in other studies. What is clear from the results is that the patterns of species succession and enzymatic potential differ. Although particular species come and go and their periods of activity differ, successive species have overlapping digestive enzymes (Fig. 5.9). The authors note that even within genera, species vary in activity period through the year (but the enzymatic potential is probably similar).

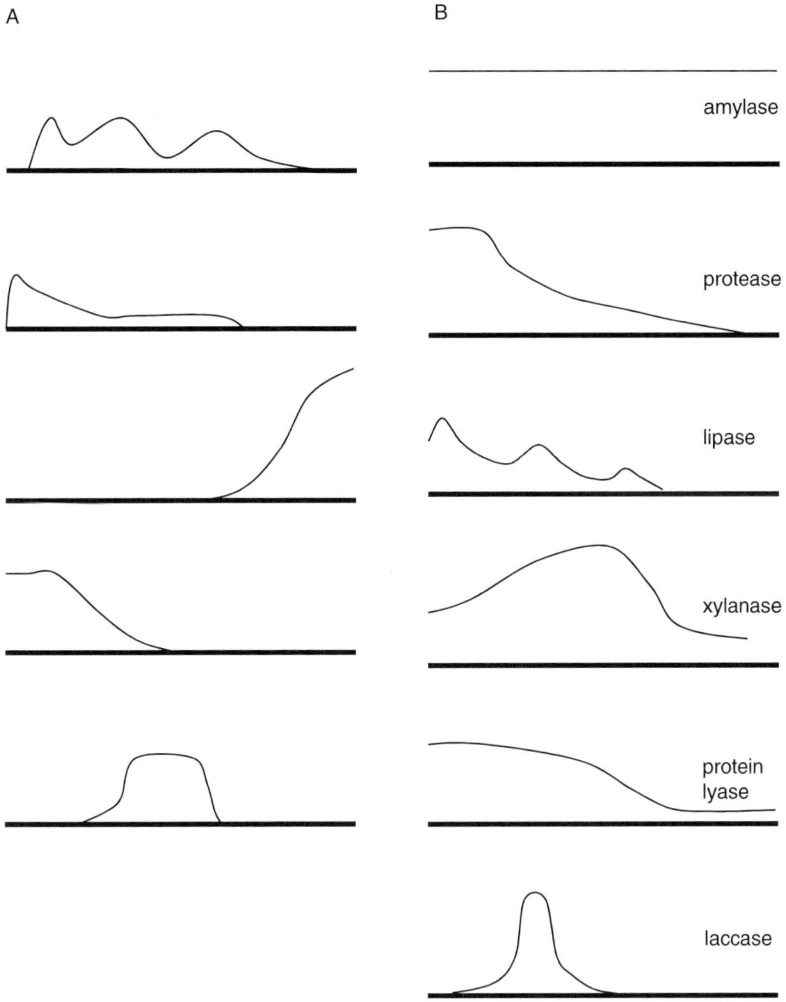

Fig. 5.9. Example of patterns in fungal species succession and enzyme activity trends over time. (A) Variations in the frequency of single species abundance over time. (B) Enzyme activity patterns over time on a substrate (adapted from Rosenbrock *et al.*, 1995).

Another important study described the spatial segregation of Basidiomycota species on decomposing *Picea abies* and *Pinus sylvestris* naturally fallen tree trunks, in a northeast Finland boreal forest (Renvall, 1995). During decomposition, the wood undergoes chemical modifications, which in turn cause structural changes in the matrix. The bark decays or falls more rapidly. The wood increases in water content with capillary water infiltration in the increasingly porous matrix. Waterlogged wood becomes anaerobic and inhibits much of the eukaryote activity. Overall, 120 species of Basidiomycota were found growing in *Picea* and 104 in *Pinus*. When considering the number of species found, one wonders whether there can be the same number of niches on a single tree trunk. The description suggested that indeed there are sufficient microhabitats to accommodate at least a majority of the species. In part, seasonal preferences in activity period can account for some of the species diversity. Succession over the period of tree decomposition and microhabitat differentiation within the tree can account for some of the remaining diversity. Several of the conclusions are outlined here. Species were restricted in substrate-volume colonized along the length of the tree. There were species that clearly preferred the base of the tree, and others that preferred the crown, or areas in between. This reflects the changing composition of the tree cell walls and tissues along the length, from old to young. Early saprotrophs give way to later colonists which tend to be more specialized and k-selected species. The growth of certain species was profuse and colonized large volumes, whereas other species were restricted to small volumes. These may reflect competition and inhibition between species. However, one also needs to consider the palatability of various hyphae to grazers inside the wood. The white rot species were more restricted in time and successional sequence than brown rot species. The brown rot fungi were active for longer durations but decreased in abundance from 25 to 7% of total species with progressing decomposition. The lignolytic Basidiomycota differed greatly in their physical matrix (substratum) requirements. The presence of lignin in itself was insufficient to describe their species composition. They also vary in their competitive ability with each other, as well as preference for the extent of decomposition prior to colonization. Co-occurring species may be competing for space on resources, or they may be utilizing different molecular components of the trunk, or different depths within the tree. Certain species tend to be followed by others. There was some sequence to species succession, but a stochastic element remains depending on which of several possible successional species arrives first. This observation was particularly striking when trees that had fallen for different reasons were compared. The author distinguishes between downed trees caused by fire, sickness, drying or live fall. In each case, the initial colonizing community was different. This affected the subsequent succession of species. There were species that were characteristic of a particular tree species, of a particular stage of decomposition, or of a particular sequence in succession. However, most species were rare, and not many were very common. These observations should be expanded in the future by determining the enzymatic functions of species in the sequence. Species composition may reflect some stochastic element in colonizing ability and competitive ability with existing colonists; however, at the functional level, the enzymatic sequence may be more similar. The author (Renvall, 1995) further points out that as we continue to destroy old growth forests, fungi that are specific saprotrophs of old mature trees with thick trunks no longer have adequate substrate in our young managed forests. The longer the spores remain inactive, they are buried in the litter and in turn decomposed. These species are probably increasingly rare as inactive spores, or are extinct.

The combative interactions of wood-decaying Basidiomycota on wood over succession on the substrate were reviewed recently (Boddy, 2000). The conclusions of the review in large part confirm what we have described here previously, regarding abiotic tolerance and preferences of fungi, and the interactions between coexisting individuals and species. New colonization of substrate by a species is followed by attempts at colonization by new arrivals. This leads to either inter-specific contact between mycelia, or antagonisms at a distance between adjacent mycelia, or consumption of one mycelium by the other (the author uses the misleading term 'mycoparasitism'). The end results of these interactions are most simply categorized as deadlock or replacement of one species by another over time. In particular, growth of one fungus around the periphery of another mycelium may completely prevent its expansion and limit its activity. Mutually tolerated growth is possible when species benefit from the functional role of enzymes and growth properties of coexisting species. As one substrate is exhausted and the matrix structure changes, species that were competitive give way to other species which are more competent on the remaining substrates. The review points out that different strains of the same species often vary in their ability to compete with other species. (There is probably an allelic polymorphism between populations.) Furthermore, the outcome between species is significantly affected by water potential, temperature, and soil or substrate air composition. For example, Basidiomycota require more water (less than –4.4 MPa) than xylariaceous Ascomycota, which can function at lower water potential. In some cases where contact between mycelia is not tolerated or inhibited, one species can destroy the other's mycelium. In most cases, the *inhibition* or cell lysis that ensues is caused by secondary metabolites, not by digestive enzymes. When it is caused by digestive enzymes, it is best described as predatory *consumption* of the mycelium. This is in contrast to *decomposition* of dead hyphae by primary saprotrophs. In general, the outcome of species interactions is not always predictable. Sometime, one species may have grown 'sufficiently' in a given volume to make it difficult for a competitor to establish a colony. The exact nature of these interactions is not always obvious, and requires experimental validation.

In the field, these direct interactions between mycelia are affected by the composition of the rest of the decomposer food web. In a review of fungal succession (Frankland, 1998), the case is made with *Mycene galopus* in northern temperate forests. It is a major decomposer of plant litter, which digests *in vitro* all major components of the plant cell wall. This species is not very effective on wood as it is susceptible to wood-inhibitory chemicals, and it requires more oxygen than is available inside the wood matrix. The fungus is seasonally active on new litter in autumn, where it effectively keeps out competing colonizers. One of its competitors is in fact more effective when it arrives first. However, its growth is limited by Collembola grazing because it has a preferred tasty hypha. That grazing preference is sufficient to cause the difference in species abundance, with *Mycene galopus* dominating.

These descriptive observations of species succession can be extended in the laboratory with functional assays. Moreover, many edaphic species do not form easily visible reproductive structures, and are more difficult to identify from field samples. They require molecular identification, especially if isolates are not easily cultured. Species baited on particular substrates can be obtained naturally and kept on agar plates in the laboratory. These can be assayed for enzymatic activity, for molecular identification, or forced to compete with other isolates on agar plates. For example, a study with sterile animal hair (wool) followed decomposition by fungi on Saboureau–dextrose agar (Ghawana *et al*., 1997). The authors observed that initial colonists did not digest keratin, but were gradually replaced by keratin-digesting species such as *Acremonium* species, *Chrysosporium* species and *Trichophyton simii*. These were more specialized and became dominant as other substrates (with more easily digested bonds) became exhausted. Similarly, Holmer and Stenlid (1997) tested the competitive ability of Basidiomycota on wood discs placed on water–agar. These types of studies, although they are simple and descriptive, open the way for extended and more detailed experimentation of parameters that control species activity periods and digestive enzyme secretion.

One example of a more elaborate description of fungal primary decomposition and succession is provided by Osono and Takeda (2001). The study was conducted along a hill slope, in a cool temperate deciduous forest in Japan. Beech leaves were selected because they harboured numerous fungal species in succession and because they decompose relatively slowly. Two sites along the hill slope were used to set leaf litter bags in 2 mm mesh: an upper moder soil and a lower mull soil 200 m away. The rate of decomposition at the two sites was about the same, with about 55% mass remaining after 35 months. The litter bags were sampled by first removing the surface spores (surface sterilization or surface washing), then making observations on the hyphae within the tissue by culture on agar plates. The extent of hyphal mesh in the litter was estimated by a grid intersect method using calcofluor white stain for the cell wall and acridine orange for the nuclei in living hyphae. The litter samples were assayed for content in lignin, holocellulose, soluble carbohydrates, phenolics, and total N, P, K, Ca, Mg. Sequential loss of cell wall components indicates indirectly the enzymes that were active in the leaf litter. The rate of loss followed the sequence: polyphenols>soluble carbohydrates>holocellulose>lignin. Both N and P increased at first, indicating colonization with living biomass and increase in living cytoplasm from saprotrophs. It is followed by a decline, as fewer living hyphae remain and nutrients are mobilized. The minerals K, Ca and Mg fluctuated up and down in abundance, but with an overall decrease over this time period. This reflects changes in the species composition in the litter as well as release of nutrients from the litter. Graphs of the frequency of occurrence of Ascomycota species over 35 months showed periods of activity that succeed each other. There was an increase in species with clamp connections near the end of the sampling period. The authors suggest that this reflects a functional shift that occurred at about 21 months. Interestingly, the percentage of living fungal biomass remained more or less steady as a fraction of the litter mass. It fluctuated between 6 and 2%. However, total hyphal biomass that includes the abandoned cell walls peaked between 7 and 12 months. Succession of species isolated from the litter bags established that there were particular fungal species associated with particular stages of primary decomposition.

Lastly, any section on fungi would be incomplete without some consideration of the yeast forms. These are rarely sampled by soil ecologists, but there is clearly some pattern that can be described (Spencer and Spencer, 1997). Many species are associated with surfaces of plants and insects where they grow by osmotrophy on the tissue secretions or leakage. Alternatively, some are intracellular symbionts of invertebrates. They find their way into the soil on death of the organisms, where they are picked up again or dispersed on to new hosts. These yeasts are transient species in the soil, not necessarily adapted for soil growth at all. Some species are true edaphic species particularly in the genera *Candida*, *Cryptococcus*, *Hansenula*, *Lipomyces* and *Pichia*. Most soil species are non-fermentative Basidiomycota. Abundances are often 1.4×10^4/g soil in surface layers, decreasing to 600–1800/g in subsurface layers. There is a seasonal increase in abundance during the summer months. Habitats with enriched yeast species are limited to fruit-fall periods and fruit wastes, where abundances increase to 10^4–10^5/g. However, after oil spills, some species increase in abundance as they can metabolize a wide range of aliphatic and aromatic organic molecules, in both soil and marine environments. Yeasts are sensitive to the presence of Actinobacteria, which inhibit their growth by antibiotic secretion. They are preyed on by grazing invertebrate larvae and some adult forms, especially of certain microarthropods. Ingestion of yeast cells by protozoa and nematodes is also common. Many bacterivorous species can be cultured on yeasts. The distribution of yeasts in the soil may not be limited by oxygen as some species are anaero-tolerant and capable of fermentation. It is probable that they are enriched in certain riparian or regularly flooded soil that is also rich in organic matter. Whether they contribute to soil food webs in any significant amount is unclear. They may be infrequently active opportunists.

Secondary Saprotrophs and Other Consumers

Patterns in soil protozoa

Very few studies have tried to describe stratification patterns or succession trends with soil protozoa species. There are, none the less, many casual observations of the phenomenon. The scarcity of data reflects primarily the lack of studies on these species. There are some observations on ciliate and testate amoebae species summarized here. A great deal of work remains to be done in this area, to correlate changes in protozoa species abundance with moisture, temperature and soil composition variations, with seasons, and over longer successional periods.

Aggregation of protozoan abundance into small patches can be visualized even in a Petri plate or in visibly homogeneous experimental small ponds. The heterogeneity develops over several hours to days as colloids and organic matter aggregate. Detection of the heterogeneity depends on the size of samples and scale of the experiment (Taylor, 1979; Taylor and Berger, 1980). It is easy to miss the spatial heterogeneity by sampling infrequently or with large sample volumes (1 ml or more of liquid culture; >0.5 g of soil) which essentially mix microhabitats. The data suggest that with ciliates, species number decreases with sample size, so that patches are probably populated by no more than one, or a very small number of species. This spatial heterogeneity is obviously far more pronounced in natural soil. One would expect individuals to remain in the vicinity where excystment occurred, as long as prey remains in sufficient quantity. The most important criteria in determining the activity and abundance of protozoa in natural samples are: (i) the frequency of conditions that trigger excystment for particular species; (ii) the duration of conditions suitable for growth on available food resources; (iii) food resource limitations during periods of activity by scarcity or competition; and (iv) predation impact on active cells or cysts.

An early study of spatial distribution of species in the interstitial pores showed considerable stratification in space for different invertebrate species (Fenchel *et al.*, 1967). The study was conducted on a sandy beach near Helsingor. They identified species of Archiannelida, Gastrotricha, Harpacticoida, Nematoda, Oligochaeta, Tardigrada and Turbellaria. The authors observed that species presence was correlated with changes in salinity, oxygen content and pore size. In a now classic set of papers, these observations were extended to ciliates and several other taxa of protists (Fenchel *et al.*, 1967; Fenchel, 1968a,b,c, 1969). The sublittoral sands were subdivided into horizons based on redox potential with depth. An upper oxidized zone contained, among other groups, nematodes, gastrotrichs, rotifers and a variety of ciliates. An upper reduced zone contained nematodes and very few other invertebrates, with a variety of ciliate species. A lower reduced zone and a sulphide horizon below it were essentially free of invertebrates, but each contained species of ciliates specific to it (Fig. 5.10). These species displayed both depth preferences and considerable variation over time with seasons. Many species were sampled only at specific months, or else in low numbers.

The diurnal and seasonal fluctuations in species abundance depend in part on soil temperature and moisture. Temperature fluctuations have a role in stimulating excystment in some families. For example, many Gymnamoebae in temperate regions can be excysted after a brief warming of samples in moist conditions. However, samples from tropical regions do not respond in the same way and may prefer a period of cooling or drying prior to excystment. This illustrates that local species or strains may show variability due to adaptations to local climatic conditions. Buitkamp (1979) and Foissner (1987) proposed that local species are adapted to grow best at the highest mean annual soil temperature or at the highest soil temperature during a particular season. That is about 20°C for many species in temperate regions, and about 25–30°C in an African savannah. The fungivorous Grossglockneridae (Colpodea) prefer cooler temperatures of 4–12°C, even in the laboratory. Many cysts require periods of desiccation prior to excystment, and are dispersed by wind or water, such as stalk-forming (sorocarp) species in *Sorogena* (Colpodea in Ciliophora), Mycetozoa and Myxogastria. Some sorocarps disperse spores after drying, while others are dispersed by rehydration. Their activity is therefore seasonal (such as Myxogastria) or more frequent with periodic moisture changes in Mycetozoa. Amoebozoa tend to encyst when water tension falls below about 0.15 MPa, but some Testacealobosea are active in much drier conditions (see Cowling, 1994).

Many species of testate amoebae are restricted to fresh litter or the O_L and O_F layers, and penetrate inside the dead tissues. The dominant surface litter species are often laterally compressed or wedge shaped, such as *Assulina*, *Corythion*, *Euglypha*, *Heleopera*, *Hyalosphenia*, *Nebela* and

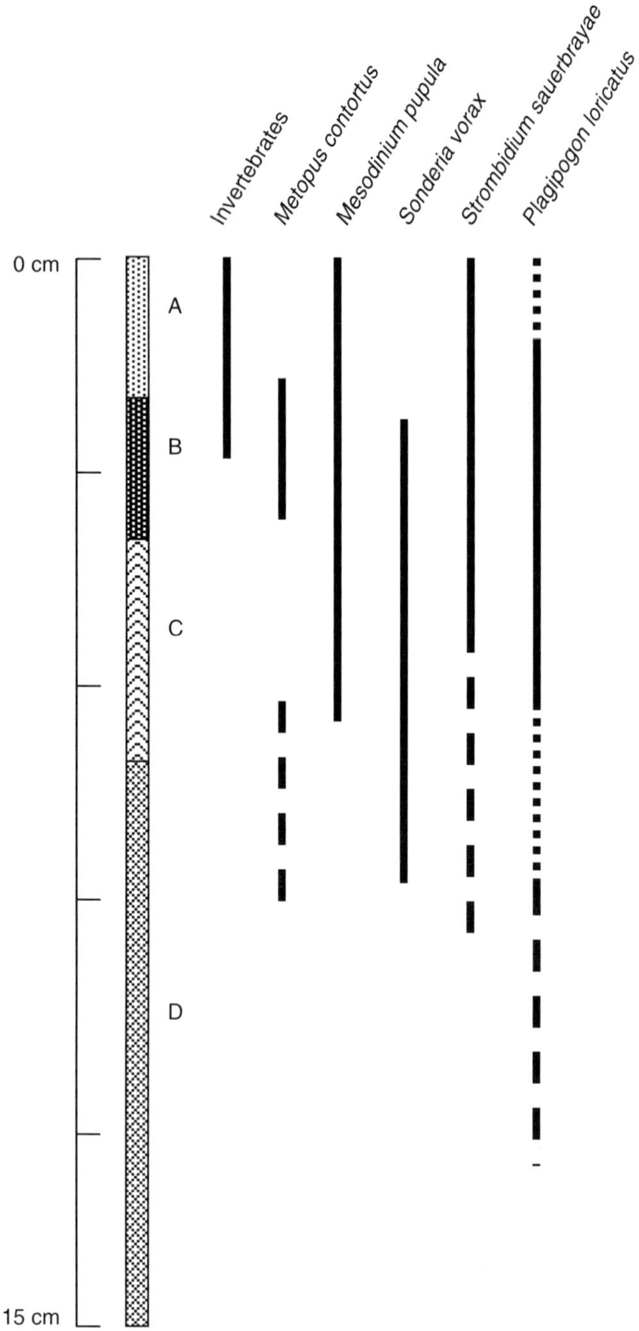

Fig. 5.10. Ciliate distribution with depth in sublittoral sands. Horizons are: A, oxidized yellow sand; B, blackish–greyish sand; C, decomposing *Zostera*; D, blackish sand (data adapted from Fenchel, 1969).

Trinema. Species that occur deeper in the organic horizons tend to have more vacuous test chambers. Depth of occurrence in the profile is limited primarily by pore size and depth of preferred food. Common genera in the soil profile include *Centropyxis*, *Cyclopyxis*, *Difflugia*, *Difflugiella*, *Phryganella*, *Plagiopyxis*, *Pseudodifflugia*, *Pseudowerintzewia*, *Schoenbornia* and *Tracheleuglypha*. A summary of horizon preferences for some species was provided by Chardez (1968) (Fig. 5.11). Their horizontal species patchiness in visually homogeneous *Sphagnum* moss was also described (Mitchell *et al*., 2000). In laboratory cultures, soil testate amoebae usually excyst and become active only after several weeks (or 2–3 months) and remain active for no more than a few days. When suitable growth conditions or food are limiting, Testacealobosea become inactive in the test. Some species form a cyst inside the test protected by a thick membranous secretion. Other species seal the test opening with a hyaline secretion (the epiphragm) or with mineral particles. However, inactive periods without encystment are common and probably suggest suitable conditions in the absence of food resources, like a stationary phase culture. Therefore, the activity of testate amoebae tends to be very seasonal and varies with litter fall. At greater depths, below 2 m, subsoil samples still contain nanoflagellates (see Cowling, 1994). Their abundance is most affected by the texture, and they are probably carried through the profile with gravitational water. They none the less remain amenable to laboratory cultivation.

Patterns in nematodes

Under conditions where food resources are available in sufficient quantity, and other abiotic factors are also favourable, the presence of active nematodes is restricted to thin water films and water-filled pores. They are excluded from dry pores and viscous films, as determined by the soil water matric potential. Nematodes tend to accumulate in moist pockets as the soil dries, providing isolated refuges. Nematodes survive dry periods by anhydrobiosis in the humid air of soil pores even when there is insufficient film water. The air-filled pores favour the locomotion of microarthropods and Collembola, which are usually reduced in numbers in wetter sites. This becomes an opportunity for microarthropod consumers of nematodes. Different nematode species tolerate different soil water matric potentials (Wallace, 1958; Neher *et al*., 1999). This was investigated in a series of experiments with intact soil cores kept under increasingly dry conditions (–3 to –50 kPa) (Neher *et al*., 1999). The cores were sampled four times over 21–58 days when conditions mimicked natural seasonal conditions. The main conclusions of this study can be summarized as follows. For some species (not all), the effect of drying was the same between seasons. Nematode abundance

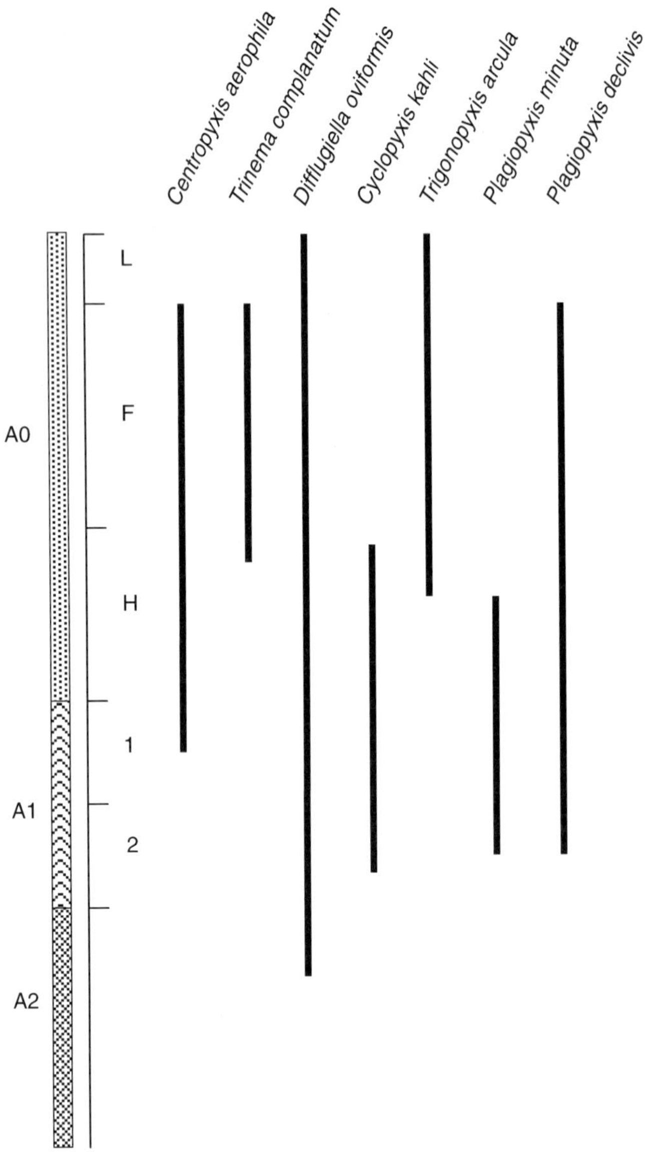

Fig. 5.11. Activity of selected testate amoebae through the profile (simplified from Chardez, 1968).

did not correlate with estimates of the fungal or bacterial biomass, under these microcosm conditions. Predacious and omnivorous nematodes increased in abundance when there were more nematodes in general. This study also observed that under wetter conditions (–3 kPa), the bacterial biomass was greater than under drier conditions. In contrast, fungal species preferred drier conditions. This is not unlike previous conclusions from similar studies with primary saprotrophs (see earlier in this chapter).

The natural heterogeneity of soil with depth and across the horizontal plane creates many distinct microsites for nematodes, which maintain species diversity at the landscape level. Within microsites and at small scales, samples tend to be dominated by very few species. For example, when comparing nematode species composition between samples within one field site, there is usually little overlap between samples (Ettema, 1998). Each soil core contains few dominant species, with several infrequent species. This reinforces the idea that the soil consists of small overlapping (three-dimensional) patches. One effect of patchy species distribution is to separate, spatially, species that would otherwise be competitors. The idea was verified by monitoring the coexistence of five species of the bacterivorous *Chronogaster* in a riparian wetland (Ettema *et al.*, 2000). A map of the occurrence and abundance of each of five species of *Chronogaster* over four seasons at this site was established using statistical methods (spatiotemporal variograms, cross-correlograms and log-normal kriging). This predicted the abundance probability of each species across the landscape, between sampled cores. The map depicts location and abundance changes of each species relative to each other, over the seasons. The results of the analysis emphasize the effect of sample size, and of distance between samples, on the resolution of the data. When the sample contains too many microsites or patches, the population spatial distributions are blurred and may appear homogeneous. However, the study identifies patches of species that are more aggregated. The data suggest that patch dynamics concepts can be applied to explain the ecology of this genus at this site. However, the authors remark that it is not clear which aspect of the species biology is responsible for the spatial segregation, since the biology appears to be very similar between these species. Differences in prey species preferences or slight variations in life history can be implicated, as well as purely stochastic explanations due to unpredictable microsite disturbances.

The parameters that affect soil nematode distribution locally are numerous, and include soil bulk density and mean pore diameter, as well as moisture, temperature, soil air CO_2 concentration, preferred prey species abundance, chemical irritants or inhibitors (see Chapter 1 and Ettema, 1998). Predation and parasitism will also impact population dynamics. The effect of these parameters can be easily demonstrated experimentally in the laboratory and in the field. For example, the entomophagic *Steinernema riobrave* (Steinernematidae) was shown to occur and chose soil depths that reflected its moisture preferences (Gouge *et al.*, 2000). In the field, individuals are encountered at depths which remain within a preferred soil moisture range over the seasons.

In another field experiment, Bakonyi and Nagy (2000) manipulated temperature and moisture content on a series of 2 × 2 m plots. The experiment was only conducted over 4 months, but the authors provide several conclusions based on this initial trend. Significant changes in nematode species composition between plots were detected near the end of the experiment. It is difficult based on short field studies to distinguish between the magnitude of the experimental effect and of the underlying seasonal effect. However, it is clear that over this time period, there were shifts in species composition that were due to moisture and temperature changes at the manipulated plots, compared with control plots. One such shift in composition occurred in the ratio of bacterivorous to fungivorous taxa. The abundance of fungivores was reduced in wetter plots, showing a preference for drier sites. It is unclear whether the fungal biomass between wet and drier plots was mostly active or inactive. That would affect grazing preferences (see Sohlenius and Wasilewska, 1984). The activity of bacterial and fungal species varies greatly with these changes (see earlier in this chapter). Other factors such as dissolved O_2 content of the soil solution at wetter sites and changes in predation on the nematodes were not considered. The authors attempted to explain the results based simply on the nematode genera, not considering the effect of the manipulations on other species in the food web. Imposing an abrupt change in soil moisture and temperature inevitably caused a decrease in nematode abundance in all manipulated plots. It would also affect species in other phyla that are competitors with or predators on nematodes. It would be interesting to see the results of follow-up studies from this experiment after three annual cycles.

These factors are important and must be considered along with the biology of other coexisting species. During normal seasonal fluctuations, the abundance of all active species in the soil is affected by these factors. There are changes due to gradual changes in abiotic conditions (such as pH, temperature and moisture), as well as litter substrate species and chemistry, and the abundance of new competitors and predators. For example, in a description of seasonal changes in the composition of species and edaphic parameters, Hanel (2000) observed that the biological factors (food and species composition) influenced the nematode populations more than abiotic parameters. Neither soil temperature nor moisture were the main drivers of the changes in species composition directly. The seasonal fluctuations in species numbers correlated well with other coexisting taxa of enchytraeids, rotifers or tardigrades, which were competitors or predators of nematodes (Fig. 5.12). Correlations were found between predacious nematodes and tardigrades ($r = 0.673$) which are both competitors for small invertebrates, fungivorous nematodes and enchytraeids ($r = 0.830$), omnivorous nematodes and enchytraeids ($r = 0.919$), and bacterivorous nematodes

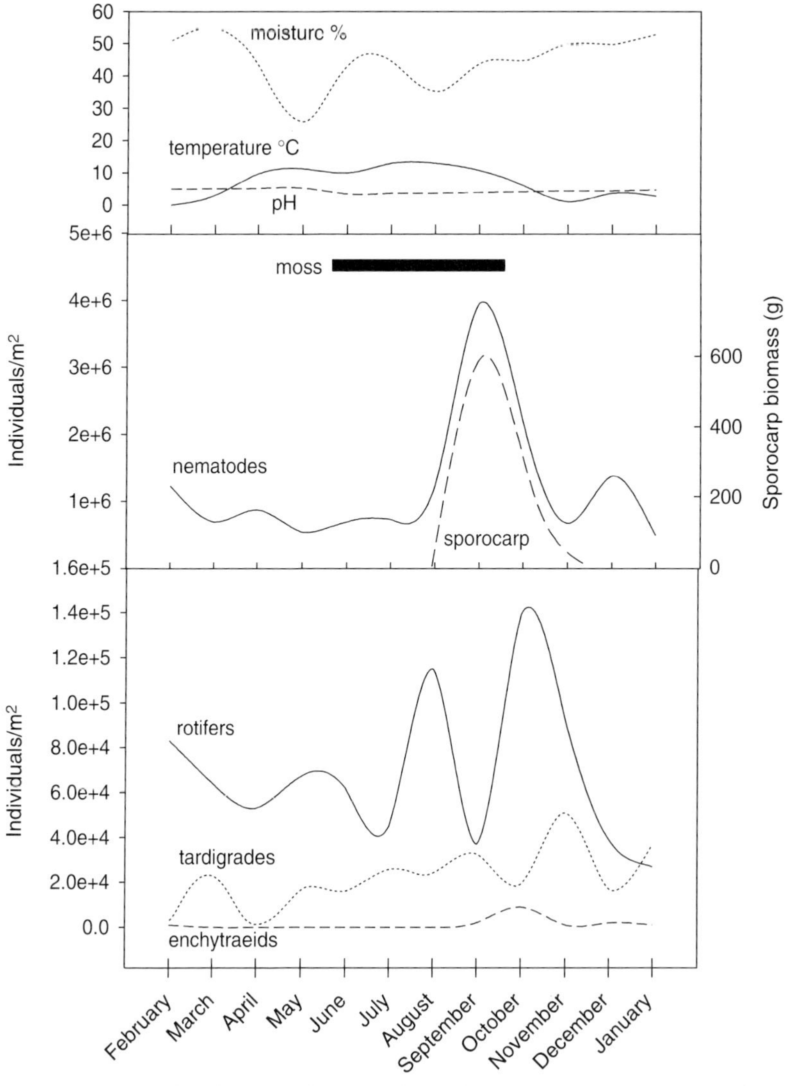

Fig. 5.12. Seasonal distribution of nematodes and co-occurring wet-extracted species. Adapted from Hanel (2000).

and rotifers ($r = 0.654$). It is not clear what proportion of rotifers were bacterivorous or predacious. A large fraction of nematodes during sporocarp formation were *Filenchus* species associated with mycorrhizae. It is very clear from this more integrative approach that population dynamics through seasons are affected by the entire food web community, across taxonomic boundaries.

Over longer time periods, in field sites at different successional stages of the plant assemblage, certain trends can be identified in nematode functional group and species abundance. In one study carried out in the Fontainebleau forest near Paris, there were trends with depth between successional stages and nematode functional groups (Armendariz and Arpin, 1997). The study site was a beech forest that had been free of management for about 300 years. Four types of areas were distinguished in this forest: (i) growing forest; (ii) mature forest; (iii) senescence phase; and (iv) clearings due to wind-throw from a 1990 storm. Each of these areas within the forest could be distinguished by the depth of the horizons in the profile (Fig. 5.13). Overall, the mean percentage of nematodes at each depth was 76% in L–F litter layers, 17.6% in the next 0–5 cm, and 6.3% in the 5–10 cm depths. However, a particular depth did not match the same organic horizon exactly between the different forest age areas. In the 5–10 cm soil, nematode numbers were higher in the mature forest areas because of the higher organic matter content. The abundance of nematodes per gram of soil was relatively steady between these successional stages, but the abundance of different functional groups differed (Fig. 5.14). It is interesting to note the extent of change in the soil profile of this old forest after the clearings were created in 1990. The storm was followed by extensive removal of damaged trees and woody litter. This caused an increase in

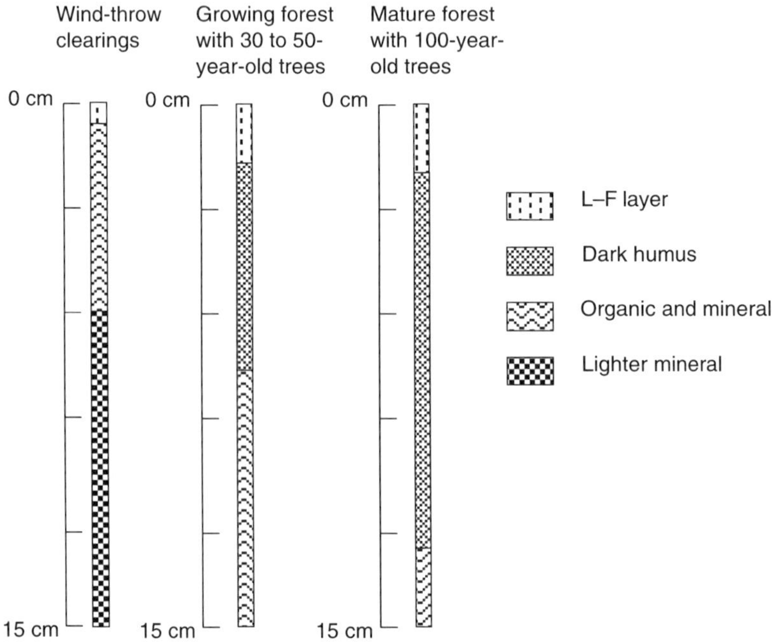

Fig. 5.13. Variations in the soil profile at sampled sites in a forest nematode study (data from Armendariz and Arpin, 1997).

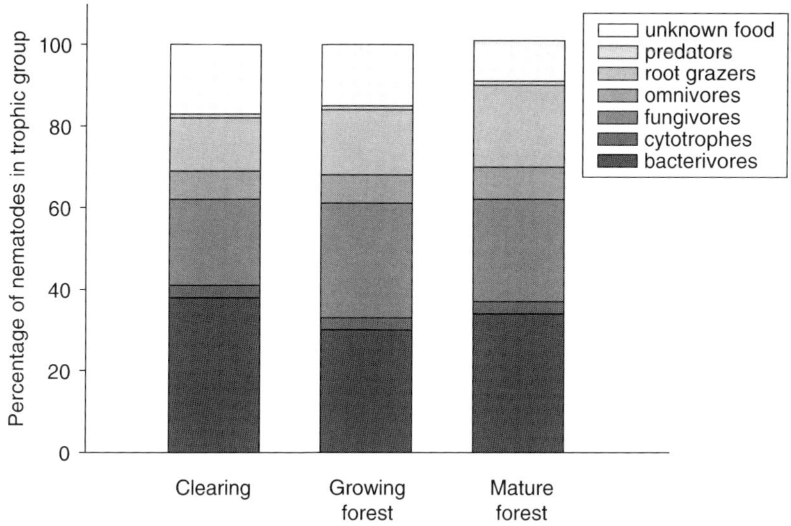

Fig. 5.14. Relative percentage of nematode functional groups at three types of sites in a forest nematode study (data from Armendariz and Arpin, 1997).

the abundance of bacterivorous and predacious nematodes at first. Then root feeders increased in abundance as many roots were dying.

The effect of plant succession stages on the soil community can be described without waiting for the succession to occur. Fields at different plant succession stages, or under different soil management regimes, can be selected within a similar geographical region and compared. For example, one study was carried out in South Bohemia, with a humid and mild temperate climate (Hanel, 1995). Samples were obtained over 2 successive years from sites representing a successional series (Table 5.2). These were an agricultural field under cultivation, a field that was recently left fallow, a meadow and an oak forest. The list of nematode species and frequency of occurrence were described. This study indicates that seasonal fluctuations in the nematode composition were mostly driven by fluctuations in weather patterns. Litter composition and temperature were secondary to more immediate weather changes. Interestingly, both the forest and meadow were very similar in nematode biomass and species abundance over 2 successive years. Forest soils are more difficult to sample for total species composition. The heterogeneity of the soil is such that to find rarer species may require a large sampling effort through several seasons. Hypothetically, the nematode species composition could have been underestimated. However, there is not a general agreement between studies with regard to nematode species number and stage in plant succession. Sometimes earlier stages, such as young forests or meadows, have more species.

Table 5.2. Nematode abundance from sites at different secondary succession stages of the plant community. Data from 2 consecutive years are shown.

Field site	Cultivated field	Fallow field	Meadow	Oak forest
Age	—	1–2 years	9–10 years	60–70 years
No. of species (successive years)	52, 65	53, 60	94, 86	92, 75
Abundance ($\times 10^6$) (per m^2 to 10 cm depth)	0.54, 0.73	0.60, 1.00	1.31, 1.13	1.03, 1.52
Biomass (g/m^2 to 10 cm depth)	0.22, 0.46	0.29, 1.18	1.57, 1.55	1.57, 1.62
Bacterivores	55%	56%	35%	56%
Fungivores	24%	18%	16%	23%
Omnivores	10%	13%	32%	18%
Root feeders	9%	11%	10%	<1%
Predacious	2%	2%	7%	3%

Data from Hanel (1995).

A second example considered the nematode assemblage in pine stands on old mining dumps in Germany, near Cottbus (Hanel, 2001). The pine stands are in a geographically small region with similar climate and a soil derived from remediated coal mining sands. The field sites had been recovering after remediation for 1–32 years. They were compared with a semi-natural 40-year-old pine stand that was not affected by mining, also on sandy soil (Table 5.3). The results reveal some variation in soil carbon and phosphorus content and availability between the sites, even though they are similar in origin. Soil pH varied between 2.70 and 5.02 at these sites. The results suggest that variations in both pH and soil humus content between sites were important variables in determining the nematode functional group composition, and species composition. Overall, 47 genera were identified, with only six commonly occurring genera, from 10 cm^2 cores sampled to 10 cm depth. Correlation coefficients and principal component analysis could distinguish between older and younger sites based on nematode species composition. Enchytraeids appear only in later succession stages at these sites. Analysis of nematode species composition indicates that there is succession (species replacement) with site maturation. However, the number of genera per site increases quickly and does not vary to the same extent. Both ‘r’ and ‘k’ strategists tend to coexist, even in the 40-year-old forest site, reflecting the persistence of opportunity for ‘r’ strategists in the soil environment. Several genera tended to be more common in the older stands (*Metateratocephalus*, *Cervidellus*, *Drilocephalobus*, *Eumonhystera*, *Teratocephalus*, *Alaimus*, *Geomonhystera*, *Eucephalobus*, *Tylolaimorphus*, *Deladenus*, *Thonus* and *Aporcelaimellus*). There was a clear tendency to increased size of individuals with soil

Table 5.3. Site recovery from mining over a successional chronosequence. Summary of results from wet-extracted invertebrates, indicating nematode functional groups.

Site	*P. sylvestris*	*P. nigra*	*P. sylvestris*				
Age (years)	1	14	17	18	32	32	40
P_{total} mg/kg	437	273	155	169	598	358	193
$P_{(aq)}$ mg/kg	2	3	5	4	4	5	4.5
$\%C_{organic}$	1.46	3.61	1.92	5.86	19.16	12.12	2.85
Enchytraeidae ($\times 10^3/m^2$)	0	3	6	7	18	21	21
Rotifera ($\times 10^3/m^2$)	22	31	22	27	23	36	114
Tardigrada ($\times 10^3/m^2$)	6	12	32	39	44	37	99
Nematoda ($\times 10^3/m^2$)	574	499	629	502	498	594	1303
Mean μg of individual nematode	0.15	0.33	0.41	0.41	0.60	0.98	0.23
Nematoda (genera)	13	24	22	23	22	27	32
Bacterivores	64.5%	54.9%	65.3%	61.8%	45.6%	41.8%	43.6%
Fungivores	34.8%	27.3%	25.8%	25.5%	23.3%	9.8%	17.4%
Hyphae/root hairs	<0.2%	11.4%	0%	0.6%	13.1%	27.8%	25.0%
Plant root	<0.2%	8.0%	<0.2%	0	0.6%	<0.2%	0
Omnivores	0.35%	4.6%	6.0%	8.6%	12.2%	12.8%	12.3%
Predacious	<0.2%	0.6%	0.3%	0.6%	5.4%	7.6%	<0.2%
Insect pathogen	0	0.4%	2.4%	3.2%	0	0	1.5%

From Hanel (2000), percentages were rounded to one decimal place.

organic carbon content. This probably reflected better nutrition, or uncharacterized site properties (such as soil porosity). As in other studies, the frequency of plant root feeding species is very low, except in disturbed (cultivated) or newly restored sites. As the soil profile develops with site successional age, the number of genera in different functional categories increases. This reflects an overall increase in species (and functional) diversity throughout the soil food webs.

Patterns in the distribution of soil mites

The dominant factors contributing to the distribution of Oribatida, the predominant soil mites, across successional time and spatial stratification were reviewed recently (Maraum, 2000). This review identifies species fecundity, longevity, sexual and asexual reproduction, microhabitat preferences and food preferences as essential to understanding species dynamics. Furthermore, the study points out that control of abiotic and biotic parameters over the life history of the species, availability of resources through competition and predation effects, as well as parasitism on the mites or their symbionts impact the reproduction potential and species dynamics. Studies that fail to account for these factors provide an overly simplistic analysis that may lead to contradictory results when compared with other studies. In reviewing 20 selected data sets, representing deciduous and coniferous forests, pastures, meadows, agricultural fields and forests with mor, moder (lower in earthworms) and mull (higher in earthworms) forest soils, the following generalities were proposed (Maraum, 2000).

1. The total abundance of individuals per area was lowest in human impacted soil (agricultural fields and pastures) and highest in forests, especially in old growth forest with moder soil. The abundance is also influenced by the depth of the organic horizon, so that soils with thicker organic horizons have more mites because the habitat is deeper.
2. Some oribatid taxa are associated with particular habitats or successional stages. For example, the Enarthronota are very small species found in deeper soil that accumulate in late succession stages. Females give rise to daughters asexually and they lay very few eggs which develop slowly. In contrast, the Poronata occur in disturbed soil and early in succession as they are better colonizers with high fecundity and fast development.
3. The transformation of moder to mull soils by earthworms is accompanied by changes in pH and a reduction of oribatid abundance. This observation is consistent with other studies, though the interaction between earthworms and mites is more complex (McLean and Parkinson, 2000).
4. The effect of soil humidity and temperature was secondary to SOM quality (humus type) across the temperate sites compared.
5. In general, longer term studies with observation over several years are required to obtain a clearer view of the response of oribatids to experimental manipulation. That is because they are slow at colonization with relatively low reproductive rates, and several years of interaction are required to discern population and species changes.

The distribution of microarthropods locally, over several metres, is affected by the type of litter and SOM accumulated. Forest floor with mixed litter consistently harbours more species than areas with single litter species. This was verified with 1 m^2 plots of mixed and single species litter at the Coweta forest research site (Hansen, 2000). The reduction in species diversity in single litter plots was cumulative over several years and led to a simplification of the microarthropod community. Species abundance and reproductive dynamics in these simplified communities were driven increasingly by boom–bust dynamics over successive years. In mixed litter plots, with decomposition rates being out of phase between leaf litter species, more continuous and steady activity of a more complex community was maintained. The feeding habits and preferences of microarthropods are somewhat opportunist and variable with seasons, as the litter quality and microhabitat change (see Norton, 1985). In mixed litter, sufficient microhabitat diversity remains to support a diverse community through seasons. That gradually disappears in single species forest stands over successive years, or in agricultural monoculture.

In agreement with this conclusion, a comparison of single versus mixed leaf litter in an oak–pine stand over 2 years (Kaneko and Salamanca, 1999) reported a greater number of microarthropods with higher species richness in mixed litter combinations than in single litter. The mixed litter with the more complex microarthropod community also supported faster decomposition rates, and this trend persisted through the second year (Table 5.4). Moreover, these types of studies also demonstrate strong plant litter-specific effects on the microarthropod community. For example, in the Kaneko and Salamanca study (1999), the slow decomposing *Sasa veitchii* supported a greater number of microarthropods in single litter trials, and the oak–*Sasa* combination could support twice as many microarthropods as the oak–pine combination (Table 5.4). Similarly, Hansen (1999) observed that single and mixed litter bags each supported a different microarthropod community, and the red oak litter accumulated endophagous oribatids which accelerated its decomposition. These endophagous mites and other fauna which burrow into detritus (leaf litter and coarse woody debris) are poorly extracted by standard Tullgren extraction (Hanson, 1999).

Another factor affecting microarthropod distribution over several metres, besides the vegetation variability, is heterogeneity of the soil itself. Along with variations in the distribution of organisms from one plant litter species to another and the litter at the base of these plants,

Table 5.4. Effect of mixed litter species on microarthropod species richness during primary decomposition.

	Quercus serrata (Q.s.) oak	*Pinus densiflora* (P.d.) red pine		*Sasa veitchii* (S.v.) shrub
N	0.81	0.37	0.67	
C:N	61.4	147.66	55.7	
Lignin	36.4	29.9	8.6	
Decomposition	Fast	Medium	Slow (20% silica)	
Microarthropods after 1 year	99.8 ± 27.7	107.0 ± 27.8	157.3 ± 11.7	
Mixed litter	Q.s. + P.d.	P.d. + S.v.	Q.s. + S.v.	Q.s. + P.d. + S.v.
Microarthropods after 1 year	119.8 ± 47.3	182.8 ± 42.6	247.8 ± 16.9	193.8 ± 23.5

Data from Kaneko and Salamanca (1999).

the soil bulk density varies somewhat over several centimetres to metres. These changes together were responsible for the varying abundance of tardigrades, rotifers and nematodes in a short grass prairie study (Anderson *et al.*, 1984). There may also be variation in extraction efficiency with changes in the soil compaction and organic matter content. This would provide false variations due to methodology.

The vertical distribution of microarthropods varies with changes in soil temperature and moisture content (Whitford *et al.*, 1981). These migrations occur seasonally or with abrupt changes in conditions. This behavioural response is exploited in microarthropod extraction by Tullgren methods (Chapter 3). In desert environments, dramatic changes in soil moisture and temperature occur daily. These changes were followed over 1 day in the Negev desert and correlated to the microarthropod migrations through the soil profile (Steinberger and Wallwork, 1985). The main input of soil moisture is through dew in the early morning, with increasing temperature and decreasing moisture through the day and into the evening. The main taxa implicated in these migrations were small mites and Prostigmata (Cheylitidae, Nanorchestidae, Stigmocidae and Tydeidae). The mites migrate deeper with an increase in surface heat, and migrate up with cooling surface temperatures and an increase in moisture. In most soils, mite species have depth preferences, and overall distribution of mites varies with the habitat. For instance, in an oak stand, a normal distribution of individuals with depth is common, with a peak at 2–5 cm (the organic horizons) and decreasing to 30 cm and below. In a prairie soil, one can expect an exponential decrease in abundance from the surface to 30 cm. This pattern reflects the lack of surface litter accumulation in grasslands. Smaller species are often assumed to burrow deeper into smaller pore spaces, increasing their range. In general, the association of soil organisms, microarthropods included, with depth in the profile is best described according to the type of horizon. The depth of each horizon in the profile provides a particular habitat for certain species. It is this profile preference and the chemistry of the habitat in that horizon that select species composition. This point is particularly important when comparing species distribution between sites, where both bulk density and horizon thickness vary (Scheu and Schultz, 1996). These authors compared the depth preferences of soil microarthropods at several sites differing in the plant assemblage. The sites were a field that had been in cultivation for 20 years; one that had been left to colonization by weeds and grasses for 4 years; another that was left fallow for 11 years and was under grass; another that was a 50-year-old ash forest with small shrubs; and lastly a 130-year-old beech wood with a dense understorey (Table 5.5). The cultivated field had the least species diversity. Through this plant succession series, there was little difference in the isopod composition between sites that were not in cultivation. The diplopods preferred

Table 5.5. Number of species found at each site along a plant successional series.

Field site (years in succession)	Cultivated (0 years)	Grasses (4 years)	Grasses (11 years)	Ash forest (50 years)	Beech wood (130 years)
Plant species	12	30	27	15	22
Diplopod species	2	7	10	3	5
Isopod species	1	4	4	2	3
Oribatid species	4	2	8	9	25
Earthworm species	2	7	5	5	8

Data from Scheu and Schultz (1996).

the sites under grasses far more than treed forests. The authors are uncertain if they obtained the deeper earthworm species, but noted that species composition was similar between sites for those found. However, based on other studies with earthworm species, it is expected that abundances increase and community structure changes with later successional stages (Pizl, 1999). The oribatid species occur in the surface organic horizons in earlier succession stages. As the profile develops and organic matter increases deeper in the profile with successional stages, more mites occur in the mineral horizons. In this study, most mites in the beech wood forest occurred in the upper mineral soil. The species composition seems to evolve with profile development and organic matter composition over successional time.

Over several weeks to months, it is possible to follow a succession of mite species in litter, as the litter is increasingly decomposed and seasons change. In an oak stand in Brittany (France), Bellido and Deleporte (1994) described the appearance of the oribatid *Ceratoppia bipilis* in fresh unfragmented leaves, along with the bacterial and fungal colonizers. This was followed by the appearance of *Hermannia gibba*, then *Opia ornata* in the more decomposed litter. However, these successions through litter can be more complicated. For example, *Ceratozetes cisalpinus* and *Adiristes ovatus* must ingest hyphae first, before burrowing into microdetritus or pine needles, presumably to benefit from the fungal enzymes (Woodring and Cook, 1962a,b; Mignolet, 1971; Lions and Gourbière, 1988).

Over longer periods, annual variations in populations and species activity occur as seasons change (Stamou and Sgardalis, 1989). Seasonal changes of the oribatid mite populations vary with species and depend on both the climate changes and the particular adaptations of each species. The climatic and temperature changes regulate life histories and reproductive activity. Therefore, the timing of reproduction, egg-lay and development rate, as well as survival through seasonal extremes, depend on the weather, and will vary somewhat from year to year. Thus, sampling time and sampling frequency are experimental variables that affect the species sampled and their abundance. One study of soil mite composition in forest litter, conducted over 3 years, describes how different species varied in abundance over the sampling period (Metz and Farrier, 1967). Samples were collected monthly from litter bags through primary decomposition stages. The study showed that despite the diversity of the soil microarthropods in the forest stands, there were, at any time, few dominant species. The proportion of males, females with eggs and females without eggs also varied through the seasons. The authors note that for some species, for example the three *Vegaia* species found, the depth profile preference was different between the adults and juvenile stages. The juveniles occurred in the more decomposed leaf litter.

Species occurrence and abundance also vary with the successional stage of the plant assemblage, as they translate into litter species diversity and SOM changes. The increase in plant species diversity with succession usually correlates with an increased microarthropod species diversity (Travé *et al.*, 1996; Petrov, 1997; Maraum, 2000). For instance, in a selection of Boreal forests in Quebec (Canada), there was an increase in micro- and macroarthropod abundance and species (Aranea were excluded from the study) with increasing forest age (Paquin and Coderre, 1997). Forests were close together and selected to represent a range of successional recovery from the previous fire event. These were a deciduous forest (age 47 years), a mixed forest (age 144 years) and a coniferous forest (age 231 years), with dominant tree composition varying in succession from *Populus tremuloides* (<100 years); *Abies balsamea*, *Betula papyrifera*, *Picea glauca* and *Populus tremuloides* (100–200 years); to *Abies balsamea* and *Thuya occidentalis* (>200 years). These successional stages were reflected in the understorey vegetation and through the soil profile, with a reduced role for arthropods in decomposition in the mature coniferous forest soil. The change in litter species and SOM chemistry modifies the composition of the soil community and affects the development of the soil profile and soil type.

Studies of microarthropod species diversity and abundance along altitude gradients, or over several hundred kilometres, generally do not reveal consistent trends. In one study at the Coweta research forest in North Carolina, five sites were selected along an elevation gradient, and monitored over 1 year (Lamoncha and Crossley, 1998). Several interesting results confirm the lack of simple trends between sites and between studies. Overall, the study collected 74.8% Acari (68% were Oribatida), 18.2% Collembola and 7% other microarthropods. The relative abundances were steady but lowest over the winter months (January–May), then increased until November frosts when the abundances declined again. With elevation, there was no observed trend in abundances, and no incremental change with altitude. However, there were overall more species at higher elevations, and two sites with lower abundances had greater species diversity. There were plant litter-specific effects, such as stands with mixed oak which had more species, and mature or near mature stands having fewer species. The authors suggest that with forest successional age, microarthropod species richness may not increase but at least species evenness increases (with fewer dominant species) and population sizes are less variable. Because of the lack of simple trends over these distances, and between study sites, one needs to place more emphasis on heterogeneity between smaller patches and environmental variables (which are not independent). For example, the single south-facing slope had the most divergent species assemblage compared with the other sites with north-east to north-west aspects. There were dominant species found at all sites, but many dominant species were site specific. It seems that microarthropod species were determined more by the local site-specific plant community (through litter species and litter chemistry), soil microsite parameters and site microclimate (which controls life histories), rather than elevation.

Similarly, in a comparison of sites along a 600 km transect in Alaska from Fairbanks (near the Arctic circle) to Prudho Bay on the Arctic Ocean, there were significant differences between sites, with no simple trend or pattern (Thomas and MacLean, 1988). Microarthropod communities, abundances and species richness were determined primarily by local site soil and climate characteristics. The authors note that with increasing latitude (going north), there was increased site heterogeneity due to increasingly severe climate, and an overall decrease in species richness. The discussion of the data is supported by a valuable and honest critique of problems associated with the sampling, enumeration and data analysis for species richness.

The biogeography of species distribution across continents reveals certain trends. There is a consistency between data sets from different regions for abundance distribution of soil oribatids. In a large data set representing 25 sites in different continents, the oribatid assemblage was very different between ecophysiological regions (Osler and Beattie, 1999). Within each ecophysiological region, species were more similar. Each site and habitat tends to be dominated by a small number of abundant species, with a large number of less abundant species. Of the thousands of species of oribatids, in about 150 families, relatively few are ubiquitous with broad geographical distribution. However, habitat preference is reflected at the family level, probably through life history characteristics and adaptive preferences. For example, habitats with trees across the world are suitable for the same families and superfamilies of oribatids. The limited geographic range of species of Oribatida is usually attributed to their poor dispersal capacity. They are generally poor colonizers that rely on accidental transport by animals, rainwater, wind or anthropogenic causes (Travé *et al.*, 1996). These authors suggest the dispersal of many species is probably limited to several metres rather than kilometres, as they are more likely to desiccate in wind and to not survive great distances. Siepel (1986) tested the hypothesis that oribatid species distribution was limited by extremes in the biotic and abiotic environment, and was restricted to regions with stable abiotic parameters within the tolerance range of each species. The experiments were conducted in the laboratory, with 23 species exposed to a variety of extreme conditions. These included food shortages, low humidity (70% relative humidity), frost (–5°C, >90% humidity, 48 h) and heat (35°C, >90% humidity, 48 h). There was a species-specific response to survival and tolerance to these environmental conditions, and three main conclusions. (i) Most species were tolerant of frost or heat, but few to both. This distinction probably reflects seasonal activity as a summer or winter species. (ii) If a species was tolerant to either heat or frost, it was also tolerant of the desiccation treatment. (iii) If a species had a narrow temperature range, it was also susceptible to death by desiccation. These results suggest that community changes associated with succession, vegetation removal, extended drought and fire involve the reduction in abundance of susceptible species. They become replaced by more tolerant and competitive species which may already be present in low numbers, as the environmental extremes vary over the years. Unfortunately, this study was not followed with an examination of long-term consequences of stress on egg-lay, juvenile survival and fertility. Obtaining these values would help predict the dynamics of adjustment after infrequent extreme events.

Collembola patterns in soil

As for other microarthropods in the forest litter, Collembola tend to be found in patches of increased abundance. The patchiness often reflects areas of higher humidity, especially in dry conditions, but can indicate sources of food that attract particular species. Therefore, the dimensions of these patches may vary, according to the size of the area covered by twigs, bark and other litter that may provide a suitable shelter. Periodically, Collembola will aggregate for mating, attracted by pheromones. This pheromone-led aggregation occurs at different times seasonally for different species, or at times of high population density (see Hopkins, 1997).

Annual fluctuations in population numbers occur according to the duration of suitable conditions for feeding and reproduction. Most species develop from egg to adults in weeks or months. Therefore, if suitable conditions, in terms of food availability, sufficient moisture and adequate temperatures, extend long enough and often enough through the year, several life cycles can be possible. On average, most species probably complete only 1–2 life cycles annually. Survival through unsuitable conditions depends in large part on finding suitable refuge, and surviving through the increased risk of predation caused by aggregation in refuges.

The three-dimensional distribution in soil is affected by temperature and moisture, but also food resources. At short distances, the distribution of Collembola and other microarthropods can be affected by odours from food resources. This results in chemotaxis which can lead to aggregation. Moreover, the same odour can be used as cues to predators of Collembola, which will converge towards the source of the odour to prey on Collembola. For example, the predatory mite *Hyposaspis aculeifer* is attracted to the odour of fungal hyphae grazed by its collembolan prey, also attracted to the odour (Hall and Hedlund, 1999). Pseudo-scorpions are attracted to Collembola pheromones in order to prey on them as they aggregate (Schlegel and Bauer, 1994).

Vertical stratification of species occurs and reflects the adaptations of species for various soil horizons (Hagvar, 1983; Hopkins, 1997) (Fig. 5.15). Most species are not actively tunnelling through soil, and are restricted in part by their dimensions, but some smaller species, such as the

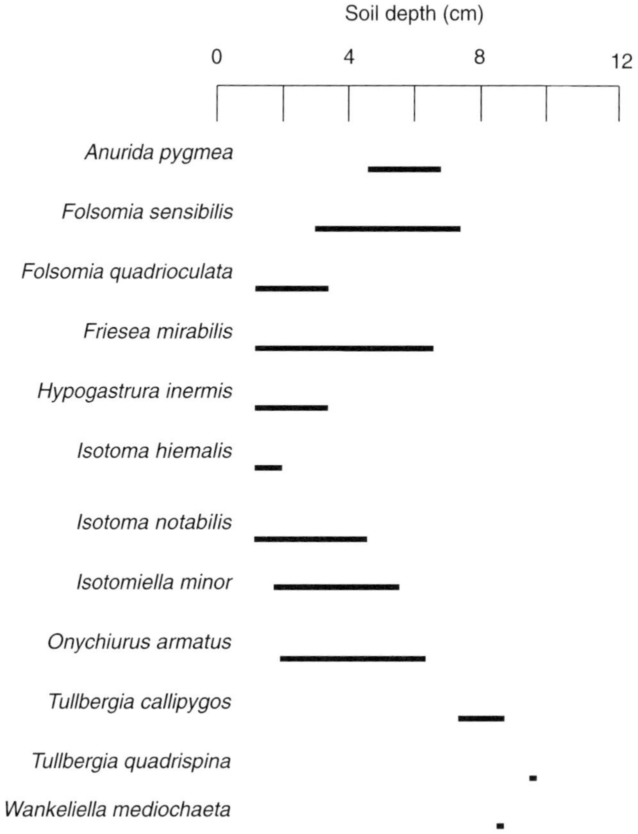

Fig. 5.15. Average range of depths for selected species of Collembola, in a Norwegian coniferous forest soil (data from Hagvar, 1983).

Tullbergiinae (Onychiuridae), do burrow into the soil. It is usually helpful to note which soil horizon, at which depth, sampled organisms are found. Smaller species found deeper in the soil tend to be missing eyes, a furca and pigmentation, but can be also found in the litter. Absence of these traits does not, however, indicate a truly interstitial species (Rusek, 1998). Taking into account profile preferences, life histories and food preferences, it becomes possible to distinguish between niches for coexisting species. Many species exploit microtunnels along roots, invertebrate and earthworm burrows to reach deeper soil. This may be for protection from desiccation or cold, or in search of food resources. There are variations through seasons in the depth preferences of Collembola, which reflect changes in the temperature, moisture and food resources (Poole, 1961; Hopkins, 1997). Similarly, there are diurnal patterns in distribution of Collembola, which affect numbers sampled depending on time of day (Frampton *et al.*, 2001). These daily migrations are not restricted to polar or arid regions, but occur in agricultural fields, trees and rooftops, and emphasize the sensitivity of Collembola to changes in moisture and temperature.

In an important study, Ponge (2000) described both depth in the profile at which various Collembola species were found, and gut content analysis of individuals sampled. This study describes a correspondence between species and depth in profile, as well as species with food preferences. Gut content analysis is very useful, but restricted to items that have cell walls, such as fungal and plant cells, or invertebrate cuticle, which remain visible in the digestive tract. Similarly, pollen, yeasts, cysts, spores, testate amoebae, diatoms and other protists with cell walls can be recognized. However, similar sized protozoa (without cell walls) which would be ingested along with the food items will not be seen as they are lysed and digested without trace.

Succession of Collembola species over long time periods, corresponding to plant succession stages, is known to occur. There are early colonizing species, as well as species restricted to intermediate stages, and some adapted to the more stable climax communities of forests (Deharveng, 1996; Lek-Ang *et al.*, 1999). Microarthropods (such as Collembola and mites), enchytraeids and other soil arthropods are also directly affected by local changes in the litter cover and corresponding microclimate. Woody debris and surface litter besides leaves are usually neglected by soil biologists. The following study tested the hypothesis that coarse woody debris (>10 cm diameter and >1 m length) provides a significant habitat for microorganisms (Marra and Edmonds, 1998). Samples were obtained from a variety of sites at three seasons over 1 year and pooled to a single statistical sample. The analysis accounted for distance from woody debris, depth in profile (0–5 or 5–10 cm), stage of decomposition and differences between clear-cut and late succession forest. The Acari and Collembola accounted for 93–97% of individuals sampled, and <2% were Coleoptera, with the total number of species distributed as follows: Acari 43%, Collembola 8% and Coleoptera 24%. There were fewer species of arthropods in the clear-cut samples, and site moisture was an important parameter. The Collembola were particularly affected, with 79% reduction in abundance in wet sites. The effect of moisture was more significant than whether the site was clear-cut or forested. Overall, the microarthropods were clearly more abundant at 0–5 cm depth than at 5–10 cm, and there was little effect of coarse woody debris at this resolution. These conclusions are consistent with other studies. The study did not consider aggregation patterns inside or under the woody debris with site drying and wetting. In an experimental study, the effect of removing or adding logging residue in a *Pinus sylvestris* forest in central Sweden was described, after 15–18 years (Bengtsson *et al.*, 1997). Although there was no significant effect of residue removal on populations of enchytraeids and Diplopoda, those of Collembola, predatory gamasid mites, spiders, dipterous larvae and predatory insects were reduced. The authors note that the relative abundance of species did not change, but the overall abundance did. This suggests a resource reduction that did not greatly modify community structure.

With respect to global biogeography, it is difficult to make predictions on variations in species number or biodiversity between ecosystems. This is largely due to enumeration of species into functional groups or families, without noting how many species. The task is complicated by the predominance of local endemic species and undescribed species (for example, see Deharveng and Lek, 1995). In temperate forests, species numbers tend to be between 100 and 200, whereas they are lower in grasslands (~30–50 species) and even lower in polar or alpine regions, which may have as few as 1–5 species (Rusek, 1998). However, most field sites are dominated by relatively few common species, and a greater number of rare and local species.

Other insects

Succession of insects occurs in decomposition of animal material. Different families of insects tend to colonize cadavers at preferred stages of decomposition. Some occur in fresh corpses, others in far more decayed matter (Table 5.6). Due to the importance of this sequence in forensic science, succession on mammal and human remains is better known (Sperling *et al.*, 1994; Byrd and Castner, 2001). For example, Cleridae (checkered beetles), Dermestidae (dermestid beetles), Nitidilidae (sap beetles) and Scarabaeidae (lamellicorn beetles) tend to dominate more decomposed and desiccated animal remains. They decrease in abundance as the cadaver desiccates. Once the animal remains are too desiccated and are fragmented, the edaphic saprotrophs and secondary saprotrophs continue the decomposition.

Table 5.6. Insect (adult) succession in animal primary decomposition stages.

Insect abundance in animal decomposition	Fresh	Bloated	Decomposing	Desiccated
Calliphoridae	Low	High	Medium	Very low
Muscidae	Low	High	Medium	Very low
Silphidae	Low	High	Medium	Very low
Sarcophagidae	Very low	High	Medium	Very low
Histeridae	Low	High	Medium	Very low
Staphylinidae	Very low	High	Medium	Very low
Nitidulidae	None	Low	High	High
Cleridae	None	Very low	Medium	High
Dermestidae	None	Low	Medium	High
Scarabaeidae	None	None	Medium	High

Note that earlier stages are relatively brief, and that the extent of decomposition is determined on qualitative properties. Larval stages tend to appear after the adults have laid eggs.
Adapted from Byrd and Castner (2001).

Similarly, animal dung provides a nutritious habitat for some insects. One study demonstrated the extent of niche differentiation between coprophagous dung beetles, on an animal safari reserve in Ireland, where a variety of dung was naturally available (Gittings and Giller, 1998). Animal dung patches were distributed in the park to trap surface invertebrates, and the dung beetles were described. The study followed the deposited dung patches over 25 days as each fresh pat desiccated. Over this time interval, there was little change in the organic matter content, but dung pats desiccated from >80% moisture to about 70% moisture. Dung beetle species showed a preference for the amount of desiccation, for the fibre content and for particular texture or nutritional qualities of dung produced by the vertebrate consumers aboveground. For example, *Aphodius depressus*, *A. erraticus* and *Sphaeridium scarabaeoides* preferred fresh dung pats; drier dung was frequented more by *S. lunatum* and *A. ater*; and further decomposed dung was frequented by *A. fimetarius*, *A. fossor* and *Geotrupes spiniger*. These dung pat preferences were in part to satisfy egg-laying and larvae requirements between different species.

Earthworm distribution in soil

Earthworms have a diurnal rhythm and tend to be more active at night, probably to avoid ultraviolet light, solar radiation and desiccation. They are active when there is sufficient moisture and become inactive as the soil dries, by dehydrating and entering dormancy. Even endogeic species emerge at night and can be trapped on the surface as they migrate. The effect of seasonal changes on cocoon production, emergence and amount of rainfall can cause masses of earthworms to migrate through the profile, or across fields. In some cases, mass migrations uphill or downhill are reported at specific days seasonally, presumably as earthworms seek more favourable conditions (in Lee, 1985). Earthworms that emerge from flooded burrows during rainfall events can be washed away to new areas. Colonization and local spread of a population are often of several metres annually.

In a review of the literature (Lee, 1985), there were no evident trends between abundances and amount of species diversity for earthworms across latitudes or altitudes. Trends seem to correlate better with site-specific parameters of quality and quantity of nutrient resources, and with soil physical properties (Curry, 1998). Overall, earthworms tend to have a patchy distribution, with individuals of one species found in patches of several metres diameter, often 20–30 m. This distribution did not correlate with soil heterogeneity over depth or distances for the endogeic *Polypheretima elongata* (Rossi *et al.*, 1997). This study found that in patches of *P. elongata*, the adults were spatially segregated from the juveniles and cocoons. There was no correlation between density of the earthworm with organic matter, depth 0–30 cm or soil physical properties. This illustrates in part that behind the simple presence and absence of individuals in a patch there are more complex demographic patterns structuring the spatial distribution.

On a smaller scale, variations in soil bulk density (compaction), root depth and food resources affect the spatial distribution of individuals. For example, in agricultural fields, there is a difference between the planted row and the spaces between the rows. When comparing earthworms along the row and between planted rows, more occur in the root zone and along planted rows (Binot *et al.*, 1997). This abundance correlated with the reduced compaction along the planted rows, and the distribution was less restricted with organic nutrient amendment between the rows.

The distribution of different species through the profile varies according to their functional grouping, such as epigeic, anecic and endogeic. These functional categories reflect food preferences and other adaptations. Comparing these adaptations and differences between species, such as reproductive strategies, depth preferences, food preferences and seasonal period of activity, it is possible to define niches for coexisting species (Lavelle *et al.*, 1980; Lee, 1985; Springett and Gray, 1997) (Fig. 5.16). These profile distributions vary somewhat with seasonal changes and periodic changes that affect soil moisture and the organic matter available, such as rainfall and temperature. In a seasonal study, Valle *et al.* (1997) reported that in drier conditions, *Hormogaster elisae* burrowed deeper, and was more difficult to find. This endogeic species adjusted its depth with rainfall, but invariably laid cocoons at

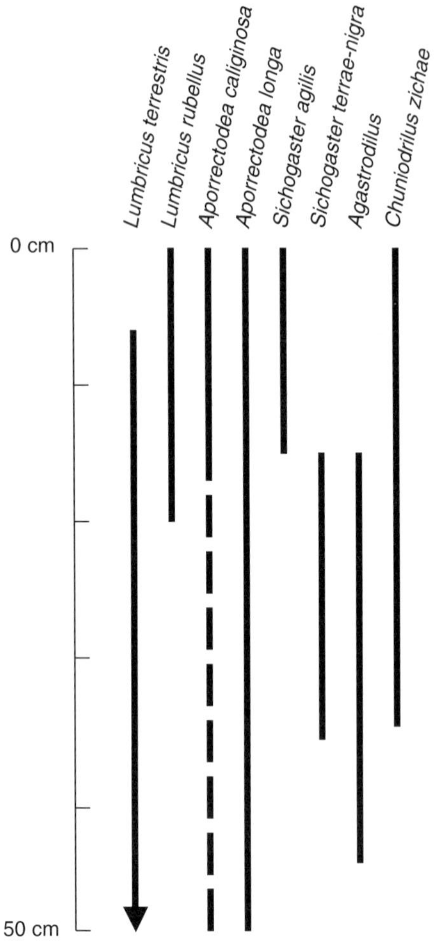

Fig. 5.16. Distribution through the profile of selected earthworms, indicating the preferred depth of residence (data from Lee, 1985; Springett *et al.*, 1997).

20–30 cm depth. There was a correlation between the abundance of cocoons, juveniles and mature individuals with rainfall and seasonal conditions that affect life history (Fig. 5.17).

Over longer periods of time, there are changes in earthworm species and abundance which accompany changes in the soil profile with plant succession. In a study of successional recovery after tree removal in Puerto Rico, several observations were noted. The typical successional trend in this tropical region is for colonization by grass into pastures over 3 years. The next phase (3–15 years) is a grass–vine–fern stage, which leads to a shrub–small tree successional stage (15–40 years). The forest stage is recognized after >40 years, at which time the soil litter

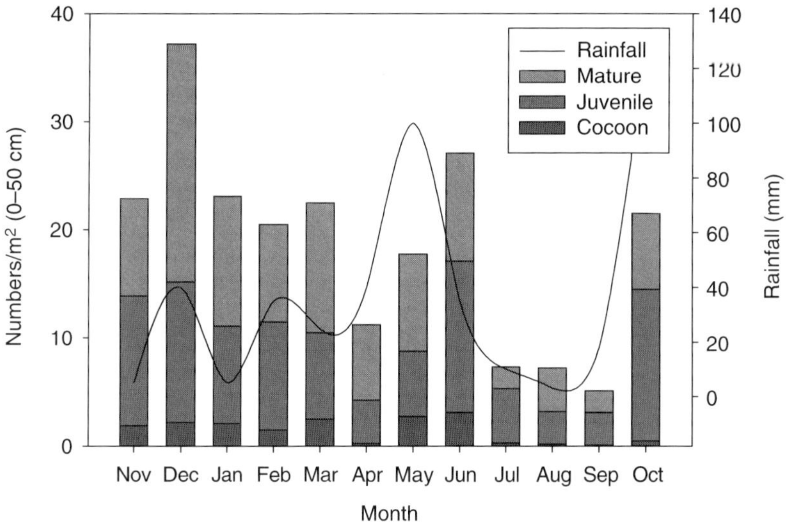

Fig. 5.17. Mean numbers of the earthworm *H. elisae* (cocoons, juveniles and mature adults) over 1 year, in relation to rainfall (adapted from Valle, 1997).

and organic matter content are 30 times more than they were in the pasture. However, the total biomass of earthworms was highest in the pastures (about nine times more) early in succession, compared with the forest stage. Some species are restricted to particular stages in succession, whereas other species occur at all stages. The reduced density of earthworms in these forests, despite the much greater amount of accumulated organic matter in the forest profiles, is suggested to be caused by the chemical quality of the organic matter (Zou and Gonzalez, 1997). However, the authors note that the functional diversity of earthworms in the forest was greater than in the pastures, measured by the increase in anecic species. There are no evident trends between abundances and species diversity for earthworms across latitudes or altitudes (Lee, 1985). Trends seem to correlate better with quality and quantity of nutrient resources, and with soil physical properties.

The direct action of earthworms and enchytraeids on soil litter, organic matter and soil burrowing changes the physical structure of soil as well as impacting the soil community. The impact of Oligochaetas on the soil structure and their stratification patterns in the soil were reviewed recently for both earthworms and enchytraeids (Hendrix, 2000; van Vliet, 2000). Despite the size difference, both have a similar diet and affect the food web in a similar manner. Earthworms and enchytraeids reduce soil protists (see Chapter 4; Day, 1950; Miles, 1963; Bonkowski and Schaeffer, 1997; Jaffee *et al.*, 1997; Bonkowski *et al.*, 2000), ingest nematodes (Dash *et al.*, 1980) and thus influence the abundance and distribution of other soil organisms. Theoretically, the fungal hyphae are also affected if there is significant disruption of the mycelia, but this should be balanced with consideration for spore dispersal through the gut and with earthworm locomotion (Dash *et al.*, 1979). Earthworms also impact the microarthropod abundances and distribution indirectly through their mechanical action on soil comminution and feeding on the mite habitat. The results between experiments sometimes seem contradictory because not all parameters are considered, such as the specific interactions of earthworm functional groups with the origin of litter species (through litter chemistry), soil stratification and local food web structure. In a forest experiment, the impact on microarthropods of new colonization by an epigeic earthworm was monitored (McLean and Parkinson, 2000). The epigeic species *Dendrobaena octaedra* mixes the litter (comminution) through L–F–H layers, and the mixing of the less decomposed L layer (or O_1) into the F–H (or O_2) soil changed its physical structure, thus modifying the habitat. The effect was to increase oribatid species richness in the L layer as *D. octaedra* numbers increased. However, oribatid numbers decreased in the F–H layers with increased *D. octaedra* individuals. This trend was true for 18 species of oribatids and other microarthropods in the F–H layers. These results were not reproduced in the enclosed environment of mesocosms, reflecting the limitations of experimental enclosure. In the mesocosms with the earthworm, oribatid abundances increased over 6 months, although it is not clear where the new individuals came from. It is assumed there was reproduction of the active species, but development of pre-existing eggs cannot be ruled out.

Synthesis and Conclusions

Succession of species and functional groups on litter species follows set trends in a particular habitat. The sequence and timing change with litter and detritus composition as a different set of species are implicated. Ponge (1991) described the sequence of species that decompose pine needles, and it remains an exemplary description of species succession in litter. The earliest colonization of pine needles in this habitat was by saprophytic Ascomycota species that appear in senescent pine needles, such as *Lophodermium*, *Lophodermella*, *Ceuthospora* and *Phacidium lacasum*. These initial stages of fungal decomposition, in the surface L layer, involve extensive colonization of plant cell walls without cell penetration. This stage is cellulolytic without evidence of ligninolysis in the needles. This was followed by the appearance of the microfungi *Verticiclodium* and *Desmazierella acicola*. This suggests that soluble organic molecules remaining in the the cell and from cell wall digestion were becoming available to osmotrophs. Then the Basidiomycota *Marasmyus* and *Collybia* appear, followed lastly by the mycorrhizal Ascomycota

Cenococcum. Lignocellulose degradation begins with these stages in the L–F layers along with penetration of vascular tissues and resin ducts. The Basidiomycota activity occurs over the winter months in the L_2 layer. It continues lignocellulose digestion, with extensive cell wall digestion in the mesophyll tissues, that produces a spongy porous pine needle. Mycorrhizae colonization occurs in the F_1 layer preferentially between the remaining hypodermis and mesophyll tissues. It is at this stage in the F_1 layer that other species also begin to colonize the pine needles, after several months of gradual digestion and perforation by saprophytic fungi. Protists with chloroplasts, protozoa and Cyanobacteria can be identified inside the needles, along with cysts of inactive species. The bacteria also colonize at this stage and succeed the fungal hyphae. The hyphae become devoid of cytoplasm, with only the cell wall remaining. It is unclear whether this is because of cytotrophic feeding on the cytoplasm by piercing fungivores, or by dying of inactive and old hyphae that have exhausted useful nutrients in the pine needle. This stage of decomposition is characterized by bacterivory, cytotrophy and fungivory inside the pine needles. Nematodes and rotifers arrive to prey on the protists and bacteria. Subsequently, enchytraeids and microarthropods arrive to feed on the remains and on the species accumulated inside. These stages also begin in the F_1 layer. These microinvertebrates penetrate by making distinct punctures in the needles and tunnel into the spongy decomposing tissue. They appear well fed, with full digestive tracts, faecal pellets accumulating behind them in the microtunnels and oviposition also occurring. Ponge (1991) comments that the enchytraeids are feeding not so much on the remaining plant cell walls as on the dead fungal hyphae cell walls. The enchytraeids are not permanent residents and burrow in and out of needles scavenging through the L–F layers for a variety of food resources. Bacteria colonize the faecal pellets of enchytraeids both inside and outside needles, and they also digest the slime layer along their microtunnels. Mycorrhizae also occur in their faecal pellets. The motility of enchytraeids in and out of the needles is probably responsible for the transport into needles of many testate amoebae, bacteria, cysts and pollen. Oribatid mites (such as Phthiacaridae) are more permanent residents, and several Oribatid species have obligate pine needle-burrowing larvae. They prefer to feed on plant cell walls infested with hyphae and contribute to some cellulose digestion. They contribute to the fragmentation of the needle and leave a trail of faecal pellets behind in the microtunnels. No bacterial growth is known to occur inside these pellets. Larvae of microinvertebrates can be found occasionally inside the needles in the F_1 layer, where they can ingest and digest remaining lysed plant cells. Epigeic lumbricids in the F_1 layer can ingest mouthfuls of decomposing needle tissue, that would contain the species in it and the faecal pellets, with the exception of the larger organisms such as enchytraeids and mites. These may have time to escape. Little recognizable plant cell wall remains in the earthworm faecal pellets (as they have extensive gut symbionts), but these faeces can be re-colonized by bacteria and mycorrhizae. In this late stage of pine needle litter decomposition, the faecal pellets, excreted detritus and needle fragments become part of the microdetritus that can be colonized by bacteria and fungi, and assimilated again by microinvertebrates. New colonization of fungal hyphae also occurs through the punctures in the needle and needle fragments that remain as macrodetritus. Pine needles can be studied through the profile and sampled over successive years (Hagvar and Edsberg, 2000). Following marked litter through the profile allows one to sample the saprotrophs that occur in increasingly decomposed litter, into secondary decomposition stages.

These local patterns of succession through primary and secondary decomposition with depth can be compared across an ecosystem, and between climatic regions. Locally, decomposition food webs change with time, as the ecosystem develops through its successional stages. These patterns are described from sampling a chronosequence of sites within a geographic region. Comparing field sites with different properties, in climate, litter species composition, soil mineralogy or successional age, contributes to our understanding of the effect of variables on decomposition food web function. Moreover, these comparisons can identify variation caused by disturbance. Past land use leaves a footprint that affects subsequent succession and food web composition. Many sites bear the mark of mining, deforestation, pollution, excessive hunting or unsustainable agricultural exploitation. The footprint of past land use remains for centuries (Verheyen *et al.*, 1999). This study was conducted near Ghent in Belgium based on land use maps between the years 1278 and 1990. The historical plant succession of the region was impacted by successive cycles of deforestation, agriculture and reforestation. The modern soil chemical properties were affected by land use history over the past several centuries, especially in some soil types (based on mineralogy). Depletion of certain minerals (magnesium and phosphorus) on some sites could be linked directly to land use history (excessive agriculture) centuries ago. The perspective from longer time scales, of historical land use which affects soil properties and decomposition today, is insufficiently integrated into soil ecology. That is particularly important if we are to manage our soils adequately today for tomorrow's generations.

Summary

The abundance of active individuals in a population varies with fluctuations in abiotic and biotic conditions. Changes in the soil physical structure through the profile (such as aeration, temperature, moisture content, organic matter quality and quantity, compaction and mean pore sizes) determine which species are active at particular depths. Food resources (from saprotroph-excreted waste inside the soil, from roots and from surface litter) determine the quality and quantity of substrate for saprotrophic species. Species are associated with horizon preferences that reflect their adaptations to physical and food resources. Variations in soil moisture affect most directly the food availability, locomotion and pore accessibility inside peds. The osmoregulation potential of species determines how much soil desiccation they can withstand. Species become active or quiescent in succession to accommodate fluctuations in growth conditions. Weather and seasonal changes modify species composition and litter deposition. For these reasons, at any one time, only a fraction of species in a sample are active. Over decades and centuries, plant species succession changes the composition of litter input. These change the decomposition food web as soil physical and chemical properties evolve. With time, as resources change, dominant species are replaced by previously rare species. This pattern is observed by comparing microhabitats over several months, or ecosystems over several decades. Comparing the evolution of food webs between habitats and ecosystems through decades will provide valuable insight into sustainable soil management practices.

Suggested Further Reading

Body, L. (2000) Interspecific combative interactions between wood decaying Basidiomycetes. *FEMS Microbiology Ecology* 31, 185–194.

Dilly, O. and Irmler, U. (1998) Succession in the food web during the decomposition of leaf litter in a black alder (*Alnus glutinosa* (Gaertn.) L.) forest. *Pedobiologia* 42, 109–123.

Ettema, C.H. (1998) Soil nematode diversity: species coexistence and ecosystem function. *Journal of Nematology* 30, 159–169.

Fenchel, T. (1969) The ecology of the marine microbenthos. IV. Structure and function of the benthic ecosystem; its chemical and physical factors and the microfauna communities with special reference to the ciliated protozoa. *Ophelia* 6, 1–182.

Ponge, J.F. (1991) Succession of fungi and fauna during decomposition of needles in a small area of Scots Pine litter. *Plant and Soil* 138, 99–113.

Renvall, P. (1995) Community structure and dynamics of wood-rotting Basidiomycetes on decomposing conifer trunks in northern Finland. *Karstenia* 35, 1–51.

Siepel, H. (1994) Life history tactics of soil microarthropods. *Biology and Fertility of Soils* 18, 263–278.

Verheyen, K., Bossuyt, B., Hermy, M. and Tack, G. (1999) The land use history (1278–1990) of a mixed hardwood forest in western Belgium and its relationship with chemical soil characteristics. *Journal of Biogeography* 26, 115–128.

Wicklow, D.T. and Carroll, G.C. (1992) *The Fungal Community: its Organisation and Role in the Ecosystem*, 2nd edn. Marcel Dekker, New York.

Integrating the Food Web

6

In previous chapters, we have seen that species within the pedon are organized through the profile, and associate with microsites of adequate substrate or prey, and with abiotic conditions through the profile. The organization of the community changes through time over several days, with soil drying, wetting, fluctuations in temperature and aeration. At a longer time scale, there are species composition changes with seasons. Over longer time scales, more significant changes occur in the structure of food webs and the abundance of species in functional groups with succession. This succession reflects changes in soil physical and chemical properties, but does not necessarily reflect plant succession. It is, however, strongly affected by changes in litter chemistry and root morphology that occur with plant succession.

The interactions of species within functional groups with the litter substrate, the soil matrix, coexisting competitor species, prey abundances, predators and abiotic conditions are complex. The complexity is caused by the large number of fluctuating variables under natural field conditions. Individuals are continuously adjusting to new conditions, so that activity and abundances of species vary. Moreover, there are numerous coexisting species within each gram of soil, so that realistic *in situ* situations need to consider the effect of changes in the activity or abundance of one species with respect to all the others. Such reconstruction of the food web at microsites or through the profile requires extensive descriptive work.

To describe decomposition from pedon sampling involves enumerating coexisting species, and obtaining estimates of activity or rates of nutrient transfer between functional groups, from species–species interactions. This reduces the food web to a set of boxes representing species in functional groups, with fluxes of nutrients between them. The difficulty in obtaining these numbers is in the methodology and determining the correct scale for the experimental studies. Since the soil is opaque and the microorganisms are only visible by microscopy, they need to be extracted. Once species are extracted, the pedon structure is destroyed. Therefore, one must reconstruct the biological processes from knowledge of species and measurements of nutrient flux rates. One method with a great deal of potential is to rely on soil cores prepared in the laboratory, or from intact samples. The cores can be maintained under steady conditions in the laboratory and replicates buried *in situ*. Each soil core represents a microcosm with one variable modified (litter species, species in functional groups, ratio of species in functional groups, or abiotic conditions) which can be subsampled at regular intervals. The cores can be constructed with isotopic tracers to help follow the rate and paths of decomposition of particular compounds through the food web. The behaviour of these cores in relation to each other can be compared further when experiments are replicated across latitudes, climatic conditions and different mineralogy.

Data from species abundances, frequency of activity of species during seasons and flux of nutrients between functional groups must be synthesized into a model of the decomposer food web. Analysis of rates of nutrient transfer between species in functional groups, of substrate modification rates, and rates of nutrient appearance in the soil solution relies on mathematical models. These models are essential to develop a dynamic and analytical tool for predicting the outcome of manipulations, and for testing the validity of food web models. There remain many opportunities to elaborate our understanding of decomposition food webs. Many analytical studies have relied on agro-ecosystems or simplified communities to develop mathematical models.

Global Impact of Decomposition

Carbon

The pool of organic matter accumulated in the surface terrestrial soil is larger than the amount of carbon (CO_2) in the atmosphere (Fig. 6.1). Only the pool of oceanic carbon is larger than that in the terrestrial soil organic horizons. The source of organic matter in the soil is estimated globally from biogeochemical cycles and data obtained from various ecosystems (Schlesinger, 1997). It is estimated that 560×10^{15} g C is held in terrestrial plant tissues, mostly as wood. A fraction is shed annually as litter from roots, leaves and woody debris for decomposition. The soil organic matter (SOM) consists of annual additions of plant-derived litter, as well as litter from organisms of the above-ground production food webs (Fig. 6.2). It also includes litter from saprotrophic organisms, as dead organisms, shed body parts and excreted wastes. The SOM is decomposed by saprotrophs releasing CO_2 and metabolic wastes, as well contributing to new biomass through growth of cells, cell divisions and sexual reproduction, yielding new individuals (Fig. 6.3). In terrestrial systems, decomposition of SOM into CO_2 is the largest source of carbon

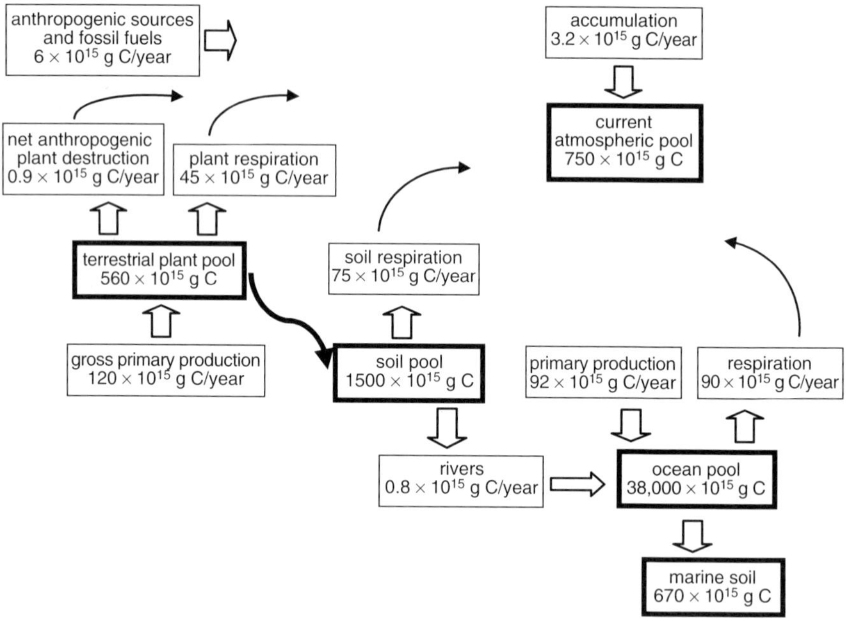

Fig. 6.1. Global carbon pools and flux rates (adapted from Schlesinger, 1997).

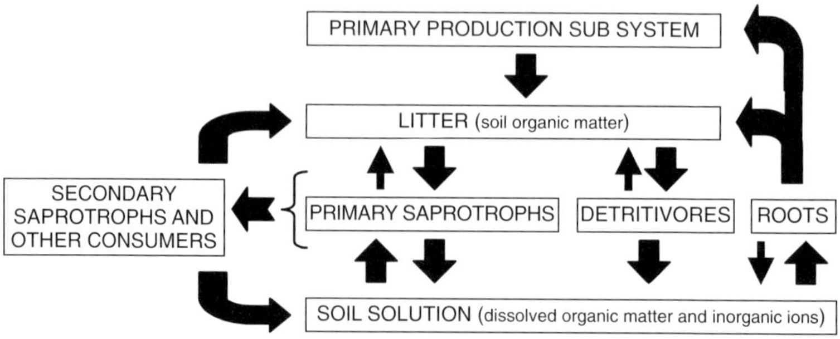

Fig. 6.2. Litter transformations and uptake through decomposition. Roots are the main sink of soil solution molecules into primary production. Metabolic wastes and excreted organic matter from saprotrophs contribute both to litter and to the soil solution. Cell walls and dead organisms also contribute to litter. Consumers are important in recycling primary saprotrophs back into litter and soil solution compartments. Respired CO_2 from decomposition (not shown) also enters primary production through photosynthesis.

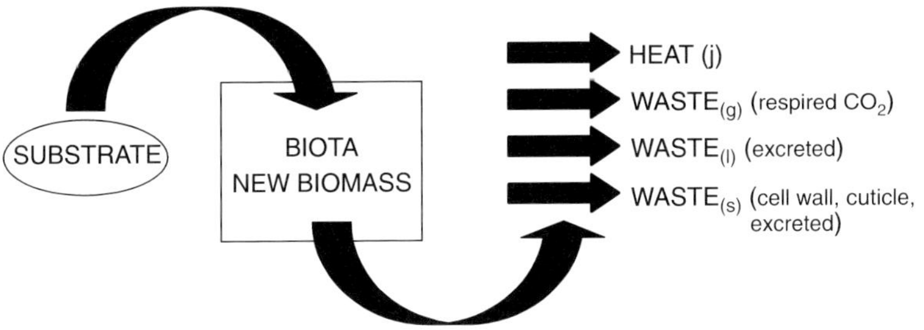

Fig. 6.3. Litter transformation by saprotrophs. Individual saprotrophs ingest food (prey and nutrients) in order to grow (new biomass) and excrete the undigested portion ($Waste_{(s)}$), the nitrogenous wastes ($W_{(l)}$) and respired gases ($Waste_{(g)}$), and release heat generated from metabolism.

recycling for photosynthesis and primary production. Until the recent impact of humans on ecosystems over the last 250 years, the release of soil CO_2 and its uptake by photosynthesis were closely balanced (Schlesinger and Jeffrey, 2000). The balance is now disrupted, with the accumulation of an additional 7×10^{15} g C annually in the equation, mostly from burning fossil fuels and destruction of habitats. There have been many discussions of the global carbon budget as a result, to predict the impact of climate change on ecosystems, and on the rate of warming caused by the rate of an increase in CO_2 (a greenhouse gas) in the atmosphere. In particular, soils are seen as important because they are more easily managed than the ocean or marine soil (Amundson, 2001; Fang *et al.*, 2001; Pacala *et al.*, 2001). The global approach to ecosystem analysis relies on two key observations: (i) that ecosystems are distributed into similar life zones based on similar mean annual precipitation and mean annual temperature (Holdridge, 1947; Whittaker, 1975); and (ii) that plant communities vary along climatic and SOM gradients (Post *et al.*, 1982). These observations allow calculations of carbon pool fluxes within life zones and comparisons between ecosystems (Table 6.1, see Amundson, 2001). The cycling of organic matter does not occur independently for each element, even though they are often represented graphically independently (Figs 6.1, 6.4 and 6.5; Schlesinger, 1997).

Phosphorus

SOM consists of litter derived from cellular tissues, which is rich in polymers of sugars and lignin, but poor in phospholipids and polymers of nucleic acids and amino acids. Therefore, the food substrate for the soil species is low in nitrogen- and phosphorus-containing molecules. The availability of nitrogen is often rate-limiting, although some soils may also be deficient in soluble phosphorus or other minerals. In grasslands, for

Table 6.1. Terrestrial life zones, showing area and amount of carbon input and turnover from litter sources.

Life zone	Area (10^{12} m^2)	Soil C (kg/m^2)	Soil C input (kg/m^2/year)	Residence time (years)
Tundra	8.8	21.8	0.102	213
Boreal desert	2	10.2	0.050	204
Boreal moist forest	4.2	11.6	0.190	61
Boreal wet forest	6.9	19.3	0.681	28
Temperate cool steppe	9	13.3	0.300	44
Temperate thorn steppe	3.9	7.6	0.462	16
Temperate cool forest	3.4	12.7	0.912	14
Temperate warm forest	8.6	7.1	0.826	9
Tropical desert bush	1.2	2	0.083	24
Tropical savannah	24	5.4	0.479	11
Tropical very dry forest	3.6	6.1	0.472	13
Tropical dry forest	2.4	9.9	0.458	22
Tropical moist forest	5.3	11.4	2.491	5
Tropical wet forest	4.1	19.1	3.732	5
Cool desert	4.2	9.9	0.214	46
Warm desert	14	1.4	0.043	33

From data presented in Amundson (2001).

example, 1–20 kg/ha/year of phosphorus input comes from plant litter Whitehead (2000). The situation is exacerbated in regularly harvested agricultural soils, where the annual input of litter is minimized by harvest and removed off-field. Obtaining nitrogen- and phosphorus-containing molecules and soluble ions for cell growth is competitive. Soluble phosphorus compounds are readily scavenged and stored in intracellular pools. The soil solution near the rhizosphere is notoriously poor in K, N and P ions, as plants are effective and most competitive at absorbing these ions, especially in mycorrhizal roots. Typically, soil organisms, particularly eukaryotes, are effective at absorbing dissolved ions through cell membranes by osmotrophy. Even though they consist of 2–6% of the total soil organic carbon, they hold up to 20% of the soil phosphorus. The soil organisms are probably the most important source of soluble and biologically available phosphorus compounds. Most of the available phosphates are as $H_2PO_4^-$, HPO_4^{2-} and polyphosphates in cells. The difficulty in tracing phosphorus is that only a small fraction is available to the soil solution, and the larger fraction is bound to clays and inactive fractions of the soil. The fraction that is available for cell membrane transport into cells is rapidly absorbed and immobilized. Much of the phosphorus is insoluble or bound to soil particles and not available for membrane transport. Some is bound to calcium phosphates including apatites, some is clay bound such as to kaolinite, and some is bound to

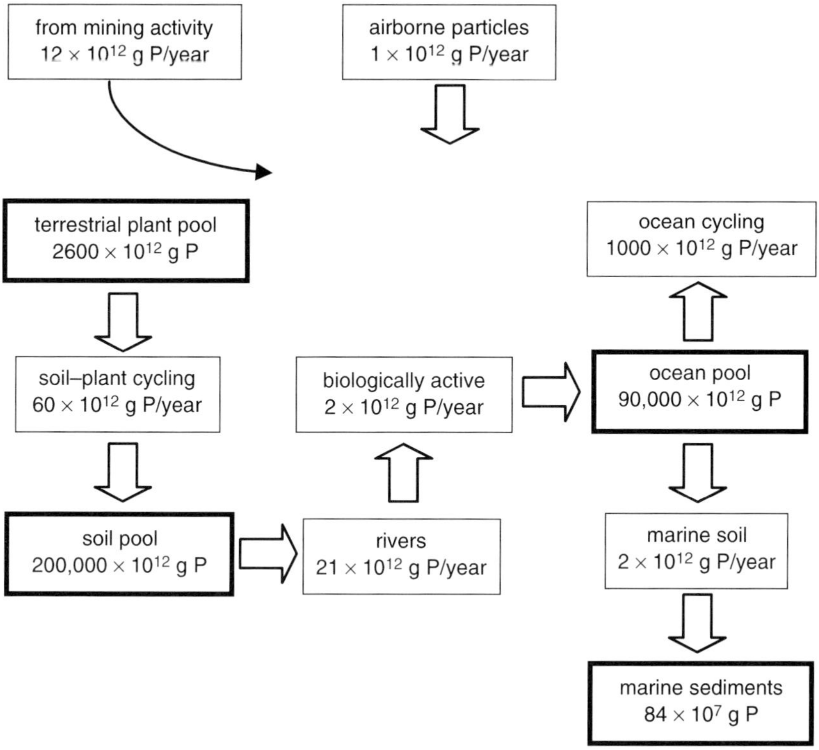

Fig. 6.4. Global phosphorus pools and flux rates (adapted from Schlesinger, 1997).

organic matter. Apatite consists of $3[Ca_3(PO_4)_2].CaX_2$, where X can be carbonate, OH^-, F^- or C^-. Sometimes, other P-containing minerals are present, such as strengite ($FePO_4 \cdot 2H_2O$), variscite ($AlPO_4 \cdot 2H_2O$) and other iron or aluminium phosphates. A fraction of the organic phosphorus pools may be liberated through secretion of *phosphatases* by protists and bacteria, releasing soluble inorganic phosphate ions. Phosphorus limitation in the decomposition food web is most limiting to ATP cytoplasmic pools, and to nucleotide synthesis which prevents DNA replication and reduces cellular activity. Reduced phosphorus availability limits cell growth and activity. In turn, reduced cellular activity decreases the rate of decomposition and the cycling of other nutrients. Therefore, the overall rate of nutrient use and uptake decreases in P-poor soils.

Nitrogen

By far the largest pool of nitrogen is in the atmosphere as $N_{2\,(g)}$, but it is not available to most biological reactions (Table 6.2, Fig. 6.5). The triple bond that holds the molecule is strong and energetically expensive to break, thus

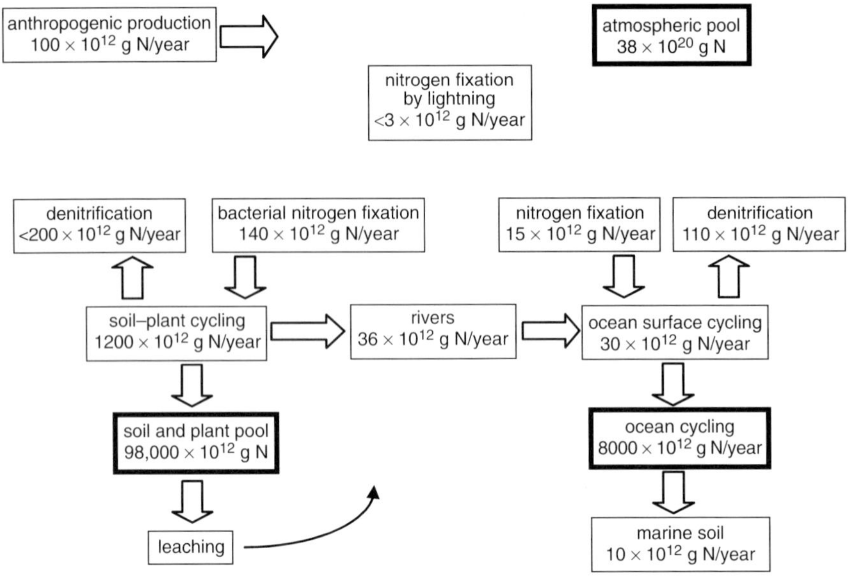

Fig. 6.5. Global nitrogen pools and flux rates (adapted from Schlesinger, 1997).

Table 6.2. Nitrogen pools in the atmosphere.

N_2	3.9×10^9
N_2O	1300
NH_4^+	1.8
Other	2.5–4

Data from Tamm (1991).

it is almost inert. 'Fixed nitrogen' refers to biologically available molecules that have been oxidized or reduced to other nitrogenous molecules. Nitrogen fixation by lightning forms nitrogen oxide molecules, at an estimated average rate of deposition on land of $<5 \times 10^{12}$ g N/year. Biological nitrogen fixation is carried out by several genera of bacteria, such as *Azospirillum*, *Azotobacter*, *Bacillus*, *Beijerinckia*, *Bradyrhizobium*, *Chromatium*, *Chlorobium*, *Clostridium*, *Desulfotomaculum*, *Desulfovibrio*, *Klebsiella*, *Pseudomonas*, *Rhizobium*, *Rhodomicrobium*, *Rhodopseudomonas*, *Rhodospirillum*, *Thiobacillus*, *Vibrio* and some *Actinobacteria*; in freshwater systems, the Cyanobacteria are the principal nitrogen fixers. On average, the total biological nitrogen fixation on land is about 10 kg N/year/ha, with about 5–9 kg N/year/ha from symbiotic associations between bacteria and plant roots. The greatest amount of biological nitrogen fixation occurs in soils poor in nitrogen but active in decomposition. The soil nitrogen is mostly held in the organic matter and in the cells of edaphic species, with about 5% (soil N dry weight) in the inorganic fraction (Table 6.3). The pool of organic nitrogen

Table 6.3. Nitrogen pools in terrestrial systems.

Source	Pool size
Plant biomass	$1.1–1.4 \times 10^4$
Animal biomass	200
Surface litter	$1.9–3.3 \times 10^3$
Soil organic matter and soil biota	3×10^5
Insoluble inorganic nitrogen	1.6×10^4

Data from Tamm (1991).

exists in chitin largely from invertebrate cuticles and fungal cell walls, mucopolysaccharides from roots and bacteria, polymers such as proteins and nucleic acids, and their monomers. Fresh litter from above-ground sources contributes an annual input of organic nitrogen and from excretion of urine and uric acid from metabolic wastes. The principal source of dissolved inorganic nitrogen is from metabolic nitrogenous waste of saprotrophs decomposing litter and from above-ground animals (Table 6.4). In a temperate grassland soil, for example, with about 10% organic matter, one can estimate that about 0.08–0.5% will be dissolved inorganic nitrogen. The availability of the dissolved nitrogenous compounds depends on several parameters. The clay mineralogy will affect cation exchange capacity and binding of both organic compounds and inorganic N ions to soil particles. The extent of complexation and colloid formation of N compounds in humus will vary with clay composition, pH and rainfall. Leaching of ions also varies with rainfall frequency and intensity. Loss of N from soil occurs through volatilization as NH_3, N_2 and nitrogen oxides, and leaching of soluble ions. Soil erosion and emigration or displacement of animals are also significant causes of reduced N input. An important source of N input is from the atmospheric pool and through anthropogenic addition, including fertilizers and planting of legumes (plants with nitrogen-fixing symbiotic bacte-

Table 6.4. Summary of main forms of nitrogenous wastes from soil organisms.

Taxon	Excreted waste	Excreted N
Oligochaetae	Cast, soil	NH_4, urea and uric acid (when soil is dry)
Onychophoran	Faecal pellets	Uric acid
Collembola	Faecal pellets	Uric acid, insoluble urates
Oribatida	Faecal pellets	Guanine crystals
Rotifera	Humus	NH_4
Nematoda	Humus, pellets	NH_4, purines
Fungi	Cell wall, vesicles	NH_4, chitin in cell wall
Protozoa	Humus	NH_4
Bacteria	Capsule, cell wall	NH_4, cell wall

ria). In modern times, the anthropogenic input on land is by far larger than natural sources. It is estimated that >80 × 10^{12} g N/year is added through agricultural fertilizer usage (NH_4^+, urea), with an additional 20 × 10^{12} g N/year from fossil fuel combustion. Natural N inputs are from litter and root decomposition, metabolic N excretion by saprotrophs and aboveground species (Tables 6.2 and 6.4). As the input N increases, the amount that is available to saprotrophs and plant roots also increases. The preferred form of nitrogen for uptake by osmotrophy is ammonia, although nitrate can be transported through cell membranes of many species. Some fungi cannot transport nitrate, but plant roots can transport ammonia and nitrate. As the N input amount increases, through biological transformations, the loss of soluble forms by leaching and gaseous forms by volatilization also increases. Because the activity of nitrogen fixation metabolic pathways depends to a large extent on the soil conditions, the activity varies with seasons and weather. It is therefore difficult to make generalization and predictions about the amount of nitrification and denitrification processes between field sites, without knowledge of the conditions at the time of sampling and of the food web composition.

Nitrogen fixation

The enzyme responsible for **nitrogen fixation** is *nitrogenase* (abbreviated *nif*) and consists of two subunits. One subunit requires iron and the other requires both iron and molybdenum prosthetic groups. The enzyme requires ATP, reduced ferredoxin and other electron acceptors, and it is extremely sensitive to the oxygen concentration. Most nitrogenase enzymes are inactivated irreversibly by oxygen, so that the enzyme must be physiologically protected. The bacteria create a low oxygen microenvironment in the cell or wait for adequate conditions for nitrogenase activity. In most cases, exposure to oxygen is the primary limiting factor in nitrogen fixation, when all other parameters are favourable. However, there are both aerobic and anaerobic species that carry out nitrogen fixation (Fig. 6.6). The end-product of nitrogen fixation is ammonia, with hydrogen gas released as a by-product. The ammonia is used in cellular anabolic metabolism to synthesize organic molecules **(ammonium assimilation)** by transport through the cell membrane. Soil bacteria capable of fixing nitrogen are common, but those that form symbiotic associations with roots are more efficient at producing fixed nitrogenous compounds. In these symbioses, the N compounds are available to plant primary production and do not become part of the soil decomposition system until the plant biomass turns to litter. Only a small portion will be returned through root exudates. Legume is a general term given to crops that permit symbiosis with nitrogen-fixing bacteria. The bacteria form nodules inside root tissues in a species- or strain-specific manner. It is sometimes necessary to inoculate a field or crop with the adequate strains to induce nodule formation.

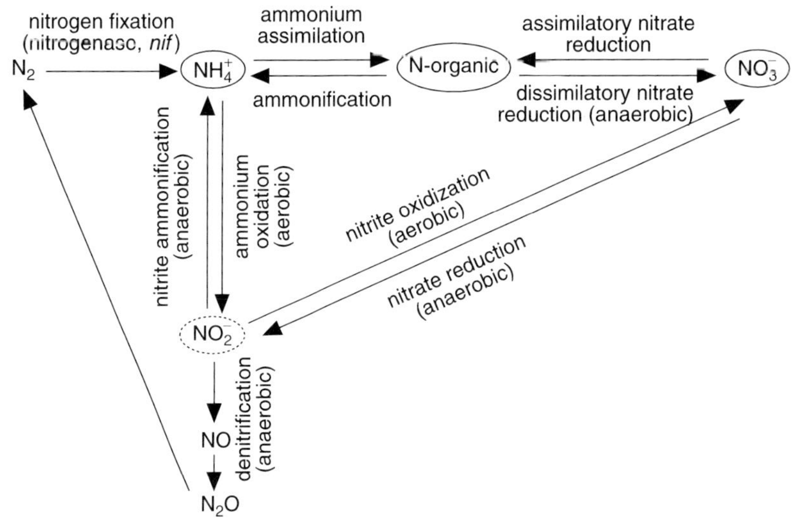

Fig. 6.6. Representation of the nitrogen cycle through the decomposition food web (see text for details).

Nitrogen-fixing bacteria, especially non-symbiotic species, are an important source of nitrogen in the decomposition food web and often represent a keystone functional group. In nitrogen-limited soils, the rate of nitrogen fixation can be the rate-limiting step in decomposition.

Ammonification refers to the release of ammonia from organic molecules. A large portion of the ammonia is released as metabolic nitrogenous waste by various organisms (Tables 6.4 and 6.5). An important source of ammonia is from urea (excreted from various Animalia) through *urease* enzyme activity produced by certain primary saprotrophs. The enzyme requires nickel as a prosthetic group. The presence of the enzyme in soils is crucial when urea-based fertilizers are used. The ammonia is more accessible than urea or organic N molecule monomers for cell membrane transport by osmotrophy.

Table 6.5. Glossary of the main biochemical transformations in the nitrogen cycle.

N fixation	Biological transformation of N_2 (g) into organic or inorganic compounds
Ammonium assimilation	Cellular absorption of NH_3 or NH_4^+ from the environment for new cell biomass
Nitrification	Oxidation of ammonia to NO_2^- and NO_3^- for chemical energy
Nitrate reduction	Reduction of NO_3^- followed by its absorption into cell biomass
Ammonification	Transformation of organic N to ammonia, e.g. from urea or uric acid
Denitrification	Reduction of NO_3^- to any N oxides or $N_{2(g)}$

Nitrification

There is a two-step process which involves the oxidation of ammonia to nitrite ions, followed by the oxidation of nitrite to nitrate ions, called **nitrification** (Table 6.5, Fig. 6.6). Each step is an energy-yielding process and occurs in a small number of autotrophic species. The two steps are often carried out by separate species of bacteria, but they occur in close proximity and simultaneously so that a net accumulation of nitrite is not usually observed. In soils, the most common species implicated are *Nitrosolobus* and *Nitrosomonas*, and in acid soils *Nitrosospira* (**ammonia oxidation**) and *Nitrobacter* (**nitrite oxidation**). The latter reaction is carried out by *nitrite oxidoreductase* and these species are capable of limited heterotrophic growth, unlike the ammonia oxidizers which are strict autotrophs. Other species of primary saprotrophs are capable of a limited amount of ammonium oxidation. On their own, they probably do not contribute significantly to ammonium oxidation; however, if their pooled numbers are great, their contribution could be significant. Known heterotrophic nitrifiers are the fungus *Aspergillus*, and the bacteria *Alcaligenes*, *Arthrobacter* and several Actinobacteria. The importance of this pathway in nitrogen transformations depends on the clay mineralogy. If the clays are positively charged, the negative nitrite and nitrate ions are more likely to be bound and less accessible to osmotrophy. In the reverse situation, it would be more beneficial to transform the positive ammonium ions into the negatively charged oxidized ions. Nitrate ions are easily transported through cell membranes and assimilated into organic molecules (**assimilatory nitrate reduction**). It is the preferred form of nitrogen uptake for many species, especially of plants.

Denitrification

Soluble forms of nitrogen in the soil are transported through cell membranes by osmotrophy for assimilation into complex organic molecules by anabolic metabolism. These soluble forms include a variety of small organic molecules, ammonium, nitrate and nitrite ions (Table 6.4, Fig. 6.6). In the absence of oxygen, several pathways of nitrogen transformation become more dominant. Dissimilatory **nitrate reduction** becomes one important mechanism. It involves the oxidation of cell organic molecules utilizing nitrate ions as terminal electron acceptors, producing nitrite ions. This pathway is more efficient than the typical fermentation pathway. Two types of nitrate reduction are distinguished. In the first type, certain facultative anaerobic species can reduce nitrate to nitrite by transferring electrons from anabolic metabolism. Under certain conditions, these species will further transform the nitrite ions excreted into ammonia by **nitrite ammonification**. Under these conditions, the nitrate can be re-utilized as ammonia by osmotrophy and assimilated. Some soil genera implicated in this process include, *Clostridium*, *Desulfovibrio*, *Enterobacter*, *Erwinia*, *Escherichia*, *Klebsiella*, *Nocardia*, *Pseudomonas*, *Spirillum* and *Vibrio*.

In the second type, nitrate reduction is followed by further reduction of nitrite ions to NO, N_2O and N_2. This pathway is called **denitrification** because it produces gaseous N molecules which are not biologically useful. The main denitrifiers in the soil are *Agrobacterium*, *Alcaligenes*, *Azospirillum*, *Bacillus*, *Bradyearhizobium*, *Cytophaga*, *Flavobacterium*, *Nitrosomonas*, *Propionibacterium*, *Pseudomonas*, *Rhizobium* and *Rhodopseudomonas*. The enzymes for this process are close together in the cell membrane. Their activity is inhibited by oxygen and they require metal atom prosthetic groups. Nitrate is first transformed to nitrite by *nitrate reductase* (*Nar*), nitrite is transformed to nitric oxide by *nitrite reductase* (*Nir*), nitric oxide is transformed to nitrous oxide by *nitric oxide reductase* (*Nor*), and nitrous oxide is converted to dinitrogen by *nitrous oxide reductase* (*Nos*). Again here, nitrate reduction, ammonification and denitrification processes occur in close proximity in the soil. Peds are more aerobic on the periphery and more anaerobic in the centre. Bacteria are dispersed throughout the peds and utilize the available soil solution under the conditions they are in. Typically, nitrate assimilation at the periphery of peds is accompanied by nitrate reduction and denitrification in the anaerobic centre of peds and at anaerobic microsites. The exact balance depends on the combination of bacteria functioning in syntrophy at microsites or in peds, under soil aeration status. The amount of N osmotrophy and bacterivory by other soil organisms further protects from loss of nitrogen from the system. In aerobic conditions, ammonium oxidation is carried out by *ammonia mono-oxygenase* which has a broad substrate specificity, including methane and several halogenated organic molecules. Under anaerobic conditions, these species can also carry out further reduction of nitrite to N_2O and thus contribute to denitrification. During a spell of anaerobic conditions, such as several days of flooded fields, a significant portion of the accumulated fixed nitrogen and dissolved N molecules can be lost to the atmosphere by denitrification reactions.

How to Trace Nutrients

The fate of nutrients in the litter, through saprotrophs and into the soil solution, is of great interest. Dissolved nutrients in the soil solution are the only source of mineral ions and inorganic ions into roots and primary production. Therefore, understanding the controls and regulators of the dynamics of the rates of litter transformation into dissolved compounds is of primary importance in predicting plant growth. There have been many attempts at describing and understanding these parameters (Kalbitz *et al.*, 2000). It is particularly clear that a fraction of dissolved molecules, released from litter polymers by saprotrophs, becomes bound in the mineral soil, and that a fraction is lost by leaching through the organic horizons. Using data sets from a variety of forests, Neff and Asner (2001) proposed a model from calculated flux values for dissolved organic carbon, through successive soil strata. Their study concludes that saprotroph activity (stated as 'microbial') regulates the rate of release of dissolved nutrients, whereas the soil mineralogy (geochemistry) and local hydrology determine the amounts of loss from the system. While dissolved nutrients are lost to deeper layers (vertical flow) and through water runoff (horizontal flow), the mineral horizons can effectively adsorb a major portion of the dissolved nutrients. This pool of adsorbed nutrients in the mineral horizons consists of more or less labile molecules that permit biological activity in the mineral horizons. The binding capacity of reduced organic and inorganic molecule in mineral soil causes increases in dissolved organic matter loss through the profile. This nutrient loss appears in primary streams and ground water. The loss of a fraction of the SOM at each rainfall event through the organic horizons is called **hydrological short-circuiting**, because it by-passes organic matter adsorption through the mineral horizons. This causes pulses of dissolved organic nutrient input into streams, and it is exacerbated by loss or absence of mineral horizons. The loss of inorganic nutrients is increased by loss or absence of roots which are normally responsible for the absorption of a major portion of ions and soil solution. This is exacerbated by clear-cutting forests and bare-fallow agricultural fields which cause soil erosion, as hydrodynamic dispersal and mass flow through the soil are increased, and mineral soil is flushed through the profile. Inorganic nutrients are also adsorbed by mineral soil rich in iron and aluminium hydroxides. These soils keep adsorbed ions out of the soil solution. In a set of field and laboratory experiments, Qualls (2000) compared the mechanisms of organic and inorganic nutrient retention in forest soils. He proposed a conceptual model for the factors controlling leaching of soluble organic nutrients, based on leaching–sorption experiments. The study concludes that most of the organic molecules are held to each other through hydrogen bonds between hydroxyl side groups. This is to be expected since it is the principal form of interaction between biological molecules in cell walls, and between proteins or nucleic acids. Chemical adsorption sites in the mineral soil bind soluble and dissolved organic and inorganic molecules. This has the effect of buffering the amount of dissolved molecules, so that only a fraction is in solution. Competition for ligand exchange sites (by sulphates, phosphates, oxalates and fluoride) facilitates desorption and suppresses adsorption, even at relatively low ion concentrations.

Most of these types of studies and reviews focus on the abiotic parameters and the chemical interactions between soil minerals and the soluble nutrients. Almost none of the laboratory measurements were obtained on samples with an intact or relatively unharmed biological component. The methodology for these experiments involves considerable handling and mixing of soil, which inevitably lead to species destruction and loss of the stratified spatial habitat. In cases where a biological component is assayed, it is usually restricted to the bacterial, spores and encysted components that were not destroyed. For this reason, the continuous contribution of saprotrophs to the solution is usually grossly underestimated in laboratory studies.

In one recent field study, the dynamics and magnitude of both dissolved organic carbon and nitrogen in forest soil were monitored every 2 weeks for 2 years. The study obtained soil solution through the profile to 90 cm depth in the Bw–Cw horizon of a spruce forest (Michalzik and Matzner, 1999). The fluxes of dissolved organic N and C decreased with depth. At 90 cm depth, the flux of dissolved organic N reached the detection limit (from 7.1 kg/ha/year in the Oi horizon) and it was 16.5 kg/ha/year for dissolved organic C (from 172.7 kg/ha/year in the Oi horizon). Nutrients released in the surface horizons percolated rapidly to deeper horizons. The nutrient flux was strongly correlated with water flux, but did not correlate with pH or electric conductivity over the range of field values. Graphs of the fortnightly results indicate the magnitude and extent of changes in the flux with seasons, which are especially high during snow melt. Within the forest organic horizons (*the forest floor*), ammonium ions, amino sugars and amino-bearing molecules, dissolved organic nitrogen and dissolved organic carbon molecules were elevated. The concentration of all these components decreased rapidly in the mineral horizons below it. Only nitrate ion concentrations remained about the same between the organic and mineral horizons, increasing a little in deeper soil. Unlike other nitrogen and carbon molecules which are derived from litter decomposition and metabolic wastes, the primary input of nitrates at this site was from bulk precipitation and through-fall. The authors were able to construct a budget for N and C nutrients for the site through the soil profile. It would be interesting to follow-up on this approach, to determine the flux of organic and inorganic molecules, in the litter and soil solution, into each saprotroph functional group and into plant roots. In particular, we do not know how much of the ammonia, nitrate and organic molecules are solubilized from litter, and how much is released as metabolic wastes from saprotrophs (Fig. 6.2, Tables 6.3 and 6.4). What is clear, however, is that a low soil solution concentration can mask a high flux rate, from litter and solid substrate, through the soil solution into saprotrophs and roots. The dynamics of dissolved organic C and dissolved organic N did not correlate very well (correlation coefficients 0.4–0.2). The authors suggested that the mechanism and rate of release could be different for each pool. There were also weak correlations between several abiotic parameters and soil solution composition, in the field. The paper illustrates the difficulty in understanding the processes that release nutrients from the litter, and of nutrient uptake and transformation by saprotrophs, without describing the principal path of substrate molecules though the food web.

Tracer studies

Litter transformation through decomposition into solid, liquid and gaseous molecules can be followed and quantified with isotopic tracer studies (Fig. 6.3). Cells of species can be isotopically labelled by feeding on substrates enriched with rare isotopes or radioactive isotopes. The labelled molecule is a molecule enriched in an otherwise rare isotope, which can then be tracked against the background presence of that isotope. Labelled substrate polymers, soluble monomers or inorganic ions can be followed through time as each labelled molecule is transported into cells. Molecular transformations can be followed by analysis of molecules that contain the isotope through time. Molecules are absorbed, assimilated and excreted by a succession of species. Translocation of the isotope through functional groups, by a succession of species, is followed by looking for the isotope in soil species extractions. The method permits identification of functional groups and key species involved at each step. These isotope tracer methods are more accurate (Kuzyakov, 2001; Kuzyakov *et al.*, 2001). For example, radioactive detection and quantification of root biomass reveals 20~60% more root than traditional soil–root washing. The method is also more sensitive and accurate at identifying changes in substrate quantity. The tracer method is also the only method of following the rate of transformations of substrate through ingestion–digestion–excretion by a succession of species. The techniques have been used primarily to follow the route of litter accumulation into soil from plant photosynthates and plant residues (Spaccini *et al.*, 2000; Kuzyakov *et al.*, 2001). The next step, that of following the labelled litter substrates through the saprotroph food web, is in its infancy. There have been elegant attempts at demonstrating nutrient transfer between plants and mycorrhizae, both in the laboratory and in the field (Bethenfalvy *et al.*, 1991; Simard *et al.*, 1997). These need to be taken to a finer resolution, to measure rates of nutrient transfer between species in functional groups, to include protozoa, nematodes, microarthropods and other organisms. There are exciting opportunities in the years ahead in following the route of substrate transformation through decomposition, instead of whole soil chemical extractions of organic matter fractions.

Two types of isotope tracers are useful. The first kind involves enrichment with radioactive or rare isotopes of elements which, when incorporated into substrates. can be traced through the food web as molecules are absorbed into cellular biomass, secreted, excreted or respired (Fig. 6.3). The radioactive isotopes are detected by scintillation counts of extracted samples if there is sufficient radioactivity in samples. Usually, more refined techniques rely on mass spectrometry to separate the isotopes. This method requires much less sample and it is more sensitive, though more expensive to perform. Alternatively, autoradiography can be used to detect cells and organisms with the radioactive isotope on microscope slides. Good resolution with small samples is also possible, using a thermal combustion-elemental analyser for stable isotope analysis, followed by gas chromatogram separation (instrument Delta-plus XL) (Werner *et al.*, 1999). The fate of the labelled atom through molecular transformations can be traced through functional groups and species, and the amount and rate of transfer can be quantified. For example, ^{33}P, ^{15}N, ^{13}C and ^{14}C enrichment into a substrate can be followed through soil fractions or species through time. Typically, in microcosms, ^{15}N, ^{14}C and ^{3}H are incorporated into substrate molecules. ^{14}C is a useful tracer for organic molecules in general. For example, the fate of a particular C in cellulose, chitin or an amino acid can be traced, to determine what fraction is respired into CO_2 and accumulated into saprotroph biomass over a time period. ^{14}C can be traced through excreted organic matter to monitor its subsequent absorption and digestion by successive species. For example, Kuzyakov (1997) followed the turnover of amino acids and nucleic acids using ^{15}N- and ^{14}C-labelled molecules. In this study, uracil, glycine and alanine labelled at different positions with ^{15}N and ^{14}C were applied to red clover root residues and followed for more than 1 year. The radioactive isotope ^{3}H is useful primarily in thymidine because it is incorporated specifically into DNA at replication. The usefulness of ^{3}H is in autoradiography, because the emitted β particle travel path is very short, and the particle remains near the cell. For primary saprotrophs, uptake is through cell membrane transport and osmotrophy. For specific incorporation of ^{3}H into DNA, the uptake by secondary saprotrophs and other consumers must be through prey ingestion, not through membrane transport (or osmotrophy) (Berger and Kimball, 1964; Berger, 1971). Its incorporation into cells is a clear indication that DNA synthesis has occurred and that the cell was active and growing.

The second type of isotopic tracers rely on natural abundances that can be assayed if differentially enriched substrates can be found (Nadelhoffer *et al.*, 1996; Gerzabek *et al.*, 2001a,b; Robinson, 2001). This method applied to natural abundances of ^{15}N is not easy to carry out, as it is difficult to obtain samples that are sufficiently different in nitrogen isotopic composition (Robinson, 2001). The technique is easier with respect to carbon isotopes (Mathers *et al.*, 2000). The organic matter C isotopic composition is similar to the litter and living species from which it originates. However, C3 and C4 plants have different ^{13}C:^{12}C ratios, so that residue from one type of plant can be used as a tracer in soil grown with the other type of plant. Another approach to tracing ^{14}C in SOM and through the profile is from 'bomb-produced radiocarbon'. This technique takes advantage of the accumulation of radioactive ^{14}C isotopes in the atmosphere as a consequence of nuclear weapon detonation above-ground. These explosions began in the 1950s and continued through most of the cold war. Accurate values for the amount accumulated in the atmosphere through time exist for reference. The differences in accumulated ^{14}C isotope are large enough to detect changes at the annual or decadal scale from natural samples.

In an important study, Gaudinski *et al.* (2000) calculated estimates for residence time and fluxes of soil litter fractions at Harvard Forest, based on the respired 'bomb-radiocarbon' content. The research site is a well-drained mixed deciduous temperate forest in a hilly area in Massachusetts. The average time from C fixation by plant photosynthesis to its return into the atmosphere through decomposition is important to determine the magnitude of net C flow. Inter-annual variations in the rate of net flux have been linked to climate (Goulden *et al.*, 1996; Fung *et al.*, 1997). Total soil respiration was partitioned into two components: one containing root respiration and microbial metabolism or photosynthates less than 1 year since C fixation, and the second containing CO_2 from decomposition of litter that resides in the soil longer than 1 year. The litter was subdivided into four fractions for modelling purposes: recognizable leaf litter; recognizable fine root litter (<2 mm); humus and decomposed litter not bound to mineral surfaces; and decomposed organic matter associated with mineral surfaces. Leaching represented a small fraction of loss from the system, so that the study assumed all loss of soil C to be from decomposition and respiration (CO_2) or transfer from one litter fraction to another. The percentage carbon isotopic allocation into litter fractions and respired fractions indicates that about one-third of the annual litter input is respired during the first year in this forest (Fig. 4.3). The results also showed that about 34–51% of the annual CO_2 was released from decomposition of low-density organic matter that resides in the soil more than 1 year. In particular, the study identified an important point about the impact of the heterogeneity of the soil profile, and of the mixed source of organic matter, when averaging results from soil samples and subfractions. For example, the authors point out that even new fine root biomass contains carbon that was fixed in the past and translocated from plant storage tissues; or obtained from new or old litter though mycorrhizae. Therefore, the developmental age of the fine root itself may be new, but dating it by 'bomb-radiocarbon' is compromised by the source of the carbon used. Moreover, the paper indicates that different parts of the same <1 mm diameter fine root can contain bomb-radiocarbon different in age by 2 years. Therefore, the tip may turn over rapidly, but the soil-extractable longer and older part with a lower turnover rate will bias the data. This introduces an ambiguity in the interpretation of data. In general, the authors warn that fast-cycling pools of organic matter, which would include the saprotroph biomass, are obscured by the much larger pools of recalcitrant or less labile organic matter. The fast-cycling modern carbon becomes isotopically diluted beyond detection in the mineral horizons. Taking into account heterogeneity through the profile, and trying to account for some of the fast-cycling but small carbon pools, corrects the short- and long-term carbon budget estimates. Without accounting for the fast-cycling pools and soil heterogeneity, the accumulated soil carbon pool in their models was overestimated by about 300 g C/m^2 over 10 years, and by 550 g C/m^2 over 100 years. It should be expected that refining estimates of turnover rate through the saprotrophs, using isotope tracer techniques, and accounting for subfractions of labile organic matter sources, should further reduce previous estimates of the carbon pool age and decrease estimates of organic matter sequestration. This point is particularly important with respect to global warming carbon sequestration predictions.

Soil Food Web Models

There have been several attempts at describing the overall structure of soil food webs. These studies were conducted mostly on agricultural fields, which generally have a reduced species diversity and are easier to describe. The saprotrophs were grouped into categories, and the amount of nutrient flux through various compartments was estimated. These estimates were used to construct models of decomposition processes based on trophic interactions between the compartments. The models were then used to determine stability or dynamic behaviour of the food web under simulated manipulations.

Regulation of population growth

Growth of a population of a species is usually described mathematically with variations of the logistic growth models (Krebs, 2001; see Chapter 5). There are numerous parameters that will regulate growth in a **density-dependent** manner. As the number of individuals increases, in a habitat-restricted situation where the organisms are confined to a particular space or volume of soil, density-dependent factors become more important. Theoretically, these include food limitation, low oxygen and crowding, but can also include increased predation or viral epidemics. There are very few examples of density-dependent population growth limitation in natural soils, primarily because the experiments were not conducted in open field settings. In microcosms, it is difficult to be convinced that species were not spatially constrained excessively, or that microcosms mimic *in situ* dynamics. That would produce artefactual results. In part, the effects of density-dependent reduction in reproduction rates or fertility become evident over longer time periods than the duration of short microcosm experiments. There are also **density-independent** factors that regulate the growth and abundance of individuals. These include soil temperature, moisture and aeration status, on a diurnal to seasonal scale. It can be argued that in soil, substrate limitation or overcrowding is probably infrequent. Density-independent parameters would then determine whether species are active or inactive according to a range of suitable conditions for activity of each species. In reality, both density-dependent and density-independent parameters must be considered. *However, they may not affect the food web at the same spatial scale or time scale.* For example, density-independent parameters are more important in regulating population sizes in the short term. Individuals respond immediately to changes in temperature, oxygen availability, substrate and food availability (feeding rate), through physiological changes in the metabolic rate and mobility. The effect of variations in food availability, and of frequency and duration of inactivity, affect different species at different time scales. Protists and bacteria population dynamics vary over days, but it would be months to years before any changes in invertebrate populations would be seen. What is more complicated with soil species is that models based on simple birth–death rates and reproduction frequency do not account for species activity–inactivity transitions. It is erroneous to assume that most species are active most of the time (see Chapters 1 and 5). Furthermore, we do not have clear data on the potential for species to migrate distances through the soil pore reticulum, between suitable microsites. The space occupied by one population of earthworms may be several hundreds of metres in area, but only a few centimetres or less for a protozoan. Some microhabitats are suitable for some species, but not for others. In part, we can obtain an indirect idea of the extent of migration by creating maps of species distribution in space. The correct spatial scale and response time for species varies and needs to be considered in models.

Regulation of nutrient flux rates

The active individuals in a population represent the organisms responsible for a particular set of trophic behaviour and litter transformation. These individuals contribute to a trophic functional group activity (Chapter 4). The rate of food intake (soluble nutrients by osmotrophy and cell membrane transport, prey by phagocytosis or predation) by active individuals varies with abiotic conditions, especially temperature, moisture and aeration. The rate is also affected by availability of food (abundance) and suitability of the food resources. The role of each cell or individual in the food web is best represented as a catalyst for decomposition, whereby food resources (the substrate) are metabolized into new biomass, and excreted wastes and energy (Fig. 6.3). In particular, the secondary saprotrophs and consumers continuously recycle biomass to litter and soluble soil solution (Fig. 6.2). As we have seen earlier (Chapter 5), the functional response curves of populations resemble enzyme kinetics. At a different spatial scale, ecosystem level biogeochemical processes also are catalysed reactions which invariably show Michaelis–Menten type kinetics. The role of individuals in these food resource transformations can be written as $dX/dt = kES/(K_m + S)$, where X is the concentration of the excreted product, k is the reaction constant, E is the concentration of the catalyst (individuals), S is the food resource concentration, and K_m is the half-saturation value (or $1/2V_{max}$). The value for kE is often constant in the short term and is expressed as V_{max}. In the soil, decomposition proceeds as the sum of a large number of species interactions in the food web. Each interaction is responsible for a small part of the overall litter transformation. These are grouped into functional categories for simplification, but the underlying mechanism is one of interactions between different species, which are active under a narrow range of conditions.

This expression can be simplified to approximate a first-order reaction under certain circumstances (see Schimmel, 2001). (i) In cases where a narrow range of food resource concentration (S) is considered, a part of the type II (or type III) functional response will appear to resemble linear first-order kinetics. (ii) In cases where it is assumed that the litter transformations are never resource limited, or limited by absence of key species or functional groups. In other words, if all functional groups are always present, or new colonization restores all functional groups rapidly, then a first-order approximation could be valid. (iii) In cases where it is assumed that saprotroph overall activity and community structure are not affected by environmental conditions. Here it is assumed that all functional groups will be active and decomposition will proceed, even when species composition changes dramatically as a result of changes in the environment. (iv) If the substrate concentration or food resources are very low, then $dX/dt = kE/(K_m)$, or $dX/dt = V_{max}/K_m$. How reasonable this assumption is has not been tested sufficiently in the field. If some species are severely resource or habitat limited, then one could observe type II or type III functional responses by resource supplements, until V_{max} is reached. As Schimmel (2001) pointed out, it is not always very clear whether the first-order kinetics assumptions used in models are correct. Moreover, since models tend to incorporate the role of saprotrophs (usually only primary saprotrophs) within the equation constants and response functions, their *implicit* role is acknowledged. This simplification allows assumptions to be modelled, to make predictions. The predictions can be tested against data for validity. However, without an *explicit* role for saprotrophs in the models, it is almost impossible to dwell on the mechanisms and underlying processes that transform litter. Therefore, without a mechanistic explicit role for decomposition, models could be an important source of error in predictions.

It is increasingly dubious from a biological perspective whether soil communities recover rapidly from stress and disturbance. Clearly some species are good colonizers, but it is unlikely that all functional groups return rapidly. It is also unclear whether sufficient early colonizers return to maintain all functional groups active throughout the seasons. In general, old mining sites, abandoned bare mineral soil, clear-cut forests and agricultural fields each require years to decades to accumulate all functional groups, and even more years to accumulate species diversity resembling undisturbed sites. These sites tend to be dominated by bacterial communities at first, with some bacterivory. The full complement of testate amoebae, ciliates, nanoflagellates and fungal saprotrophs requires much longer in the succession sequence to accumulate. The majority of species of microarthropods, enchytraeids and earthworms accumulate slowly because they are poor at dispersal. In particular, some testate amoebae, ciliates and predatory nematodes and microarthropods are sensitive to disturbance. They are not resilient and are poor colonizers. Therefore, it is unlikely that some of the first-order kinetics assumptions above are valid, unless a narrow range of food resource abundance or environmental disturbance is considered. However, when those assumptions are valid, then the litter decomposition rate could be a simple function of litter chemistry, and actual evapotranspiration (moisture and temperature), at least for some of the primary saprotrophic functions.

Effect of food web structure on decomposition

A mechanistic understanding of decomposition food webs needs a good understanding of the underlying species–species interactions that structure communities. Traditionally, soil food webs were regarded as 'donor controlled', with litter input controlling the primary saprotrophs, with little consideration for the rest of the food web. Many experimental studies showed that overall food web structure or single species of consumers could affect rates of decomposition or nutrient cycling. More recently, decomposition food webs have been regarded as similar to other food webs, if all the functional groups are considered.

Based on these observations, Zheng *et al.* (1997) described a model that tried to link population dynamics with ecosystem level carbon cycling. The paper assumes that at steady state, $X_o = I/k$, where X_o is the steady-state litter mass, with input rate I and decomposition rate k which can be measured. The model was tested with fixed substrate quality and abiotic parameters. The decomposition rate *k* is sensitive to changes in the structure of the food web. One can imagine that a chain of three trophic levels is more rigid than a linked food web with several interacting functional groups. In reality, there are also changes in the quantity, quality and type of substrate that becomes available through time. Therefore, with time, the composition of the community changes, so that different species may be more or less efficient than previous ones, and the food web can be structured differently. Therefore, *k* becomes a function of the food web structure. The model also incorporates a 'recycling index' to account for the amount of litter that becomes available to consumers through primary saprotrophs. The consumer compartments account for predation, mortality, respired carbon, excreted carbon and new biomass from growth. The Lotka–Volterra model was more versatile and could even describe food web structures where increased litter input did not increase decomposition rate. Decomposition rate increased or decreased depending on whether an odd or even number of trophic levels existed. It could also predict biomass cascades between trophic levels. Both of these have been observed in field studies (see Zheng *et al.*, 1997). Donor control models only predict pyramid-shaped biomass at successive trophic levels, with increasing decomposition rate as more trophic levels are added. The authors suggest that field studies should describe in more detail the forms of interactions between functional groups. In particular, irrespective of biomass, the impact of a functional group on the rate of decomposition should vary with its species composition. Staying within the idea of individuals behaving as enzymes that process food or prey, a small number of individuals process a large amount of 'substrate' through time. Species that process litter or prey at a high rate would have a greater effect on the overall decomposition rate, irrespective of the biomass. In fact, bacterivorous protozoa are notoriously low in biomass, but high in impact on decomposition rate (Paustian *et al.*, 1990; Sohlenius, 1990; Griffiths *et al.*, 1999b).

One recent study attempted to apply donor control models by assuming that forest decomposition was a litter-based food chain (Ponsard *et al.*, 2000). The study focused on macroinvertebrates in the litter and their predators. Data on populations were obtained along a natural gradient of litter input in mixed deciduous mature forest sites near Orsay in France. The surface litter detritivores were assumed to be largely isolated from the interstitial food web. The seasonal dynamics observed in abundances varied in part with temperature and weather-related activity changes. They were also caused in part by life history succession in different species. Therefore, true predator–prey dynamics were not observed in this forest experiment. The models showed unrealistic interference between predators, so that the share of food for each predator decreases faster than the amount of increase in the number of competitors. Both predators and detritivores increased in abundance with increased litter input. This argues that the system was not strictly predator controlled (or strict top-down control). Theoretically, a strict top-down-controlled detritivore population would be insensitive to changes in its resources. At least some bottom-up control (or donor control) on the macroinvertebrate abundances must be postulated to explain the results. The study underlines the need to include more realistic trophic interactions. Death of detritivores was assumed to be by macroinvertebrate predation alone. However, eggs and juvenile stages are preyed on by microinvertebrates and are susceptible to fungal infections. Similarly, juvenile stages could be in competition with other soil fauna that were not considered. Development and survival of eggs and juveniles are affected by weather and abiotic conditions, and the overall fitness of the populations is affected by past conditions.

A different and more integrative approach to modelling decomposition food webs was proposed by Moore and de Ruiter (1991, and see de Ruiter *et al.*, 1995, 1998). In this approach, species are placed into functional groups based on food preferences and taxonomy. Then the flow of nutrients and energy between compartments is analysed. This approach aims to project the food web on to food, habitat and time as the principal niche dimensions. Community structure is measured by identifying functional groups, their biomass, activity and rate of nutrient flow between compartments. The connectedness and interaction strengths between compartments are analysed through simulations. The same model structure is applied to different sites, with adjustments for local abiotic and physiological variables. The first kind of model relies on observed population dynamics to calculate mineralization rates. Then the calculations are compared with observed estimates. In a second kind of model, the population dynamics are simulated to analyse the role of particular trophic interactions. These models were applied largely to various agro-ecosystems with respect to field management practices. Again, these models indicate that secondary saprotrophs and consumers have a large impact on nutrient cycling, despite their relatively small biomass. In particular, the protozoa dominate the faunal mineralization (de Ruiter *et al.*, 1995). The consumers are responsible for the high turnover rate of nutrients and recycling of the primary saprotrophs into litter and soil solution (Fig. 6.2). The effect of indirect interactions through food web population dynamics are significant. Population dynamics effects of perturbations were modelled using Lotka–Volterra equations by varying the predation coefficient of a specific trophic interaction. Further analysis of these models indicates the simultaneous co-occurrence of strong top-down effects at lower trophic levels, and strong bottom-up effects at higher trophic levels (de Ruiter *et al.*, 1998; Moore and de Ruiter, 2000). Some trophic interactions were shown to have strong effects on community food web stability, that did not correlate with interaction strength nor with feeding rates. Therefore, certain keystone interactions from apparently small interactions can only be identified through food web structure analysis and simulations.

In these studies, it becomes clear that a lack of detailed physiological understanding of species functioning interferes with data collection and analysis. There is a paucity of data on food preferences and feeding rates for many taxa. Available field data are dominated by overly simplified models of trophic interactions. Many taxa are rarely enumerated from samples, such as nanoflagellates, amoebae, testate amoebae, enchytraeids, rotifers and tardigrades. For example, the models used by Moore and de Ruiter do not account for the testate amoebae, rotifers or tardigrades. Fungivory in general is underestimated because it is assumed that only certain microinvertebrates are implicated. There remain exciting prospects for field sampling and microcosm tracer studies to elaborate on modelling decomposition food webs. Evaluation of model predictions and further development of food web analysis will require field experiments to test simulations and to verify predicted rates of activity and population dynamics. It will require a more complete description of food web components, and refining trophic functional groups. An ecosystem approach is essential to understanding nutrient flow through food web communities. However, estimates of nutrient transfer rates and individual trophic interactions require an understanding of the species implicated, and of their behavioural response to the fluctuating environment.

Summary

Decomposition is the sum of a large number of species interactions that occur in the soil food web. These can be simplified into a set of interactions among functional groups. An ecosystem approach is necessary to understand both food web dynamics and nutrient transformations. Tracer studies using stable or radioactive isotopes are sensitive tools that provide adequate resolution from whole soil samples to cell autoradiography. The appropriate spatial scale and response time in food web dynamics vary between species. Both abiotic conditions and density-dependent variables affect the dynamics of species activity through time. Species in decomposition food webs resemble enzyme processes that transform substrate: a small number of individuals can process a large amount of nutrients or prey. Models can simulate and predict *in situ* observations. Mechanistic models need an explicit role for species, and an understanding of the behaviour of species in the community. Without mechanistic models, we cannot understand the underlying processes that transform litter.

Suggested Further Reading

Coleman, D.C. and Hendrix, P.F. (2000) *Invertebrates as Webmasters in Ecosystems*. CAB International, Wallingford, UK.

Wall, D.H. and Moore, J.C. (1999) Interactions underground; soil biodiversity, mutualism, and ecosystem processes. *BioScience* 49, 109–117.

Wardle, D.A. (2002) *Communities and Ecosystems; Linking the Aboveground and Belowground Components*. Princeton University Press, Princeton, New Jersey.

References

Adl, M.S. (1998) Regulation of cell cycle duration, cell size and life history in the ciliate *Sterkiella histriomuscorum*. PhD thesis, University of British Columbia, Vancouver, Canada.

Adl, M.S. and Berger, J.D. (1996) Commitment to division in ciliate cell cycles. *Journal of Eukaryotic Microbiology* 43, 77–86.

Aerts, R. (1997) Climate, leaf litter chemistry and leaf litter decomposition in terrestrial ecosystems: a triangular relationship. *Oikos* 79, 439–449.

Agerer, R. (1987–1993) *Colour Atlas of Ectomycorrhizae*. Einhrn-Verlag, Schwabisch Gmund, Germany.

Alexander, C. and Hadley, G. (1984) The effect of mycorrhizal infection of *Goodyera repens* and its control by fungicide. *New Phytologist* 97, 391–400.

Alexander, C. and Hadley, G. (1985) Carbon movement between host and mycorrhizal endophyte during development of the orchid *Goodyera repens*. *New Phytologist* 101, 657–665.

Alm, E.W., Zheng, D. and Raskin, L. (2000) The presence of humic substances and DNA in RNA extracts affects hybridization results. *Applied and Environmental Microbiology* 66, 4547–4554.

Alphei, J., Bonkowski, M. and Scheu, S. (1996) Protozoa, Nematoda and Lumbricidae in the rhizosphere of *Hordelymus euroaeus* (Poaceae): faunal interactions, response of microorganisms and effects on plant growth. *Oecologia* 106, 111–126.

Alvim, P. de T. (1978) Perspectivas de producão agrícola na região amazônica. *Interciencia* 3, 243–249.

Alvim, P. de T. and Cabala, F.P. (1974) Um novo sistema de representacão gráfica da fertilidade dos solos para cacao. *Cacau Atualidades (Ilhéus/Bahia)* 11, 2–6.

Amundson, R. (2001) The carbon budget in soils. *Annual Reviews in Earth and Planetary Science* 29, 535–562.

Anderson, J.M. (1988) Invertebrate-mediated transport processes in soils. *Agriculture, Ecosystems and Environment* 24, 5–19.

Anderson, O.R. (1989) Some observations of feeding behaviour, growth and test particle morphlogy of a silica-secreting testate amoeba, *Netzelia tuberculata* Wallich (Rhizopoda, Testacea) grown in laboratory culture. *Archiv für Protistenkunde* 137, 211–221.

Anderson, O.R. and Rogerson, A. (1995) Annual abundances and growth potential of Gymnamoebae in the Hudson Estuary with comparative data from the Firth of Clyde. *European Journal of Protistology* 31, 223–233.

Anderson, R.V., Ingham, R.E., Trofymoy, J.A. and Coleman, D.C. (1984) Soil mesofaunal distribution in relation to habitat types in a shortgrass prairie. *Pedobiologia* 26, 257–261.

Anderson, T.R. and Patrick, Z.A. (1980) Soil vampyrellid amoebae that cause small perforations in conidia of *Cochliobolus sativus*. *Soil Biology and Biochemistry* 12, 159–167.

André, M. (1949) Ordre des Acariens. In: Grassé, P.-P. (ed.) *Traité de Zoologie: Anatomie, Systématique, Biologie*, Vol VI. Publ. Masson et Co., Paris, pp. 794–892.

Angle, J.S. (1994) Viruses. In: Weaver, R.W., Angle, S., Bottomley, P., Bezdicek, D., Smith, S., Tabatabai, A. and Wllum, A. (eds) *Methods of Soil Analysis, Part 2. Microbiological and Biochemical Properties*. Soil Science Society of America, Madison, Wisconsin, pp. 107–118.

Anstett, M. (1951) Sur l'activation macrobiologique des phénomènes d'humification. *Comptes Rendues Hebdomadaire des Séances de l'Académie Agricole de France* p. 230.

Armendariz, I. and Arpin, P. (1997) Nematodes and their relationship to forest dynamics: I. Species and trophic groups. *Biology and Fertility of Soils* 23, 405–413.

Arnolds, E. (1992) The analysis and classification of fungal communities with special reference to macrofungi. *Handbook of Vegetation Science* 19, 7–48.

Augé, R.M. (2001) Water relations, drought, and vesicular–arbuscular mycorrhizal symbiosis. *Mycorrhiza* 11, 3–42.

Avel, M. (1949) Classe des annélides oligochaetes. In: Grassé, P.-P. (ed.) *Traité de Zoologie: Anatomie, Systématique, Biologie*, Vol. V. Publ. Masson et Co., Paris, pp. 224–470.

Bachelier, G. (1963) *La Vie Animale dans les Sols*. Orston, Paris.

Bago, B., Pfeffer, P.E. and Shachar-Hill, Y. (2000) Carbon metabolism and transport in arbuscular mycorrhizas. *Plant Physiology* 124, 949–957.

Bakonyi, G. (1989) Effects of *Folsomia candida* on the microbial biomass in a grassland soil. *Biology and Fertility of Soils* 7, 138–141.

Bakonyi, G. and Nagy, P. (2000) Temperature and moisture-induced changes in the structure of the nematode fauna of a semiarid grassland—patterns and mechanisms. *Global Change Biology* 6, 697–707.

Balczon, J.M. and Pratt, J.R. (1996) The functional responses of two benthic algivorous ciliated protozoa with differing feeding strategies. *Microbial Ecology* 31, 209–224.

Bardele, C.F., Foissner, W. and Blanton, R.L. (1991) Morphology, morphogenesis and systematic position of the sorocarp forming ciliate *Sorogena stoinaovitchae* (Bradbury and Olive 1980). *Journal of Protozoology* 38, 7–17.

Barker, K.R. (1978) Determining population responses to control agents. In: Zehr, E.I. (ed.) *Methods for Evaluating Plant Fungicide, Nematicide and Bactericides*. American Phytopathological Society, St Paul, Minnesota.

Barker, S.J. and Tagu, D. (2000) The roles of auxins and cytokinins in mycorrhizal symbiosis. *Journal of Plant Growth Regulation* 19, 144–154.

Barr, D.J. (1983) Zoosporic grouping of plant pathogens. In: Buczacki, S.T. (ed.) *Zoosporic Plant Pathogens. A Modern Perspective*. Academic Press, London, pp. 43–83.

Barron, G.L. (1977) *The Nematode Destroying Fungi*. Lancaster Press Inc., Lancaster, Pennsylvania.

Barron, G.L. (1988) Microcolonies of bacteria as a nutrient source for lignicolus and other fungi. *Canadian Journal of Botany* 66, 2505–2510.

Bauhus, J. and Messier, C. (1999) Soil exploitation strategies of fine roots in different tree species of the southern boreal forest of eastern Canada. *Canadian Journal of Forest Research* 29, 260–273.

Beare, M.H., Coleman, D.C., Crossley, D.A., Hendrix, P.F. and Odum, E.P. (1995) A hierarchical approach to evaluating the significance of soil biodiversity to biogeochemical cycling. *Plant and Soil* 170, 5–22.

Beatty, G.A. and Lindow, S.E. (1999) Bacterial colonization of leaves: a spectrum of strategies. *Phytopathology* 89, 353–359.

Behan, M.V. and Hill, S.B. (1978) Feeding habits and spore dispersal of Oribatid mites in the North American arctic. *Revue d'Ecologie et Biologie du Sol* 15, 497–516.

Bellido, A. and Deleporte, S. (1994) Oribatid mites/Diptera interactions in a deciduous forest leaf-litter: an application of multivariate analysis under linear constraints. *Pedobiologia* 38, 429–447.

Bengtsson, J., Persson, T. and Lundkvist, H. (1997) Long-term effects of logging residue addition and removal on macroarthropods and enchytraeids. *Journal of Applied Ecology* 34, 1014–1022.

Berg, B. and Ekbohm, G. (1991) Litter mass-loss rates and decomposition patterns in some needle and leaf litter types. Litter decomposition in a Scots pine forest. VII. *Canadian Journal of Botany* 69, 1449–1456.

Berger, J.D. (1971) Kinetics of incorporation of DNA precursors from ingested bacteria into macronuclear DNA of *Paramecium tetraurelia*. *Journal of Protozoology* 18, 419–429.

Berger, J.D. and Kimball, R.F. (1964) Specific incorporation of precursors into DNA by feeding labelled bacteria to *Paramecium tetraurelia*. *Journal of Protozoology* 11, 534–537.

Berthold, A. and Palzenberger, M. (1995) Comparison between direct counts of active soil ciliates and most probable number estimates obtained by Singh's dilution culture method. *Biology and Fertility of Soils* 19, 348–356.

Bethenfalvy, G.J. *et al.* (1991) Nutrient transfer between the root zones of soybean and maize plants connected by a common mycorrhizal mycelium. *Physiologia Plantarum* 82, 423–432.

Beyrle, H. (1995) The role of phytohormones in the function and biology of mycorrhizas. In: Varma, A. and Hock, B. (eds) *Mycorrhiza: Structure, Function, Molecular Biology and Biotechnology*. Springer-Verlag, New York, pp. 365–390.

Biagini, G.A., Kirk, K., Schofield, P.J. and Edwards, M.R. (2000) Role of K^+ and amino acids in osmoregulation by the free-living microaerophilic protozoon *Hexamita inflata*. *Microbiology* 146, 427–433.

Binot, F., Hallaire, V. and Curmi, P. (1997) Agricultural practises and the spatial distribution of earthworms in maize fields. *Soil Biology and Biochemistry* 29, 577–583.

Biswal, B. and Biswal, V.C. (1999) Leaf senescence: physiology and molecular biology. *Current Science* 77, 775–782.

Black, C.A., Evans, D.D., White, J.I., Ensminger, L.E. and Clark, F.E. (1982) *Methods of Soil Analysis. Part 1.* American Society of Agronomy and Soil Science Society of America, Madison, Wisconsin.

Blackshaw, R.P. (1997) Life cycle of the earthworm predator *Artioposthia triangulata* (Dendy) in Northern Ireland. *Soil Biology and Biochemistry* 29, 245–249.

Blair, J.M., Parmelee, R.W. and Beare, M.H. (1990) Decay rates, nitrogen fluxes, and decomposer communities of single- and mixed species foliar litter. *Ecology* 71, 1976–1985.

Blanchette, R.A. (1991) Delignification by wood decay fungi. *Annual Review of Phytopathology* 29, 381–398.

Blanchette, R.A. (1995) Degradation of the ligninocellulose complex in wood. *Canadian Journal of Botany* 73 (Supplement 1), S999–S1010.

Block (1968) *Revue de l' Ecologie et Biologie du Sol* 5, 515–521.

Bobrov, A.A., Yazvenko, S.B. and Warner, B.G. (1995) Taxonomic and ecological implications of shell morphology of three testaceans (Protozoa: Rhizopoda) in Russia and Canada. *Archiv für Protistenkunde* 145, 119–126.

Boddy, L. (2000) Interspecific combative interactions between wood decaying Basidiomycetes. *FEMS Microbiology Ecology* 31, 185–194.

Boddy, L. and Watkinson, S.C. (1995) Wood decomposition, higher fungi, and their role in nutrient redistribution. *Canadian Journal of Botany* 73 (Supplement 1), S1377–S1383.

Bodenheimer, F.S. and Reich, K. (1933) Studies on soil protozoa. *Soil Science* 38, 259–265.

Boenigk, J. and Arndt, H. (2000) Particle handling during interception feeding by four species of heterotrophic nanoflagellates. *Journal of Eukaryotic Microbiology* 47, 350–358.

Boenigk, J., Matz, C., Jurgens, K. and Arndt, H. (2001) Confusing selective feeding with differential digestion in bacterivorous nanoflagellates. *Journal of Eukaryotic Microbiology* 48, 425–432.

Bongers, T. and Bongers, M. (1998) Functional diversity of nematodes. *Applied Soil Ecology* 10, 239–251.

Bonkowski, M. and Schaefer, M. (1997) Interactions between earthworms and soil protozoa: a trophic component in the soil food web. *Soil Biology and Biochemistry* 29, 499–502.

Bonkowski, M., Cheng, W., Griffiths, B.S., Alphei, J. and Scheu, S. (2000a) Microbial faunal interactions in the rhizosphere and effects on plant growth. *European Journal of Soil Biology* 36, 135–147.

Bonkowski, M., Griffiths, B.S. and Ritz, K. (2000b) Food preferences of earthworms for soil fungi. *Pedobiologia* 44, 666–676.

Bonnet, L. (1964) Le peuplement thecamoebien des sols. *Revue d'Ecologie et Biologie du Sol* 1, 123–408.

Bouché, M.B. (1977) Stratégies lombricienne. *Ecological Bulletin (Stockholm)* 25, 122–132.

Boyd, R.F. (1995) *Basic Medical Microbiology*, 5th edn. Little Brown and Co., Boston.

Breznak, J.A. and Brune, A. (1994) Role of microorganisms in the digestion of lignocellulose by termites. *Annual Reviews in Entomology* 39, 453–487.

Bruin, J., Geest, L.P.S. and Sabelis, M.W. (1999) *Ecology and Evolution of the Acari*. Kluwer Academic Publishers, Dordrecht, The Netherlands.

Buitkamp, U. (1979) Vergleichende Untersuchungen zur Temperaturadaptation von Bodenciliaten aus klimatisch verschiedenen Reigionen. *Pedobiologia* 19, 221–236.

Burgmann, H., Pesaro, M., Widmer, F. and Zeyer, J. (2001) A strategy for optimizing quality and quantity of DNA extracted from soil. *Journal of Microbiological Methods* 45, 7–20.

Byrd, J.H. and Castner, J.L. (2001) Insects of forensic importance. In: Byrd, J.H. and Castner, J.L. (eds) *The Utility of Arthropods in Legal Investigations*. CRC Press, Boca Raton, Florida, pp. 45–79.

Cadish, G. and Giller, K.E. (1997) *Driven by Nature: Plant Litter Quality and Decomposition*. CAB International, Wallingford, UK.

Carlile, M.J. (1996) The discovery of fungal sex hormones. II. Antheridiol. *Mycologist* 10, 113–117.

Carlile, M.J., Watkinson, S.C. and Gooday, G.W. (2001) *The Fungi*, 2nd edn. Academic Press, New York.

Casper, B. and Jackson, R.B. (1997) Plant competition underground. *Annual Review of Ecology and Systematics* 28, 545–570.

Cavalier-Smith, T. (1998) A revised six-kingdom system of life. *Biology Reviews (Cambridge)* 73, 203–266.

Cayrol, J.-C. (1989) Les toxines nematicides des champignons. *Revue Horticole* 293, 53–57.

Chakraborty, S. (1984) Mycophagous Amoebae in a Pasture Soil in Relation to Take-all Disease of Wheat. PhD thesis, Department of Plant Pathology, University of Adelaide.

Chakraborty, S. (1985) Survival of wheat take-all fungus in suppressive and non-suppressive soils. *Pedobiologia* 28, 13–18.

Chakraborty, S. and Old, K.M. (1982) Mycophagous soil amoeba: interactions with three plant pathogenic fungi. *Soil Biology and Biochemistry* 14, 247–255.

Chakraborty, S. and Warcup, J.H. (1985) Reduction of take-all by mycophagous amoebae in pot bioassays. In: Parker, C.A., Moore, K.S., Wong, P.T.W., Rovira, A.D. and Kollmorgan, J.F. (eds) *The Ecology and Management of Soil Borne Plant Pathogens*. The American Phytopathological Society, St Paul, Minnesota, pp. 107–109.

Chakraborty, S., Theodorou, C. and Bowen, G.D. (1985) The reduction of root colonization by mycorrhizal fungi by mycophagous amoebae. *Canadian Journal of Microbiology* 31, 295–297.

Chapman, K., Whittaker, J.B. and Heal, O.W. (1988) Metabolic and faunal activity in litters of tree mixtures compared with pure stands. *Agriculture, Ecosystems and Environment* 24, 33–40.

Chardez, D. (1968) Etudes statistiques sur l'écologie et la morphologie des thécamoebiens (Protozoa, Rhizopoda, Testacea). *Hydrobiologia* 32, 271–287.

Chardez, D. (1985) Protozoaires prédateurs de thécamoebiens. *Protistologica* 21, 187–194.

Cheshire, M.V., Dumat, C., Fraser, A.R., Hillier, S. and Staunton, S. (2000) The interaction between soil organic matter and soil clay minerals by selective removal and controlled addition of organic matter. *European Journal of Soil Science* 51, 497–509.

Chidthaisong, A. and Conrad, R. (2000) Turnover of glucose and acetate coupled to reduction of nitrate, ferric iron and sulphate to methanogenesis in anoxic rice field soil. *FEMS Microbiology Ecology* 31, 73–86.

Choo-Smith, L.P., Maquelin, K., van Vreeswijk, T., Bruining, H.A., Puppels, G.J., Thi, N.A.G., Kirschner, C., Naumann, D., Ami, D., Villa, A.M., Doglia, S.M., Lamfarraj, H., Sockalingum, G.D., Manfait, M., Allouch, P. and Endtz, H.P. (2001) Investigating microbial (micro)colony heterogeneity by vibrational spectroscopy. *Applied Environmental Microbiology* 67, 1461–1469.

Chotte, J.L., Ladd, J.N. and Amato, M. (1998) Sites of microbial assimilation and turnover of soluble and particulate 14C-labelled substrates decomposing in a clay soil. *Soil Biology and Biochemistry* 30, 205–218.

Christiansen, K., Doyle, M., Kahlert, M. and Golabeza, D. (1992) Interspecific interactions between collembolan populations in culture. *Pedobiologia* 36, 274–286.

Clark, F.E. (1969) Ecological associations among soil micro-organisms. Soil Biology, *UNESCO – Natural Resources Research* IX, 129–131.

Cleveland, L.R. and Grimstone, A.V. (1964) The fine structure of the flagellate of *Mixotricha paradoxa* and its associated microorganisms. *Proceedings of the Royal Society, London, Series B*, 159, 668–686.

Coineau, Y., Haupt, J., Delamere-Debouteville, C. and Théron, P. (1978) Un remarquable example de convergence écologique: l'adaptation de *Gordialycus tuzetae* (Nematalycidae, Acariens) à la vie dans les interstices des sables fins. *Comptes Rendues de l'Académie des Sciences* 287, 883–886.

Coleman, D.C. (1976) A review of root production processes and their influence on soil biota in terrestrial ecosystems. In: Anderson, J.M. and Macfadyen, A. (eds) *The Role of Terrestrial and Aquatic Organisms in Decomposition Processes*. Blackwell Scientific Publishing, Oxford, pp. 417–434.

Coleman, D.C. (1985) Through a ped darkly – an ecological assessment of root–soil–microbial–faunal interactions. In: Fitter, A.H., Atkinson, D., Read, D.J. and Usher, M.B. (eds) *Plants, Microbes and Animals*. British Ecological Society Special Publication 4. Blackwells, Oxford, pp. 1–21.

Coleman, D.C. and Crossley, D.A. (1991) Modern techniques in soil ecology. *Agriculture, Ecosystem and Environment* (Special issue) 34, 1–507.

Coleman, D.C. and Crossley, D.A. (1996) *Fundamentals of Soil Ecology*. Academic Press, New York.

Coleman, D.C. and Hendrix, P.F. (2000) *Invertebrates as Webmasters in Ecosystems*. CAB International, Wallingford, UK.

Coleman, D.C., Reid, C.P.P. and Cole, C.V. (1983) Biological strategies of nutrient cycling in soil systems. *Advances in Ecological Research* 10, 1–55.

Coleman, J.S., McConnaughay, K.D.M. and Ackerly, D.D. (1994) Interpreting phenotypic variation in plants. *Trends in Ecology and Evolution* 9, 187–191.

Coleman, D.C., Blair, J.M., Elliott, E.T. and Wall, D.H. (1999) Soil invertebrates. In: Robertson, G.P., Coleman, D.C., Bledsoe, C.S. and Sollins, P. (eds) *Standard Soil Methods for Long Term Ecological Research*. Oxford University Press, Oxford, pp. 349–377.

Colgate, S.M. and Dorling, P.R. (1994) *Plant Associated Toxins – Agricultural, Phytochemical and Ecological Aspects*. CAB International, Wallingford, UK.

Cooke, A. (1983) The effects of fungi on food selection by *Lumbricus terrestris*. In: Satchell, J.E. (ed.) *Earthworm Ecology, from Darwin to Vermiculture*. Chapman and Hall, London, pp. 365–373.

Couteaux, M.M. (1967) Une technique d'observation des thécamebiens du sol pour l'estimation de leur densité absolue. *Revue d'Ecologie et de Biologie du Sol* 4, 593–596.

Cowling, A.J. (1994) Protozoan distribution and adaptation. In: Darbyshire, J.F. (ed.) *Soil Protozoa*. CAB International, Wallingford, UK, pp. 5–42.

Csonka, L.N. (1989) Physical and genetic responses of bacteria to osmotic stress. *Microbiology Reviews* 53, 121–147.

Csonka, L.N. and Hanson, A.D. (1991) Procaryote osmoregulation: genetics and physiology. *Annual Reviews Microbiology* 45, 569–606.

Cuénot, L. (1949a) Les Onychophores. In: Grassé, P.-P. *Traité de Zoologie: Anatomie, Systématique, Biologie*, Vol. VI. Masson, Paris, pp. 3–37.

Cuénot, L. (1949b) Les Tardigrades. In: Grassé, P.-P. (ed.) *Traité de Zoologie: Anatomie, Systématique, Biologie*, Vol. VI. Masson, Paris, pp. 40–59.

Curry, J.P. (1994) *Grassland Invertebrates, Ecology, Influence on Soil Fertility and Effects on Plant Growth*. Chapman and Hall, London.

Curry, J.P. (1998) Factors affecting earthworm abundance in soils. In: Edwards C.A. (ed.) *Earthworm Ecology*. St Lucie Press, London, pp. 37–64.

Dash, M.C., Mishra, P.C. and Bechera, N. (1979) Fungal feeding by a tropical earthworm. *Tropical Ecology* 20, 9–12.

Dash, M.C., Senapati, B.K. and Mishra, C.C. (1980) Nematode feeding by tropical earthworms. *Oikos* 34, 322–325.

Dawid, W. (2000) Biology and global distribution of myxobacteria in soils. *FEMS Microbiology Reviews* 24, 403–427.

Day, G.M. (1950) The influence of earthworms on soil micro-organisms. *Soil Science* 69, 175–184.

Dearnaley, J.D.W. and Hardham, A.R. (1994) Golgi apparatus of *Phytophthora cinnamomi* makes three types of secretory or storage vesicles. *Protoplasma* 182, 75–79.

Deharveng, L. (1996) Soil Collembola diversity, endemism, and reforestation: a case study in the Pyrenees (France). *Conservation Biology* 10, 74–84.

Deharveng, L. and Lek, S. (1995) High diversity and community permeability: the riparian Collembola (Insecta) of a Pyrenean massif. *Hydrobiologia* 312, 59–74.

de Lorenza, G., Castaria, R., Bellinconpi, D. and Cervone, F. (1997) Fungal invasion enzymes and their inhibitors. In: Esser, E. and Lemke, P.A. (eds) *The Mycota VII*. Springer Verlag, Berlin.

Denis, R. (1949) Sous-classe des Apterigotes, ordre des Collemboles. In: Grassé, P.-P. (ed.) *Traité de Zoologie: Anatomie, Systématique, Biologie*, Vol. VI. Publ. Masson et Co., Paris, pp. 113–159.

de Ruiter, P.C., Ouborg, N.J. and Ernsting, D.C.O. (1988) Density dependent mortality in the springtail species *Orchesella cincta* due to predation by the carabid beetle *Notiophilus biguttatus*. *Entomologia Experimentalis et Applicata* 48, 25–30.

de Ruiter, P.C., Neutel, A.-M. and Moore, J.C. (1995) Modelling food webs and nutrient cycling in agro-ecosystems. *Trends in Ecology and Evolution* 9, 378–383.

de Ruiter, P.C., Neutel, A.-M. and Moore, J.C. (1998) Biodiversity in soil ecosystems: the role of energy flow and community stability. *Applied Soil Ecology* 10, 217–228.

Dictor, M.-C. Tessier, L. and Soulas, G. (1998) Reassessment of the Kec coefficient of the fumigation-extraction method in a soil profile. *Soil Biology and Biochemistry* 30, 119–127.

Didden, W.A.M. (1993) Ecology of Enchytraeidae. *Pedobiologia* 37, 2–29.

Dilly, O. and Irmler, U. (1998) Succession in the food web during the decomposition of leaf litter in a black alder (*Alnus glutinosa* (Gaertn.) L.) forest. *Pedobiologia* 42, 109–123.

Dilly, O., Bartsch, S., Rosenbrock, P., Buscot, F. and Munch, J.C. (2001) Shifts in physiological capabilities of the microbiota during the decomposition of leaf litter in a black alder (*Alnus glutinosa* (Gaertn.) L.) forest. *Soil Biology and Biochemistry* 33, 921–930.

Dindal, D.L. (1990) *Soil Biology Guide*. John Wiley & Sons, New York.

Dodd, J.C., Boddington, C.L., Rodriguez, A., Gonzalez-Chavez, C. and Mansur, I. (2000) Mycelium of arbuscular mycorrhizal fungi (AMF) from different genera: form, function and detection. *Plant and Soil* 226, 131–151.

Donisthorpe, H., St J.K. (1927) *The Guests of British Ants: Their Habits and Life Histories*. Routledge, London, pp. 202–217.

Donner, J. (1949–1950) Rotatorien des Humusböden I and II. *Oesterreich Zoologisches Zeits*, Vols 2 and 3.

Dosza-Farkas, K. (1982) Konsum verschiedener Laubarten durch Enchytraeiden (Oligochaeta). *Pedobiologia* 23, 251–255.

Dragesco, J. (1960) Ciliés mésopsammiques littoraux. Systématique, morphologie, écologie. *Traveaux de la Station Biologique de Roscoff* 12, 1–356.

Dragesco, J. (1962) On the biology of sand-dwelling ciliates. *Science Progress Series* 50, 353–363.

Duponnois, R. (1992) Les Bactéries Auxiliaires de la Mycorrhization du Douglas (*Pseudotsuga menziesii* Mirb.) Franco) par *Laccaria laccata* souche S238. Thèse de l'Université de Nancy 1, France.

Dusenbery, D.B. (1989) A simple animal can use a complex stimulus pattern, to find a location: nematode thermotaxis in soil. *Biology and Cybernetics* 60, 431–437.

Edsburg, E. and Hagvar, S. (1999) Vertical distribution, abundance, and biology of oribatid mites (Acari) developing inside decomposing spruce needles in a podsol soil profile. *Pedobiologia* 43, 413–421.

Edwards, C.A. (1998) *Earthworm Ecology*. CRC Press, Boca Raton, Florida.

Edwards, C.A. and Bohlen, P.J. (1996) *Biology and Ecology of Earthworms*, 3rd edn. Chapman and Hall, London.

Edwards, C.A. and Heath, G.W. (1963) The role of soil animals in breakdown of leaf material. In: Doekson, J. and van der Drift, J. (eds) *Soil Organisms*. North Holland, Amsterdam, pp. 76–80.

Ekelund, F. (1998) Enumeration and abundance of mycophagous protozoa in soil, with special emphasis on heterotrophic flagellates. *Soil Biology and Biochemistry* 30, 1343–1347.

Elliott, E.C., Coleman, D.C. and Cole, C.V. (1979) The influence of amoeba on the uptake of nitrogen by plants in gnotobiotic soil. In: Harley, J.L. and Scott, R.R. (eds) *The Soil–Root Interface*. Academic Press, London, pp. 221–229.

Elliott, J.M. (1977) *Some Methods for the Statistical Analysis of Samples of Benthic Invertebrates*. Freshwater Biological Station Association, Scientific Publication No. 25.
Elton, C. (1927) *Animal Ecology*. Macmillan, New York.
Eriksson, M., Dalhammar, G. and Borg-Karlson, A.-K. (1999) Aerobic degradation of a hydrocarbon mixture in natural uncontaminated potting soil by indigenous microorganisms at 20°C and 6°C. *Applied Microbiology and Biotechnology* 51, 532–535.
Ernsting, C. and van der Werf, D.C. (1988) Hunger, partial consumption of prey and prey size preference in a carabid beetle. *Ecological Entomology* 13, 155–164.
Espeleta, J.F. and Eissenstat, D.M. (1998) Responses of citrus fine roots to localized soil drying: a comparison of seedlings with adult fruiting trees. *Tree Physiology* 18, 113–119.
Ettema, C.H. (1998) Soil nematode diversity: species coexistence and ecosystem function. *Journal of Nematology* 30, 159–169.
Ettema, C.H., Rathbun, S.L. and Coleman, D.C. (2000) On spatiotemporal patchiness and the coexistence of five species of *Chronogaster* (Nematoda: Chronogasteridae) in a riparian wetland. *Oecologia* 125, 444–452.
Evans, G.O. (1992) *Principles of Acarology*. CAB International, Wallingford UK, pp. 563.
Fahey, T.J. and Hughes, J.W. (1994) Fine root dynamics in a northern hardwood forest ecosystem, Hubbard Brook Experimental Forest, NH. *Journal of Ecology* 82, 533–548.
Fang, J., Chen, A., Peng, C., Zhao, S. and Ci, L. (2001) Changes in forest biomass carbon storage in China between 1949 and 1998. *Science* 292, 2320–2322.
Farrar, F.P., Jr and Crossley, D.A. (1983) Detection of soil microarthropods in soybean fields, using a modified Tullgren extractor. *Environmental Entomology* 12, 1303–1309.
Felske, A. and Akkermans, D.L. (1998) Spatial homogeneity of abundant bacterial 16S rRNA molecules in grassland soils. *Microbial Ecology* 36, 31–36.
Fenchel, T. (1968a) The ecology of the marine microbenthos. I. The quantitative importance of ciliates as compared with metazoans in various types of sediments. *Ophelia* 4, 121–137.
Fenchel, T. (1968b) The ecology of the marine microbenthos. II. The food of marine benthic ciliates. *Ophelia* 5, 73–121.
Fenchel, T. (1968c) The ecology of the marine microbenthos. III. The reproductive potential of ciliates. *Ophelia* 5, 123–136.
Fenchel, T. (1969) The ecology of the marine microbenthos. IV. Structure and function of the benthic ecosystem; its chemical and physical factors and the microfauna communities with special reference to the ciliated protozoa. *Ophelia* 6, 1–182.
Fenchel, T. (1986) Protozoan filter feeding. *Progress in Protistology* 1, 65–114.
Fenchel, T., Jansson, B.O. and von Thun, W. (1967) Vertical and horizontal distribution of the metazoan microfauna and of some physical factors in a sandy beach in the northern part of the Øresund. *Ophelia* 4, 227–243.
Filip, Z., Claus, H. and Dippell, G. (1998) Abbau von Huminstoffen durch Bodemikroorganismen – eine Übersicht. *Zeitschrift für Pflanzenernährung und Bodenkunde* 161, 605–612.

Filip, Z., Pecher, W. and Berthelin, J. (1999) Microbial utilization and transformation of humic acids extracted from different soils. *Journal of Plant Nutrition and Soil Science* 162, 215–222.

Fitter, A.H. and Garbaye, J. (1994) Interactions between mycorrhizal fungi and other soil organisms. *Plant and Soil* 159, 123–132.

Focht, D.D. (1992) Diffusional constraints on microbial processes in soil. *Soil Science* 154, 300–307.

Fog, K. (1988) The effect of added nitrogen on the rate of decomposition of organic matter. *Biology Reviews* 63, 433–462.

Fogel, R. (1985) Roots as primary producers in below-ground ecosystems. In: Fitter, A.H., Atkinson, D., Read, D.J. and Usher, M.B. (eds) *Ecological Interactions in Soil: Plants, Microbes and Animals*. Blackwell Scientific Publishers, Oxford, pp. 23–36.

Foissner, W. (1987) Soil protozoa: fundamental problems, ecological significance, adaptations in ciliates and testacean, bioindicators and guide to the literature. *Progress in Protistology* 2, 69–212.

Foissner, W. (1991) Basic light and scanning electron microscopic methods for taxonomic studies of ciliated protozoa. *European Journal of Protistology* 27, 313–330.

Foissner, W. (1992) On the biology and ecology of mycophagous soil protozoa. *European Journal of Protistology* 28, 340.

Foissner, W. (1993) Colpodea (Ciliophora). In: Matthes, D. (ed.) *Protozoenfauna*. Gustav Fischer, Stuttgart.

Foissner, W. and Foissner, A. (1984) First record of an ectoparasitic flagellate on ciliates: an ultrastructural investigation of the morphology and the mode of attachment of *Spiromonas gonderi* nov. sp. (Zoomastigophora, Spiromonadidae) invading the pellicle of ciliates of the genus *Colpoda* (Ciliophora, Colpodidae). *Protistologica* 20, 635–648.

Foissner, W. and Korganova, G.A. (1995) Redescription of three testate amoebae (Protozoa, Rhizopoda) from a Caucasian soil: *Centropyxis plagiostoma* Bonnet and Thomas, *Cyclopyxis kahli* (Deflandre) and *C. intermedia* Kufferath. *Archiv für Protistenkunde* 146, 13–28.

Frampton, G.K., Van den Brink, P.J. and Wratten, S.D. (2001) Diel activity pattern in an arable collembolan community. *Applied Soil Ecology* 17, 63–80.

Frankland, J.C. (1998) Fungal succession – unravelling the unpredictable. *Mycological Research* 102, 1–15.

Freckman, D.W. and Baldwin, J.G. (1990) Soil Nematoda. In: Dindal, D.L. (ed.) *Soil Biology Guide*. John Wiley & Sons, New York, pp. 155–200.

Fredslund, L., Ekelund, F., Jacobsen, C.S. and Johnsen, K. (2001) Development and application of a most-probable-number-PCR assay to quantify flagellate populations in soil samples. *Applied and Environmental Microbiology* 67, 1613–1618.

Fuller, M.S. and Jaworski, A. (1987) *Zoosporic Fungi in Teaching and Research*. Southeastern Publishing Corp., Athens, Georgia.

Fung, I., Field, C.B., Berry, J.A., Thompson, M.V., Randerson, M.V., Malmstrom, C.M., Vitousek, P.M., Collatz, G.J., Sellers, P.J., Randall, D.A., Denning, A.S., Badeck, F. and John, J. (1997) Carbon 13 exchanges between the atmosphere and biosphere. *Global Biogeochemical Cycles* 11, 535–560.

Gagliardi, J.V., Angle, J.S., Germida, J.J., Wyndham, R.C., Chanway, C.P., Watson, R.J., Greer, C.W., McIntyre, T., Yu, H.H., Levin, M.A., Russek-Cohen, E., Rosolen, S., Nairn, J., Seib, A., Martin-Heller, T. and Wisse, G. (2001) Intact soil-core microcosms compared with multi-site field releases for pre-release testing of microbes in diverse soils and climates. *Canadian Journal of Microbiology* 47, 237–252.

Gaillard, V., Chenu, C., Recous, S. and Richard, G. (1999) Carbon, nitrogen, and microbial gradients induced by plant residues decomposing in soil. *European Journal of Soil Science* 50, 567–578.

Garbaye, J. (1994) Helper bacteria: a new dimension to the mycorrhizal symbiosis. *New Phytologist* 128, 197–210.

Garbaye, J. and Bowen, G.D. (1987) Effect of different microflora on the success of ectomycorrhizal inoculation of *Pinus radiata*. *Canadian Journal of Forest Research* 17, 941–943.

Gaudinski, J.B., Trumbore, S.E., Davidson, E.A. and Zheng, S. (2000) Soil carbon cycling in a temperate forest: radiocarbon based estimates of residence times, sequestration rates and partitioning of fluxes. *Biogeochemistry* 51, 33–69.

Gerzabek, M.H., Haberhauer, G. and Kirchmann, H. (2001a) Soil organic matter pools and carbon-13 natural abundances in particle size fractions of a long-term agricultural field experiment receiving organic amendments. *Soil Science of America Journal* 65, 352–358.

Gerzabek, M.H., Haberhauer, G. and Kirchmann, H. (2001b) Nitrogen distribution and ^{15}N natural abundances in particle size fractions of a long-term agricultural field experiment. *Journal of Plant Nutrition and Soil Science* 164, 475–481.

Ghabbour, S.I. (1966) Earthworms in agriculture: a modern evaluation. *Revue d'Ecologie et Biologie du Sol* 3 259–271.

Ghawana, V.K., Shrivastava, J.N. and Kushwaha, R.K.S. (1997) Some observations on fungal succession during decomposition of wool in soil. *Mycoscience* 38, 79–81.

Giblin, A.E., Laundre, J.A., Nadelhoffer, K.J. and Shaver, G.R. (1994) Measuring nutrient availability in arctic soils using ion exchange resins: a field test. *Soil Science Society of America Journal* 58, 1154–1162.

Giovannetti, M., Sbrana, C. and Logi, C. (2000) Microchambers and video-enhanced light microscopy for monitoring cellular events in living hyphae of arbuscular mycorrhizal fungi. *Plant and Soil* 226, 153–159.

Gittings, T. and Giller, P.S. (1998) Resource quality and the colonization and succession of coprophagous dung beetles. *Ecography* 21, 581–592.

Gochenaur, S.E. (1987) Evidence suggests that grazing regulates ascospore density in soil. *Mycologia* 79, 445–450.

Gonzalez, G. and Seastedt, T.R. (2001) Soil fauna and plant litter decomposition in tropical and subalpine forests. *Ecology* 82, 955–964.

Gouge, D.H., Smith, K.A., Lee, L.L. and Henneberry, T.J. (2000) Effect of soil depth and moisture on the vertical distribution of *Steinernema riobrave* (Nematoda: Steinernematidae). *Journal of Nematology* 32, 223–228.

Goulden, M.L., Munger, J.W., Song-Miao, F., Daube, B.C. and Wofsy, F.C. (1996) Measurements of carbon sequestration by long-term eddy covariance: methods and a critical evaluation of accuracy. *Global Change Biology* 2, 169–182.

Grassé, P.-P. (1949–1995) *Traité de Zoologie: Anatomie, Systématique, Biologie*. Publ. Masson et Co., Paris.

Grassé, P.-P. (1965) *Traité de Zoologie: Anatomie, Systématique, Biologie*, Vol. IV, fasc. II and III. Publ. Masson et Co., Paris.

Grayston, S.J., Vaughan, D. and Jones, D. (1996) Rhizosphere carbon flow in trees, in comparison with annual plants: the importance of root exudation and its impact on microbial activity and nutrient availability. *Applied Soil Ecology* 5, 29–56.

Grayston, S.J., Griffith, G.S., Mawdsley, J.L., Campbell, C.D. and Bardgett, R.D. (2001) Accounting for variability in soil microbial communities of temperate upland grassland ecosystems. *Soil Biology and Biochemistry* 33, 533–551.

Gregorich, E.G. and Carter, M.R. (1997) *Soil Quality for Crop Production and Ecosystem Health*. Elsevier, Amsterdam.

Gregory, P.J. and Hinsinger, P. (1999) New approaches to studying chemical and physical changes in the rhizosphere: an overview. *Plant and Soil* 211, 1–9.

Griffin, D.M. (1969) Soil water in the ecology of fungi. *Annual Review of Phytopathology* 7, 289–310.

Griffiths, B.S. (1994) Soil nutrient flow. In: Darbyshire, J.F. (ed.) *Soil Protozoa*. CAB International, Wallingford, UK.

Griffiths, B.S., Ritz, K., Ebblewhite, N. and Dobson, G. (1999a) Soil microbial community structure: effects of substrate loading rates. *Soil Biology and Biochemistry* 31, 145–153.

Giffiths, B.S., Bonkowski, M., Dobson, G. and Caull, S. (1999) Changes in soil microbial community structure in the presence of microbial-feeding nematodes and protozoa. *Pedobiologia* 43, 297–304.

Grundmann, G.L. and Normand, P. (2000) Microscale diversity of the genus *Nitrobacter* in soil on the basis of analysis of genes encoding rRNA. *Applied and Environmental Microbiology* 66, 4543–4546.

Hagvar, S. (1983) Collembola in Norwegian coniferous forest soils. II. Vertical distribution. *Pedobiologia* 25, 383–401.

Hagvar, S. and Edsburg, E. (2000) Vertical transport of decomposing spruce needles during nine years in a raw humus soil profile in southern Norway. *Pedobiologia* 44, 119–131.

Haines, B.L., Waide, J.B. and Todd, R.L. (1982) Soil solution nutrient concentrations sampled with tension and zero-tension lysimeters: report of discrepancies. *Soil Science Society of America Journal* 46, 658–661.

Hall, M. and Hedlund, K. (1999) The predatory mite *Hyoaspis aculeifer* is attracted to food of its fungivorous prey. *Pedobiologia* 43, 11–17.

Hanel, L. (1995) Secondary successional stages of soil nematodes in cambisols of South Bohemia. *Nematologica* 41, 197–218.

Hanel, L. (2000) Seasonal changes of soil nematodes, other soil microfauna and fungus fruiting bodies in a spruce forest near Ceske Budejovice, Czech Republic. *Biologia, Bratislava* 55, 435–443.

Hanel, L. (2001) Succession of soil nematodes in pine forests on coal-mining sands near Cottbus, Germany. *Applied Soil Ecology* 16, 23–34.

Hansen, R.A. (1999) Red oak litter promotes a microarthropod functional group that accelerates its decomposition. *Plant and Soil* 209, 37–45.

Hansen, R.A. (2000) Effect of habitat complexity and composition on a diverse litter microarthropod assemblage. *Ecology* 81, 1120–1132.

Harold, F.M. (1999) In pursuit of the whole hypha. *Fungal Genetics and Biology* 27, 128–133.

Harper, S.M., Kerven, G.L., Edwards, D.G. and Ostatek-Boczynski, Z. (2000) Characterization of fulvic and humic acids from leaves of *Eucalyptus camaldulensis* and from decomposed hay. *Soil Biology and Biochemistry* 32, 1331–1336.

Harris, R.F. (1981) Effect of water potential on microbial growth and activity. In: Parr, J.F., Gardner, W.R. and Elliott, L.F. (eds) *Water Potential Relations in Soil Microbiology: Proceedings of a Symposium.* SSSA special publication 9, Soil Science Society of America, Madison, Wisconsin.

Harrison, A.F. (1971) The inhibitory effect of oak leaf litter tannins on the growth of fungi in relation to litter decomposition. *Soil Biology and Biochemistry* 3, 167–172.

Harry, M., Gambier, B. and Garnier-Sillam, E. (2000) Soil conservation for DNA preservation for bacterial molecular studies. *European Journal of Soil Biology* 36, 51–55.

Hattori, T. (1994) Soil microenvironment. In: Darbyshire, J.F. (ed.) *Soil Protozoa.* CAB International, Wallingford, UK, pp. 43–64.

Hausmann, K. and Hullsmann, N. (1996) *Protozoology*, 2nd edn. Thieme Medical Publishers, New York.

Hawes, M.C. (1998) Function of root border cells in plant health: pioneers in the rhizosphere. *Annual Review of Phytopathology* 36, 311–327.

Heal, O.W. (1963) Morphological variation in certain Testacea (Protozoa: Rhizopoda). *Archiv für Protistenkunde* 106, 351–368.

Heal, O.W. (1964) The use of cultures for studying Testacea (Protozoa: Rhizopoda) in soil. *Pedobiologia* 4, 1–7.

Hedlund, K., Boddy, L. and Preston, C.M. (1991) Mycelial responses of the soil fungus *Mortierella isabellina* to grazing by *Onychiurus armatus* (Collembola). *Soil Biology and Biochemistry* 23, 361–366.

Hendrick, R.L. and Pregitzer, K.S. (1992) The demography of fine roots in a northern hardwood forest. *Ecology* 73, 1094–1104.

Hendrix, P.F. (2000) Earthworms. In: Sumner, M. (ed.) *The Handbook of Soil Science.* CRC Press, Boca Raton, Florida, pp. C77–C85.

Hill, N.M. and Patriquin, D.G. (1992) Interactions between fungi and nitrogen-fixing bacteria during decomposition. In: Carroll, G.C. and Wicklow, D.T. (eds) *The Fungal Community: Its Organization and Role in the Ecosystem.* Marcel Dekker Inc., Madison, Wisconsin, pp. 783–796.

Hobbie, S. (2000) Interactions between litter, lignin and soil nitrogen availability during leaf litter decomposition in a Hawaiian montane forest. *Ecosystems* 3, 484–494.

Hodge, A. (2000) Microbial ecology of the arbuscular mycorrhiza. *FEMS Microbiology Ecology* 32, 91–96.

Holdridge, L.R. (1947) Determination of world plant formations from simple climatic data. *Science* 105, 367–368.

Holling, C.S. (1959) Some characteristics of simple types of predation and parasitism. *Canadian Journal of Entomology* 91, 385–398.

Holmer, L. and Stenlid, J. (1997) Competitive hierarchies of wood decomposing Basidiomycetes in artificial systems based on variable inoculum sizes. *Oikos* 79, 77–84.

Homma, Y. (1984) Perforation and lysis of hyphae of *Rhizoctonia solaris* and conidia of *Cochliobolus miyabeanus* by soil myxobacteria. *Phytopathology* 74, 1234–1239.

Homma, Y. and Cook, R.J. (1985) Influence of matric and osmotic water potentials and soil pH on the activity of the giant vampyrellid amoebae. *Phytopathology* 75, 243–246.

Homma, Y. and Kegasawa, K. (1984) Predation on plant parasitic nematodes by soil vampyrellid amoebae. *Japanese Journal of Nematology* 14, 1–7.

Hopkins, S.P. (1997) *Biology of the Springtails*. Oxford University Press, New York.

Huang, S., Pollack, H.N. and Shen, P.-Y. (2000) Temperature trends over the past five centuries reconstructed from borehole temperature. *Nature* 403, 756–758.

Hubert, J. and Lukesova, A. (2001) Feeding of the panphytophagous oribatid mite *Scheloribates laevigatus* (Acari: Oribatida) on cyanobacterial and algal diets in laboratory experiments. *Applied Soil Ecology* 16, 77–83.

Hubert, J., Sustr, V. and Smrz, J. (1999) Feeding of the oribatid mite *Scheloribates laevigatus* (Acari: Oribatida) in laboratory experiments. *Pedobiologia* 43, 328–339.

Hubert, J., Kubatova, A. and Sarova, J. (2000) Feeding of *Scheloribates laevigatus* (Acari: Oribatida) on different stadia of decomposing grass litter (*Holcus lanatus*). *Pedobiologia* 44, 627–639.

Ishikawa, M.C. and Bledsoe, C.S. (2000) Seasonal and diurnal patterns of soil water potential in the rhizosphere of blue oaks: evidence for hydraulic lift. *Oecologia* 125, 459–465.

Jaeger, C.H., Lindow, S.E., Miller, W., Clark, E. and Firestone, M.K. (1999) Mapping of sugar and amino acid availability in soil around roots with bacterial sensors of sucrose and tryptophan. *Applied and Environmental Microbiology* 65, 2685–2690.

Jaffee, B.A., Santos, P.F. and Muldoon, A.E. (1997) Suppression of nematophagous fungi by enchytraeid worms. *Oecologia* 112, 412–423.

Jakobsen, I., Gazey, C. and Abbott, L.K. (2001) Phosphate transport by communities of arbuscular mycorrhizal fungi in intact cores. *New Phytologist* 149, 95–103.

Jamieson, B.G.M. (1981) *The Ultrastructure of the Oligochaetae*. Academic Press, Sydney.

Johnsen, K., Enger, O., Jacobsen, C.S., Thirup, L. and Torsvik, V. (1999) Quantitative selective PCR of 16S ribosomal DNA correlates well with selective agar plating in describing population dynamics of indigenous *Pseudomonas* spp. In soil hot spots. *Applied and Environmental Microbiology* 65, 1786–1789.

Johnson, N.C., O'Dell, T.E. and Bledsoe, C.S. (1999) Methods for ecological studies of mycorrhizae. In: Robertson, G.P., Coleman, D.C., Bledsoe, C.S. and Sollins, P. (eds) *Standard Soil Methods for Long Term Ecological Research*. Oxford University Press, Oxford, pp. 374–412.

Johnson, S.R., Ferris, V.R. and Ferris, J.M. (1972) Nematode community structure of forest woodlots. I. Relationships based on similarity coefficients of nematode species. *Journal of Nematology* 4, 175–182.

Kaiser, K., Kaupenjohann, M. and Zech, W. (2001) Sorption of dissolved organic carbon in soils: effects of soil sample storage, soil-to-solution ratio, and temperature. *Geoderma* 99, 317–328.

Kalbitz, K., Solinger, S., Park, J.-H. and Matzner, E. (2000) Controls on the dynamics of dissolved organic matter in soils: a review. *Soil Science* 165, 277–304.

Kaneko, N. and Salamanca, E.F. (1999) Mixed leaf litter effects on decomposition rates and soil microarthropod communities in an oak–pine stand in Japan. *Ecological Research* 14, 131–138.

Kaneko, N., McLean, M.A. and Parkinson, D. (1995) Grazing preferences of *Onychiurus subtenuis* (Collembola) and *Oppiella nova* (Oribatei) for fungal species inoculated on pine needles. *Pedobiologia* 39, 538–546.

Karling, J.S. (1977) *Chytridiomycetarum Iconographia.* Lubrecht and Kramer, Monticello, New York.

Kasai, K., Morinaga, T. and Horikoshi, T. (1996) Fungal succession in the early decomposition process of pine cones on the floor of *Pinus densiflora* forests. *Mycoscience* 36, 325–334.

Kasprzak, K. (1982) Review of enchytraeids (Ologochaeta, Enchytraeidae) community structure and function in agricultural ecosystems. *Pedobiologia* 23, 217–232.

Keller, S. and Zimmermann, G. (1989) Mycopathogens of soil insects. In: Wilding, N., Collins, N.M., Hammond, P.M. and Weber, J.F. (eds) *Insect–Fungus Interactions.* Academic Press, New York, pp. 239–270.

Kernaghan, G., Currah, R.S. and Bayer, R.J. (1997) Russulaceous ectomycorrhizae of *Abies lasiocarpa* and *Picea engelmannii. Canadian Journal of Botany* 75, 1843–1850.

Kessin, R.H. (2001) *Dictyostelium. Evolution, Cell Biology, and the Development of Multicellularity.* Development and Cell Biology Series, Cambridge University Press, Cambridge.

Kessler, K.J.J. (1990) Destruction of *Gnomonia leptostyla* perithecia on *Juglans nigra* leaves by microarthropods associated with *Elaeagnus umbellate* litter. *Mycologia* 82, 387–390.

Kethley, J. (1990) Acarina: Prostigmata (Actinedida). In: Dindal, D.L. (ed.) *Soil Biology Guide.* John Wiley & Sons, New York, pp. 667–756.

Keyser, P., Kirk, T.K. and Ziekus, J.G. (1978) Ligninolytic enzyme system of *Phanerochaete chrysosporium:* synthesized in the absence of lignin in response to nitrogen starvation. *Journal of Bacteriology* 135, 790–797.

Kilbertus, G. (1980) Etudes des microhabitats contenus dans les agrégats du sol, leur relation avec la biomasse bactérienne et la taille des prokaryotes présents. *Revue d'Ecologie et de Biologie du Sol* 17, 543–557.

Kirby, H. (1950) *Materials and Methods in the Study of Protozoa.* University of California Press, Berkeley.

Kjelleberg, S. (1993) *Starvation in Bacteria.* Plenum Press, New York.

Klironomos, J.N. and Hart, M.M. (2001) Animal nitrogen swap for plant carbon. *Nature* 410, 651–652.

Klironomos, J.N. and Kendrick, B. (1996) Palatability of microfungi to soil microarthropods in relation to the functioning of arbuscular mycorrhizae. *Biology and Fertility of Soils* 21, 43–52.

Klironomos, J.N. and Moutoglis, P. (1999) Colonization of non-mycorrhizal plants by mycorrhizal neighbours as influenced by the collembolan, *Folsomia candida. Biology and Fertility of Soils* 29, 277–281.

Klironomos, J.N., Bednarczuk, E.M. and Neville, J. (1999) Reproductive significance of feeding on saprobic and arbuscular mycorrhizal fungi by the collembolan *Folsomia candida*. *Functional Ecology* 13, 756–761.

Kobatake, M. (1954) The antibacterial substances extracted from lower animals. I. The earthworm. *Kekkabu (Tuberculosis)* 29, 60–63.

Korganova, G.A. and Geltser, J.G. (1977) Stained smears for the study of soil Testacida (Protozoa, Rhizooda). *Pedobiologia* 17, 222–225.

Krebs, C.J. (2000) *Ecological Methodology*, 2nd edn. Benjamin Cummings, San Francisco, California.

Krebs, C. (2001) *Ecology*, 6th edn. Benjamin Cummings, San Francisco, California.

Kurtzman, C.P. and Fell, J.W. (1998) *The Yeasts – a Taxonomic Study*, 4th edn. Elsevier, New York.

Kuzyakov, Y.V. (1997) The role of amino acids and nucleic bases in turnover of nitrogen and carbon in soil humic fractions. *European Journal of Soil Science* 48, 121–130.

Kuzyakov, Y.V. (2001) Tracer studies of carbon translocation by plants from the atmosphere into the soil (a review). *Eurasian Journal of Soil Science* 34, 28–42.

Kuzyakov, Y., Ehrensberger, H. and Stahr, K. (2001) Carbon partitioning and below-ground translocation by *Lolium perenne*. *Soil Biology and Biochemistry* 33, 61–74.

Lagerloef, J., Andren, O. and Paustian, K. (1989) Dynamics and contribution to carbon flows of Enchytraeidae (Oligochaeta) under four cropping systems. *Journal of Applied Ecology* 26, 183–199.

Lajtha, K. (1988) The use of ion exchange resin bags to measure nutrient availability in an arid ecosystem. *Plant and Soil* 105, 105–111.

Lajtha, K., Seeley, B. and Valiela, I. (1995) Retention and leaching losses of atmospherically-derived nitrogen in the aggrading coastal watershed of Waquoit Bay, MA. *Biogeochemistry* 28, 33–54.

Lajtha, K., Jarrell, W.M., Johnson, D.W. and Sollins, P. (1999) Collection of soil solution. In: Robertson, G.P., Coleman, D.C., Bledsoe, C.S. and Sollins, P. (eds) *Standard Soil Methods for Long Term Ecological Research*. Oxford University Press, Oxford, pp. 168–182.

Lal, V.B., Singh, J. and Prasad, B. (1981) A comparison of different size of samplers for collection of Enchytraeidae (Oligochaeta) in tropical soils. *Journal of Soil Biology and Ecology* 1, 21–26.

Laminger, H. (1978) The effects of soil moisture fluctuations on the testacean species *Trinema enchelys* (Ehrenberg) Leidy in a high mountain brown earth podsol and its feeding behaviour. *Archiv für Protistenkunde* 120, 446–454.

Laminger, H. and Bucher, M. (1984) Fressverhalten einiger terrestischer Testaceen (Protozoa, Rhizopoda). *Pedobiologia* 27, 313–322.

Lamoncha, K.L. and Crossley, D.A. (1998) Oribatid mite diversity along an elevation gradient in a southeastern Appalachia forest. *Pedobiologia* 42, 43–55.

Lancini, G. and Demain, A.L. (1999) Secondary metabolism in bacteria: antibiotic pathways, regulation and function. In: Lengeler, J.W., Drews, G. and Schelgel, H.G. (eds) *Biology of the Prokaryotes*. Blackwell Science, New York, pp. 627–647.

Larcher, W. (1973) Temperature resistance and survival. In: Precht, H. (ed.) *Temperature and Life*. Springer-Verlag, Berlin, pp. 203–231.

Lavelle, P. (1983a) Agastrodilus Omodeo and vaillaut, a genus of carnivorous earthworms from the Ivory Coast. In: Satchell, J.E. (ed.) *Earthworm Ecology, from Darwin to Vermiculture*. Chapman and Hall, London, pp. 425–429.

Lavelle, P. (1983b) The structure of earthworm communities. In: Satchell, J.E. (ed.) *Earthworm Ecology, from Darwin to Vermiculture*. Chapman and Hall, London, pp. 449–456.

Lavelle, P., Saw, B. and Schaeffer, R. (1980) The geophagous earthworm community in the Lante savana (Ivory Coast): niche partitioning and utilization of soil nutritive resources. In: Dindal D. (ed.) *Soil Biology, as Related to Land Use Practise*. EPA, Washington, DC, pp. 653–672.

Laverack, M.S. (1963) *The Physiology of Earthworms*. Pergamon Press, London.

Leander, B.S. and Farmer, M.A. (2001) Comparative morphology of the euglenid pellicle. II. Diversity of strip substructure. *Journal of Eukaryotic Microbiology* 48, 202–217.

Lee, J.J. and Soldo, A.T. (1992) *Protocols in Protozoology*. Society of Protozoologists, Allen Press, Lawrence, Kansas.

Lee, J.J., Leedale, G.F. and Bradbury, P. (2000) *The Illustrated Guide to the Protozoa*, 2nd edn. Society of Protozoologists, Allen Press, Lawrence, Kansas.

Lee, K.E. (1985) *Earthworms*. Academic Press, New York.

Lee, N., Nielsen, P.H., Andreasen, K.H., Juretschko, S., Nielsen, J.L., Schleiffer, K.-H. and Wagner, M. (1999) Combination of fluorescent *in-situ* hybridization and microautoradiography – a new tool for structure–function analyses in microbial ecology. *Applied and Environmental Microbiology* 65, 1289–1297.

Lek-Ang, S., Deharveng, L. and Lek, S. (1999) Predictive models of collembolan diversity and abundance in a riparian habitat. *Ecological Modelling* 120, 247–260.

Lemaire, M. (1996) The cellulosome – an exocellular multiprotein complex specialized in cellulose degradation. *Critical Reviews in Biochemistry and Molecular Biology* 31, 201–236.

Lengeler, J.W., Drews, G. and Schelgel, H.G. (1999) *Biology of the Prokaryotes*. Blackwell Science, New York.

Leopold, A.C. (1961) Senescence in plant development. The death of plants or plant parts may be of positive ecological or physiological value. *Science* 134, 1727–1732.

Leuschner, C. (1998) Water extraction by tree fine roots in the forest floor of a temperate Fagus–Quercus forest. *Annales des Sciences Forestières (Paris)* 55, 141–157.

Levina, N.N., Heath, I.B. and Lew, R.R. (2000) Rapid wound responses of *Saprolegnia ferax* hyphae depend upon actin and Ca^{2+}-involving deposition of callose plugs. *Protoplasma* 214, 199–209.

Lions, J.Cl. and Gourbière, F. (1988) Populations adultes et immatures d'*Adoristes ovatus* (Acarien, Oribate) dans les aiguilles de la littière d'*Abies Alba*. *Revue de l' Ecologie et Biologie du Sol* 25, 343–352.

Lipson, S.M. and Stotzky, G. (1987) Interactions between viruses and clay minerals. In: Rao, V.C. and Melnick, J.L. (eds) *Human Viruses in Sediments, Sludges and Soils*. CRC Press, Boca Raton, Florida, pp. 197–230.

Lloyd, F.E. (1927) The behaviour of *Vampyrella lateritia* with special reference to the work of Professor Chr. Gobi. *Archiv für Protistenkunde* 67, 219–236.

Lopez, B., Sabate, S. and Gracia, C. (1998) Fine roots dynamics in a Mediterranean forest: effects of drought and stem density. *Tree Physiology* 18, 601–606.

Lou, Y., Meyerhoff, P.A. and Loomis, R.S. (1995) Seasonal patterns and vertical distributions of fine roots of alfalfa (*Medicago sativa* L.). *Field Crops Research* 40, 119–127.

Lüftenagger, G., Petz, W., Foissner, W. and Adam, H. (1988) The efficiency of a direct counting method in estimating the number of microscopic soil organisms. *Pedobiologia* 31, 95–101.

Lussenhop, J. (1992) Mechanisms of microarthropod–microbial interactions in soil. *Advances in Ecological Research* 23, 1–33.

Lynn, D.H. (1991) The implications of recent descriptions of kinetid structure to the systematics of the ciliated protists. *Protoplasma* 164, 123–142.

Lynn, D.H. and Small, E.B. (2000) Phylum Ciliophora. In: Lee, J.J., Leedale, G.F. and Bradbury, P. (eds) *The Illustrated Guide to the Protozoa.* Allen Press, Society of Protozoologists, pp. 371–656.

MacDonald, D.W. (1983) Predation on earthworm by terrestrial vertebrates. In: Satchell, J.E. (ed.) *Earthworm Ecology, from Darwin to Vermiculture.* Chapman and Hall, London, pp. 425–429.

Macrae, A. (2000) The use of 16S rDNA methods in soil microbial ecology. *Brazilian Journal of Microbiology* 31, 77–82.

Magill, A.H. and Aber, J.D. (1998) Long-term effects of experimental nitrogen additions on foliar litter decay and humus formation in forest ecosystems. *Plant and Soil* 203, 301–311.

Majdi, H., Damm, E. and Nylund, J.-E. (2001) Longevity of mycorrhizal roots depends on branching order and nutrient availability. *New Phytologist* 150, 195–202.

Mangold, O. (1953) Experiments zur Analyse des chemischen Sinns des Ragenwurms. 2. Versuche mit Chinin, Sauren und Susstoffen. *Zoologische Jahrbücher, Abteilung für Allgemeine Zoologie und Physiologie der Tiere* 63, 501–557.

Maraum, M. (2000) The structure of oribatid mite communities (Acari, Oribatida). Patterns, mechanisms and implications for future research. *Ecography* 23, 374–383.

Marques, R., Ranger J., Gelhaye, D., Pollier, B., Ponette, Q. and Goedert, O. (1996) Comparison of chemical composition of soil solution collected by zero-tension plate lysimeters with those from ceramic-cup lysimeters in a forest soil. *European Journal of Soil Science* 47, 407–417.

Marra, J.L. and Edmonds, R.L. (1998) Effects of coarse woody debris and soil depth on the density and diversity of soil invertebrates on clearcut and forested sites on the Olympic Peninsula, Washington. *Environmental Entomology* 27, 1111–1124.

Martens, R. (2001) Estimation of ATP in soil: extraction methods and calculation of extraction efficiency. *Soil Biology and Biochemistry* 33, 973–982.

Martin, C.H. and Lewin, K.R. (1915) Notes on some methods for the examination of soil protozoa. *Journal of Agricultural Science (Cambridge)* 7, 106–119.

Martin, T.L., Trevors, J.T. and Kaushik, N.K. (1999) Soil microbial diversity, community structure and denitrification in a temperate riparian zone. *Biodiversity and Conservation* 8, 1057–1078.

Mathers, N.J., Mao, Z.H., Xu, Z.H., Saffigna, P.G., Berners-Price, S.J. and Perera, M.C.S. (2000) Recent advances in the application of ^{13}C and ^{15}N NMR spectroscopy to soil organic matter studies. *Australian Journal of Soil Research* 38, 769–787.

Maupas, E. (1919) Essais d'hybridation chez les nématodes. *Bulletin Biologique de la France et de la Belgique* 52, 466–498.

Maupas, E. (1900) Modes et formes de reproduction des nematodes. *Archives de Zoologie Experimentale* 8, 463–624.

Mayr, C., Winding, A. and Hendriksen, N.B. (1999) Community level physiological profile of soil bacteria unaffected by extraction method. *Journal of Microbiological Methods* 36, 29–33.

McConnaughay, K.D.M. and Coleman, J.S. (1999) Biomass allocation in plants: ontogeny or optimality? A test along three resource gradients. *Ecology* 80, 2581–2593.

McLean, M.A. and Parkinson, D. (2000) Introduction of the epigeic earthworm changes the oribatid community and microarthropod abundances in a pine forest. *Soil Biology and Biochemistry* 32, 1671–1681.

McLean, M.A., Kaneko, N. and Parkinson D. (1996) Does selective grazing by mites and collembola affect litter fungal community structure? *Pedobiologia* 40, 97–105.

Meentemeyer, V. (1978) Macroclimate and lignin control of decomposition. *Ecology* 59, 465–472.

Meglitsch, P.A. and Schram, F.R. (1991) *Invertebrate Zoology*, 3rd edn. Oxford University Press, Oxford.

Merbach, W., Mirus, E., Kent, G., Remus, R., Ruppel, S. and Russow, R. (1999) Release of carbon and nitrogen compounds by plant roots and their possible ecological importance. *Journal of Plant Nutrition and Soil Science* 162, 373–383.

Metz, L.J. and Farrier, M.H. (1967) Acarina associated with decomposing forest litter in the North Carolina piedmont. In: Evans, G.O. (ed.) *Proceedings of the 2nd International Congress of Acarology*. Akademiai Kiado, Budapest, pp. 43–53.

Michaelsen, W. (1903) *Die Geographische Verbreitung der Oligochaeten*. Berlin.

Michalzik, B. and Matzner, E. (1999) Dynamics of dissolved organic nitrogen and carbon in a Central European Norway spruce ecosystem. *European Journal of Soil Science* 50, 579–590.

Mignolet, R. (1971) Etat actuel des connaissances sur les relations entre la microfaune et microflore édaphiques. *Revue de l' Ecologie et Biologie du Sol* 9, 655–670.

Miles, H.B. (1963) Soil protozoa and earthworm nutrition. *Soil Science* 93, 407–409.

Miller, D.N., Bryant, J.E., Madsen, E.L. and Ghiorse, W.C. (1999) Evaluation and optimization of DNA extraction and purification procedures for soil and sediment samples. *Applied and Environmental Microbiology* 65, 4715–4724.

Mitchell, E.A.D., Borcard, D., Buttler, A.J., Grosvernier, Ph., Gilbert, D. and Gobat, J.-M. (2000) Horizontal distribution patterns of testate amoebae (Protozoa) in a *Sphagnum magellanicum* carpet. *Microbial Ecology* 39, 290–300.

Mitchell, H.J. and Hardham, A.R. (1999) Characterization of the water expulsion vacuole in *Phytophthora nicotiniae* zoospores. *Protoplasma* 206, 118–130.

Money, N.P. (1997) Wishful thinking of turgor revisited. The mechanics of fungal growth. *Fungal Genetics and Biology* 21, 173–187.

Montagnes, D.J.S. and Lynn, D.H. (1987a) A quantitative protargol stain (QPS) for ciliates: method and description and test of its quantitative nature. *Marine Microbial Food Webs* 2, 83–93.

Montagnes, D.J.S. and Lynn, D.H. (1987b) Agar embedding on cellulose filters: an improved method of mounting protists for protargol and Chatton–Lwoff staining. *Transactions of the American Microscopical Society* 106, 183–186.

Moore, D. (1998) *Fungal Morphogenesis*. Cambridge University Press, Cambridge.

Moore, J.C. and de Ruiter, P.C. (1991) Temporal and spatial heterogeneity of trophic interactions within below-ground food webs. *Agriculture, Ecosystems and Environment* 34, 371–397.

Moore, J.C. and de Ruiter, P.C. (2000) Invertebrates in detrital food webs along gradients of productivity. In: Coleman, D.C. and Hendrix, P.F. (eds) *Invertebrates as Webmasters in Ecosystems*. CAB International, Wallingford, UK, pp.161–184.

Moore, J.C., St John, T.V. and Coleman, D.C. (1985) Ingestion of vesicular–arbuscular mycorrhizal hyphae and spores by soil microarthropods. *Ecology* 66, 1979–1981.

Muyzer, G., deWaal, E.C. and Uitterlinden, A.G. (1993) Profiling of complex microbial populations by denaturing gradient gel electrophoresis analysis of polymerase chain reaction amplified genes encoding for 16S rRNA. *Applied and Environmental Microbiology* 59, 695–700.

Nadelhoffer, K.J. and Raich, J.W. (1992) Fine root production estimates and below-ground carbon allocation in forest ecosystems. *Ecology* 73, 1139–1147.

Nadelhoffer, K., Shaver, G., Fry, B., Gibblin, A., Johnson, L. and McKane, R. (1996) 15N Natural abundances and N use by tundra plants. *Oecologia* 107, 386–394.

Nagy, P. (1996) A comparison of extraction methods of free-living terrestrial nematodes. *Acta Zoologica Academiae Scientiarum Hungaricae* 42, 281–287.

Nakatsu, C., Torsvik, V., Ovreas, L. (2000) Soil community analysis using DGGE of 16S rDNA polymerase chain reaction products. *Soil Science Society of America Journal* 64, 1382–1388.

Neff, J.C. and Asner, G.P. (2001) Dissolved organic carbon in terrestrial ecosystems: synthesis and a model. *Ecosystems* 4, 29–48.

Neher, D.A., Weicht, T.R., Savia, M., Gorres, J.H. and Amador, J.A. (1999) Grazing in a porous environment. 2. Nematode community structure. *Plant and Soil* 212, 85–99.

Nevoigt, E. and Stahl, U. (1997) Osmoregulation and glycerol metabolism in the yeast *Saccharomyces cerevisiae*. *FEMS Microbiology Reviews* 21, 231–241.

Newell, K. (1984a) Interactions between two decomposer Basidiomycetes and a collembolan under Sitka spruce: distribution, abundance and selective grazing. *Soil Biology and Biochemistry* 16, 227–233.

Newell, K. (1984b) Interactions between two decomposer Basidiomycetes and a collembolan under Sitka spruce: grazing and its potential effects on fungal distribution and litter decomposition. *Soil Biology and Biochemistry* 16, 235–239.

Ngassam, P., de Puytorac, P. and Grain, J. (1994) On *Paraptychostomum alame* N.G., N.Sp., a commensal ciliate for the digestive tract of Oligochaetes of the Cameroons, in a new subclass Hysterocinetia. *Journal of Eukaryotic Microbiology* 41, 155–162.

Nishide, H., Toyota, K. and Kimura, M. (1999) Effects of soil temperature and anaerobiosis on degradation of biodegradable plastics in soil and their degrading microorganisms. *Soil Science and Plant Nutrition* 45, 963–972.

Noborio, K. (2001) Measurement of soil water content by time domain reflectometry: a review. *Computers and Electronics in Agriculture* 31, 213–237.

Nogrady, T. (1993) Plankton regulation dynamics: experiments and models in Rotifer continous cultures. *Ecological Studies 98*, Springer Verlag, New York.

Norris, J.R., Read, D. and Varma, A.K. (1994) *Techniques for Mycorrhizal Research.* Academic Press, New York.

Norton, R.A. (1985) Aspects of the biology and systematics of soil arachnids, particularly saprophagous and mycophagous mites. *Quest Entomologica* 21, 523–541.

Norton, R.A. (1990) Acarina: Oribatida. In: Dindal, D.L. (ed.) *Soil Biology Guide.* Wiley Interscience, New York, pp. 779–803.

Norton, R.A., Kethley, J.B., Johnston, D.E. and O'Connor, B.M. (1993) Phylogenetic perspectives on genetic systems and reproductive modes of mites. In: Wrench, D.L. and Ebbert, M.A. (eds) *Evolution and Diversity of Sex Ratio in Insects and Mites.* Chapman and Hall, New York, pp. 8–99.

O'Connor, B.M. (1994) Life-history modifications in astigmatid mites, In: Houck, M.A. (ed.) *Mites: Ecological and Evolutionary Analysis of Life-history Patterns*, Chapman and Hall, New York, pp. 136–159.

Oka, N., Hartel, P.G., Finlay-Moore, O., Gagliardi, J., Zuberer, D.A., Fuhrmann, J.J., Angle, J.S. and Skipper, H.D. (2000) Misidentification of soil bacteria by fatty acid methyl ester (FAME) and BIOLOG analyses. *Biology and Fertility of Soils* 32, 256–258.

Old, K.M. and Chakraborty, S. (1986) Mycophagous soil amoebae: their biology and significance in the ecology of soil-borne plant pathogens. *Progress in Protistology* 1, 163–194.

Old, K.M. and Darbyshire, J.F. (1978) Soil fungi as food for giant amoebae. *Soil Biology and Biochemistry* 10, 93–100.

Orlov, D.S. (1999) Soil fulvic acids: a history of study, importance, and reality. *Eurasian Soil Science* 32, 1044–1049.

Osche, G. (1952) Systematik und Phylogenie der Gattung Rhabditis (Nematoda). *Zoologische Jahrbuch (Systematik)* 81, 190–280.

Osler, G.H. and Beattie, A.J. (1999) Taxonomic and structural similarities in soil oribatid communities. *Ecography* 22, 567–574.

Osono, T. and Takeda, H. (2001) Organic chemical and nutrient dynamics in decomposing beech leaf litter in relation to fungal ingrowth and succession during 3-year decomposition process in a cool temperate deciduous forest in Japan. *Ecological Research* 16, 649–670.

Page, A.L., Miller, R.H. and Keeney, D.R. (1982) *Methods of Soil Analysis. Part II.* American Society of Agronomy and Soil Science Society of America, Madison, Wisconsin.

Page, F.C. (1976) *Illustrated Key to Freshwater and Soil Amoebae*, 2nd edn. Freshwater Biology Association 34, John Wiley & Sons, New York.

Pacala, S.W., Hurtt, G.C., Baker, D., Peylin, P., Houghton, R.A., Birdsey, R.A., Heath, L., Sundquist, E.T., Stallard, R.F., Ciais, P., Moorcroft, P., Caspersen,

J.P., Shevliakova, E., Moore, B., Kohlmaier, G., Holland, E., Gloor, M., Harmon, M.E., Fan, S.M., Sarmiento, J.L., Goodale, C.L., Schimel, D. and Field, C.B. (2001) Consistent land- and atmosphere-based US carbon sink estimates. *Science* 292, 2316–2320.

Paquin, P. and Coderre, D. (1997) Changes in soil macroarthropod communities in relation to forest maturation through three successional stages in the Canadian boreal forest. *Oecologia* 112, 104–111.

Patterson, D.J. (1980) Contractile vacuoles and associated structures: their organization and function. *Biology Reviews (Cambridge)* 55, 1–46.

Patterson, D.J. (1996) *Free-living Freshwater Protozoa – a Colour Guide*. UNSW Press, Sydney.

Patricola, E., Villa, L. and Arizzi, M. (2001) TEM observations on symbionts of *Joennia annectens*. *Journal of Natural History* 35, 471–480.

Paul, E.A. and Clark, F.E. (1996) *Soil Microbiology and Biochemistry*, 2nd edn. Academic Press, New York.

Paul, E.A., Harris, D., Klug, M.J. and Ruess, R.G. (1999) The determination of microbial biomass. In: Robertson, G.P., Coleman, D.C., Bledsoe, C.S. and Sollins, P. (eds) *Standard Soil Methods for Long Term Ecological Research*. Oxford University Press, Oxford, pp. 291–317.

Paustian, K., Andren, O., Clarholm, M., Hansson, A.-C., Johansson, G., Lagerlof, J., Lindberg, T., Petterson, R. and Sohlenius, B. (1990) Carbon and nitrogen budgets of four agro-ecosystems with annual and perennial crops, with and without N fertilization. *Journal of Applied Ecology* 27, 60–84.

Pawluk, S. (1987) Faunal micromorphological features in moder humus of some western Canadian soils. *Geoderma* 40, 3–16.

Pérez-Uz, B. (1996) Bacterial preferences of growth kinetic variation in *Uronema marinum* and *Uronema nigricans* (Ciliophora: Scuticociliatida). *Microbial Ecology* 31, 189–198.

Perez-Vargas, J., Poggi-Varaldo, H.M., Calva-Calva, G., Rios-Leal, E., Rodrigez-Vazquez, R., Ferrera-Cerrato, R. and Esparza-Garcia, F. (2000) Nitrogen fixing bacteria capable of utilising kerosene hydrocarbons as a sole carbon source. *Water Science and Technology* 42, 407–410.

Perry, R.N. and Wright, D.J. (1998) *The Physiology and Biochemistry of Free-living and Plant Parasitic Nematodes*. CAB International, Wallingford, UK.

Peter, H., Weil, B., Burkovski, A., Kramer, R. and Morbach, S. (1998) *Corynebacterium glutamicum* is equipped with four secondary carriers for compatible solutes. *Journal of Bacteriology* 180, 6005–6012.

Peters, S., Koschinsky, S., Schwieger, F. and Tebbe, C.C. (2000) Succession of microbial communities during hot composting as detected by PCR-single-strand-conformation polymorphism-based genetic profiles of small subunit rRNA genes. *Applied and Environmental Microbiology* 66, 930–936.

Petrov, P. (1997) The reactions of communities of oribatid mites to plant succession on meadows. *Ekologia Polska* 45, 781–793.

Pfister, G. and Arndt, H. (1998) Food selectivity and feeding behaviour in omnivorous filter-feeding ciliates: a case study for *Stylonychia*. *European Journal of Protistology* 34, 446–457.

Piearce, T.G. and Phillips, M.J. (1980) The fate of ciliates in the earthworm gut: an *in-vitro* study. *Microbial Ecology* 5, 313–320.

Pimm, S.L. (1982) *Food Webs*. Chapman and Hall, London.

Pizl, V. (1999) Earthworm succession in abandoned fields: a comparison of deductive and sequential approaches to study. *Pedobiologia* 43, 705–712.

Pline, M., Diez, J.A. and Dusenbery, D.B. (1988) Extremely sensitive thermotaxis of the nematode *Moloidogyne incognita*. *Journal of Nematology* 20, 605–608.

Ponsard, S., Arditi, R. and Jost, C. (2000) Assessing top-down and bottom-up control in a litter-based soil macro-invertebrate food chain. *Oikos* 89, 524–540.

Pourriot, R. (1965) *Recherches sur l'écologie des Rotifères*. Masson, Paris.

Polis, G.A. (1991) Complex trophic interactions in deserts: an empirical critique of food web theory. *American Naturalist* 138, 123–155.

Ponge, J.F. (1991) Succession of fungi and fauna during decomposition of needles in a small area of Scots Pine litter. *Plant and Soil* 138, 99–113.

Ponge, J.-F. (2000) Vertical distribution of Collembola (Hexapoda) and their food resources in organic horizons of beech forest. *Biology and Fertility of Soils* 32, 508–522.

Poole, T.B. (1961) An ecological study of the Collembola in a coniferous forest soil. *Pedobiologia* 1, 113–137.

Post, W.M., Emanuel, W.R., Zinke, P.J. and Stangenberger, A.G. (1982) Soil carbon polls and world life zones. *Nature* 298, 156–159.

Potts, M. (1994) Desiccation tolerance of prokaryotes. *Microbiology Reviews* 58, 1755–1805.

Powell, M.J. (1984) Fine structure of the unwalled thallus of *Rozella polyphagii* in its host *Polyphagus euglenae*. *Mycologia* 76, 1039–1048.

Powell, M.J. (1993) Looking at mycology with a janus face: a glimpse at Chytridiomycetes active in the environment. *Mycologia* 85, 1–20.

Pregitzer, K.S., King, J.S., Burton, A.J. and Brown, S.E. (2000) Responses of tree fine roots to temperature. *New Phytologist* 147, 105–115.

Prescott, C.E., Zabek, L.M., Staley, C.L. and Kabzems, R. (2000) Decomposition of broadleaf and needle litter in forests of British Columbia: influences of litter type, forest type, and litter mixtures. *Canadian Journal of Forest Research* 30, 1742–1750.

Price, P.W. and Benham, G.S.J. (1976) Vertical distribution of Pomerantziid mites (Acarina, Pomeranziidae). *Proceedings of the Entomological Society of Washington*. 78, 309–313.

Prot, J.C. and Ferris, H. (1992) Sampling approaches for extensive surveys in nematology. *Journal of Nematology (Supplement)* 24S, 757–764.

Pussard, M., Alabouvette, C. and Levrat, P. (1994) Protozoan interactions with the soil microflora and possibilities for biocontrol of plant pathogens. In: Darbyshire, J.F. (ed.) *Soil Protozoa*, CAB International, Wallingford, UK, pp. 122–145.

Qualls, R.G. (2000) Comparison of the behaviour of soluble organic and inorganic nutrients in forest soils. *Forest Ecology and Management* 138, 29–50.

Renvall, P. (1995) Community structure and dynamics of wood-rotting Basidiomycetes on decomposing conifer trunks in northern Finland. *Karstenia* 35, 1–51.

Richard, K.S. and Arme, C. (1982) Integumentary uptake of dissolved organic materials by earthworms. *Pedobiologia* 23, 358–366.

Robertson, G.P., Coleman, D.C., Bledsoe, C.S. and Sollins, P. (1999) *Standard Soil Methods for Long Term Ecological Research*. Oxford University Press, Oxford.

Robinson, D. (2001) Delta 15N as an integrator of the nitrogen cycle. *Trends in Ecology and Evolution* 16, 153–162.

Rodriguez, G.G., Phipps, D., Ishiguro, K. and Ridgway, H.F. (1992) Use of a fluorescent redox probe for direct visualization of actively respiring bacteria. *Applied and Environmental Microbiology* 58, 1801–1808.

Rogers, B.F. and Tate, R.L., III (2001) Temporal analysis of the soil microbial community along a toposequence in Pineland soils. *Soil Biology and Biochemistry* 33, 1389–1401.

Rogerson, A. and Berger, J. (1981) Effect of crude oil and petroleum-degrading micro-organisms on the growth of freshwater and soil Protozoa. *Journal of General Microbiology* 124, 53–59.

Rogerson, A. and Berger, J. (1982) Ultrastructural modification of the ciliated protozoan, *Colpidium colpoda*, following chronic exposure to partially degraded crude oil. *Transcations of the American Microscopical Society* 101, 27–35.

Rogerson, A., Shiu, W.Y., Huang, G.L., Mackay, D. and Berger, J. (1983) Determination and interpretation of hydrocarbon toxicity to ciliated protozoa. *Aquatic Toxicology* 3, 215–228.

Ronn, R., Grunert, J. and Ekelund, F. (2001) Protozoan response to addition of the bacteria *Mycobacterium chlorophenolicum* and *Pseudomonas chlororaphis* to soil microcosms. *Biology and Fertility of Soils* 33, 126–131.

Rosenbrock, P., Buscot, F. and Munuch, J.-C. (1995) Fungal succession and changes in the fungal degradation potential during the initial stage of litter decomposition in a black alder forest (*Alnus glutinosa* (L.) Gaertn.). *European Journal of Soil Biology* 31, 1–11.

Rossi, J.P., Lavelle, P. and Albrecht, A. (1997) Relationship between spatial pattern of the endogeic earthworm *Polypheretima elongata* and soil heterogeneity. *Soil Biology and Biochemistry* 29, 485–488.

Rother, A., Radek, R. and Hausmann, K. (1999) Characterization of surface structures covering termite flagellates of the family Oxymonadidae and ultrastructure of two Oxymonad species, *Microrhapalodina multinucleata* and *Oxymonas* sp. *European Journal Protistology* 35, 1–16.

Rouelle, J. (1983) Introduction of amoeba and *Rhizobium japonicum* into the gut of *Eisenia fetida* (Sav.) and *Lumbricus terrestris* L. In: Satchell, J.E. (ed.) *Earthworm Ecology, from Darwin to Vermiculture.* Chapman and Hall, London, pp. 375–381.

Rowan, A., McCully, M.E. and Canny, M.J. (2000) The origin of the exudates from cut maize roots. *Plant Physiology and Biochemistry* 38, 957–967.

Ruess, L. and Dighton, J. (1996) Cultural studies on soil nematodes and their fungal hosts. *Nematologica* 42, 330–346.

Ruess, L., Zapata, E.J.G. and Dighton, J. (2000) Food preferences of a fungal-feeding *Aphelenchoides* species. *Nematology* 2, 223–230.

Ruppert, E.E. and Barnes, R.D. (1994) *Invertebrate Zoology*, 6th edn. Saunders College Publishing, New York.

Rusek, J. (1998) Biodiversity of Collembola and their functional role in the ecosystem. *Biodiversity and Conservation* 7, 1207–1219.

Rytter, R. (1999) Fine root production and turnover in a willow plantation estimated by different calculation methods. *Scandinavian Journal of Forest Research* 14, 526–537.

Saetre, P. (1999) Spatial patterns of ground vegetation, soil microbial biomass and activity in a mixed spruce–birch stand. *Ecography* 22, 183–192.

Salisbury, F.B. and Ross, C.W. (1999) *Plant Physiology*, 4th edn. Brooks/Cole Publishing Company, New York.

Salonius, P.O. (1981) Metabolic capabilities of forest soil microbial populations with reduced species diversity. *Soil Biology and Biochemistry* 13, 1–10.

Satchell, J.E. (1967) *Lumbricidae.* In: Burgess, A. and Raw, F. (eds) *Soil Biology.* Academic Press, London, pp. 259–322.

Sattelmacher, B. (2001) The apoplast and its significance for plant mineral nutrition. *New Phytologist* 149, 167–192.

Scheu, S. and Schultz, E. (1996) Secondary succession, soil formation and development of a diverse community of oribatids and saprophagous soil macroinvertebrates. *Biodiversity and Conservation* 5, 235–250.

Schimmel, J. (2001) Biogeochemical models: implicit versus explicit microbiology. In: Schulze, E.D., Heimann, M., Harrison, S., Holland, E., Lloyd, J., Prentice, I.C. and Schimel, D. (eds) *Global Biogeochemical Cycles in the Climate System.* Academic Press, New York, pp. 177–183.

Schlegel, D. and Bauer, T. (1994) Capture of prey by two pseudoscorpion species. *Pedobiologia* 38, 361–373.

Schlegel, H.G. (1993) *General Microbiology*, 2nd edn. Cambridge University Press, Cambridge.

Schlesinger, W.H. (1997) *Biogeochemistry: an Analysis of Global Change*, 2nd edn. Academic Press, San Diego.

Schlesinger, W.H. and Jeffrey, A.A. (2000) Soil respiration and the global carbon cycle. *Biogeochemistry* 48, 7–20.

Schonborn, W. (1965) Untersuchungen uber die Ernahrung bodenbewohnender Testaceen. *Pedobiologia* 5, 205–210.

Schonborn, W. (1983) Beziehungen zwischen Produktion, Mortalität, und Abundanz terrestrischer Testacea-Gemeinschaften. *Pedobiologia* 25, 403–412.

Schonborn, W. (1992) Adaptive polymorphism in soil-inhabiting testate amoebae (Rhizopoda): its importance for delimitation and evolution of asexual species. *Archiv für Protistenkunde* 142, 139–155.

Schubart, H.O.R. (1973) The occurrence of Nematalycidae (Acari, Prostigmata) in central Amazonia with a description of a new genus and species. *Acta Amazonica* 3, 53–57.

Schulte, H. (1954) Beitrage sur Oekologie und systematic des Boden-Rotatorien. *Zoologische Jahrbuch Systematik* Vol. 82.

Schuster, R. and Murphy, P.W. (1991) *The Acari.* Chapman and Hall, New York.

Seastedt, T.R. (1984) The role of microarthropods in decomposition and mineralisation processes. *Annual Review of Entomology* 29, 25–46.

Sherman, J.M. (1914) The number and growth of protozoa in soil. *Zentralblatt für Bakteriologie Parasitenkunde* 41, 625–630.

Sherr, B.F., Sherr, E.B. and McDaniel, J. (1992) Effect of protistan grazing on the frequency of dividing cells in bacterioplankton assemblages. *Applied and Environmental Microbiology* 58, 2381–2385.

Siepel, H. (1986) The importance of unpredictable and short term environmental extremes for biodiversity in oribatid mites. *Biodiversity Letters* 3, 26–34.

Siepel, H. (1994) Life history tactics of soil microarthropods. *Biology and Fertility of Soils* 18, 263–278.

Simard, S.W., Perry, D.A., Jones, M.D., Myrold, D.D., Durall, D.M. and Molina, R. (1997) Net transfer of carbon between ectomycorrhizal tree species in the field. *Nature* 388, 579–582.

Singh, B.N. (1941) Selectivity in bacterial food by soil amoebae in pure and mixed culture and in sterilised soil. *Annals of Applied Biology* 28, 52–65.

Singh, B.N. (1942) Selection of bacterial food by soil flagellates and amoebae. *Annals of Applied Biology* 29, 18–22.

Slavikova, E. and Vadkertiva, R. (2000) The occurrence of yeasts in the forest soils. *Journal of Basic Microbiology* 40, 207–212.

Small, E.B. and Lynn, D.H. (1988) Ciliophora. In: Lee J.J. (ed.) *Illustrated Guide to the Protozoa*. The Society of Protozoologists, Allen Press. pp. 393–575.

Smart, C. (1994) Gene expression during leaf senescence. *New Phytologist* 126, 419–448.

Smethurst, P.J. (2000) Soil solution and other soil analyses as indicators of nutrient supply: a review. *Forest Ecology and Management* 138, 397–411.

Smith, P., Andren, O., Brussaard, L., Dangerfield, M., Ekschmitt, K., Lavelle, P. and Tate, K. (1998) Soil biota and global change at the ecosystem level: describing soil biota in mathematical models. *Global Change Biology* 4, 773–784.

Smith, S.E. and Read, D.J. (1997) *Mycorrhizal Symbiosis*, 2nd edn. Academic Press, New York.

Smrz, J. and Trelova, M. (1995) The association of bacteria and some soil mites (Acari: Oribatida and Acaridida). *Acta Zoologica Fennica* 196, 120–123.

Snider, R.M. and Snider, R.J. (1997) Efficiency of arthropod extraction from soil cores. *Entomological News* 108, 203–208.

Söderström, B.E. (1979) Some problems in assessing the fluorescein diacetate-active fungal biomass in the soil. *Soil Biology and Biochemistry* 11, 147–148.

Sohlenius, B. (1990) Influence of cropping system and nitrogen input on soil fauna and microorganisms in a Swedish arable soil. *Biology and Fertility of Soils* 9, 168–173.

Sohlenius, B. and Wasilewska, L. (1984) Influence of irrigation and fertilization on the nematode community in a Swedish pine forest soil. *Journal of Applied Ecology* 21, 327–342.

Soil Survey Staff (1988) Keys of soil taxonomy (fourth printing). *SMSS Technical Monograph* 6. Cornell University, Ithaca, New York.

Sollins, P., Hofmann, P. and Caldwell, B.A. (1996) Stabilization and destabilization of soil organic matter: mechanisms and controls. *Geoderma* 74, 65–105.

Sonneborn, T.M. (1970) Methods in *Paramecium* research. *Methods in Cell Physiology* 4, 242–335.

Spaccini, R., Piccolo, A., Haberhauer, G. and Gerzabek, M.H. (2000) Transformation of organic matter from maize residues into labile and humic fractions of three European soils as revealed by ^{13}C distribution and CPMAS-NMR spectra. *European Journal of Soil Science* 51, 583–594.

Sparks, D.L. (1999) *Soil Physical Chemistry*, 2nd edn. CRC Press, Boca Raton, Florida.

Sparrow, F.K. (1960) *Aquatic Phycomycetes*, 2nd revised edn. University of Michigan Press, Ann Arbor, Michigan.

Sparrow, F.K. (1973) Hyphochytrids. In: Ainsworth, G.C., Sparrow, F.K. and Sussman, A.S. (eds) *The Fungi*, Vol. IVb. Academic Press, New York, pp. 61–110.

Spencer, J.F.T. and Spencer, D.M. (1997) Ecology. In: Spencer, J.F.T. and Spencer, D.M. (eds) *Yeasts in Natural and Artificial Habitats*. Springer-Verlag, New York, pp. 33–58.

Sperling, F.A.H., Anderson, G.S. and Hickey, D.A. (1994) A DNA based approach to the identification of insect species used for post-mortem interval estimation. *Journal of Forensic Sciences* 39, 418–427.

Sposito, G. (1989) *Soil Particle Surface*. Oxford University Press, New York.

Springett, J. and Gray, R. (1997) The interaction between plant roots and earthworm burrows in pastures. *Soil Biology and Biochemistry* 29, 621–625.

Stamou, G.P. and Sgardalis, S.P. (1989) Seasonal distribution patterns of oribatid mites (Acari, Cryptostigmata) in a forest ecosystem. *Journal of Animal Ecology* 58, 893–904.

Steinberger, Y. and Wallwork, J.A. (1985) Composition and vertical distribution patterns of the microarthropod fauna in a Negev desert soil. *Journal of Zoology (London) Series A* 206, 329–339.

Steinberger, Y., Zelles, L., Bai, Q.Y., von Lutzow, M. and Munch, J.C. (1999) Phospholipid fatty acid profiles as indicators for the microbial community structure in soils along a climatic transect in the Judean Desert. *Biology and Fertility of Soils* 28, 292–300.

Stevenson, F.J. (1994) *Humus Chemistry: Genesis, Composition, Reactions*, 2nd edn. John Wiley & Sons, New York.

Stotzky, G. (2000) Persistence and biological activity in soil of insecticidal proteins from *Bacillus thuringiensis* and of bacterial DNA bound on clays and humic acids. *Journal of Environmental Quality* 29, 691–705.

Sumner, M. (2000) *Handbook of Soil Science*. CRC Press, Boca Raton, Florida.

Swift, M.J., Heal, O.W. and Anderson, J.M. (1979) *Decomposition in Terrestrial Ecosystems*. University of California Press, Berkeley, California.

Tamm, C.O. (1991) Nitrogen in terrestrial ecosystems. *Ecological Studies* 81, 1–100.

Tate, R.L. (1987) *Soil Organic Matter: Biological and Ecological Effects*. Wiley Interscience Press, New York.

Tate, R.L. (2000) *Soil Microbiology*, 2nd edn. John Wiley & Sons Inc., New York.

Taylor, W.D. (1979) Sampling data on the bacterivorous ciliates of a small pond compared to neutral models of community structure. *Ecology* 60, 876–883.

Taylor, W.D. and Berger, J. (1980) Microspatial heterogeneity in the distribution of ciliates in a small pond. *Microbial Ecology* 6, 27–34.

Tholen, A., Schink, B. and Brune, A. (1997) The gut microflora of *Reticulitermes flavipes*, its relation to oxygen, and evidence for oxygen dependent acetogenesis by the most abundant *Enterococcus* species. *FEMS Microbiology Ecology* 35, 27–36.

Thomas, R.H. and MacLean, S.F. (1988) Community structure in soil Acari along a latitudinal transect of Tundra sites in northern Alaska. *Pedobiologia* 31, 113–198.

Thorn, R.G., Moncalvo, J.M., Reddy, C.A. and Vilgalys, R. (2000) Phylogenetic analysis and the distribution of nematotrophy support a monophyletic Pleurotaceae within the polyphyletic pleurotoid–lentinoid fungi. *Mycologia* 92, 241–252.

Thornthwaite, C.W. and Mather, J.R. (1957) Instructions and tables for computing potential evapotranspiration and the water balance. *Publications in Climatology* 10, 185–311.

Tokumasu, S. (1996) Mycofloral succession on *Pinus densiflora* needles on a moder site. *Mycoscience* 37, 313–321.
Tokumasu, S. (1998) Fungal successions on pine needles fallen at different seasons: the succession of surface colonizers. *Mycoscience* 39, 417–423.
Toutain, F., Vilemin, G., Albrecht, A. and Reisinger, O. (1982) Etude ultrastructurale des processus de biodegradation. II. Modèles enchytraeids–litières de feuillus. *Pedobiologia* 23, 145–156.
Towne, E.G. (2000) Prairie vegetation and soil nutrient responses to ungulate carcasses. *Oecologia* 122, 232–239.
Tracey, M.V. (1955) Cellulase and chitinase in soil amoebae. *Nature* 175, 815.
Travé, J., André, H.M., Taberly, G.T. and Bernini, F. (1996) *Les Acariens Oribates*. AGAR Publishers, Wavre, Belgium.
Triemer, R.E. and Farmer, M.A. (1991) The ultrastructural organization of the heterotrophic euglenids and its evolutionary implications. In: Patterson, D.J. and Larsen, J. (eds) *The Biology of Free-living Heterotrophic Flagellates*. Clarendon Press, Oxford, pp. 185–204.
Tyler, G. and Olsson, T. (2001) Concentration of 60 elements in the soil solution as related to the soil acidity. *European Journal of Soil Science* 52, 151–165.
Valle, J.V., Moro, R.P., Garvin, M.H., Trigo, D. and Diaz-Cosin, D.J. (1997) Annual dynamics of the earthworm *Hormogaster elisae* (Oligochaeta, Hormogastridae) in central Spain. *Soil Biology and Biochemistry* 29, 309–312.
van Elsas, J.D., Duarte, G.F., Keijzer-Wolters, A. and Smit, E. (2000) Analysis of the dynamic of fungal communities in soil via fungal-specific PCR of soil DNA followed by denaturing gradient gel electrophoresis. *Journal of Microbiological Methods* 43, 133–151.
Vanette, R.C. and Ferris, H. (1998) Influence of bacterial type and density on population growth of bacteria-feeding nematodes. *Soil Biology and Biochemistry* 30, 949–960.
Van Hanen, E.J., Agterveld, M.P., Gons, H.J. and Laanbroek, H.J. (1998) Revealing genetic diversity of eukaryotic microorganisms in aquatic environments by denaturing gradient gel electrophoresis. *Journal of Phycology* 34, 206–213.
Van Tichelen, K.K. and Colpaert, J.V. (2000) Kinetics of phosphate absorption by mycorrhizal and non-mycorrhizal Scots pine seedlings. *Physiologia Plantarum* 110, 96–103.
Van Vliet, P.C.J. (2000) Enchytraeids. In: Sumner, M. (ed.) *The Handbook of Soil Science*. CRC Press, Boca Raton, Florida, pp. C70–C75.
Van Wely, K.H.M., Swaving, J., Freudlard, R. and Driessen, A.J.M. (2001) Translocation of proteins across the cell envelope of Gram positive bacteria. *FEMS Microbiology Reviews* 25, 437–454.
Varga, X. (1959) Untersuchungen uber die Mikrofauna der Waldstreu einigen Waldtypen von Bukkgeliegerge (Ungarn). *Acta Zoologishes Academie Science. Hungarien*, Vol. 4.
Vargas, R. and Hattori, T. (1986) Protozoan predation of bacterial cells in soil aggregates. *FEMS Microbiology Ecology* 38, 233–242.
Vegter, J.J. (1987) Phenology and seasonal resource partitioning in forest floor Collembola. *Oikos* 48, 175–185.
Verheyen, K., Bossuyt, B., Hermy, M. and Tack, G. (1999) The land use history (1278–1990) of a mixed hardwood forest in western Belgium and its rela-

tionship with chemical soil characteristics. *Journal of Biogeography* 26, 115–128.

Viaud, M., Pasquier, A. and Brygoo, Y. (2000) Diversity of soil fungi studied by PCR-RFLP of ITS. *Mycological Research* 104, 1027–1032.

Vinolas, L.C., Healey, J.R. and Jones, D.L. (2001) Kinetics of soil microbial uptake of free amino acids. *Biology and Fertility of Soils* 33, 67–74.

Vitousek, P.M. and Howarth, R.W. (1991) Nitrogen limitation on land and in the sea: how can it occur? *Biogeochemistry* 13, 87–115.

Vitousek, P.M. and Sanford, R.L. (1986) Nutrient cycling in moist tropical forest. *Annual Reviews of Ecology and Systematics* 17, 137–167.

Vossbrinck, C.R., Coleman D.C. and Woolley, T.A. (1979) Abiotic and biotic factors in litter decomposition in a semiarid grassland. *Ecology* 60, 265–271.

Walker, G., Simpson, A.G.B., Edgcomg, V., Sogin, M.L. and Patterson, D.J. (2001) Ultrastructural identities of *Mastigamoeba punctachora, M. simplex* and *M. commutans* and assessment of hypothesis of relatedness of the pelobionts (Protista). *European Journal of Protistology* 37, 25–49.

Wall, D.H. and Moore, J.C. (1999) Interactions underground; soil biodiversity, mutualism, and ecosystem processes. *BioScience* 49, 109–117.

Wallace, H.R. (1958) Movement of eelworms. I. The effect of pore size and moisture content of the soil on the migration of the beet eelworm, *Heterodera schachterii* Scmidt. *Annals of Applied Biology* 46, 74–85.

Walter, D.E. (1987) Trophic behaviour of 'mycophagous' microarthropods. *Ecology* 68, 226–229.

Walter, D.E. (1988) Predation and mycophagy by endeostigmatid mites (Acariformes: Prostigmata). *Experimental and Applied Acarology* 4, 159–166.

Walter, D.E. and Kaplan, D.T. (1990) Feeding observations on two astigmatic mites, *Schiewebea rocketti* Woodring (Acaridae) and *Histriostoma bakeri* Hughes and Jackson, associated with citrus feeder roots. *Pedobiologia* 34, 281–286.

Walter, D.E. and Proctor, H.C. (1999) *Mites: Ecology Evolution and Behaviour*. CAB International, Wallingford, UK.

Walter, D.E., Kethley, J. and Moore, J.C. (1987) A heptane flotation method for recovering microarthropods from semiarid soils, with comparison to the Merchant–Crossley high-gradient extraction method and estimates of microarthropod biomass. *Pedobiologia* 30, 221–232.

Wanner, M. (1995) Biometrical investigations of terrestrial testate amoebae (Protozoa: Rhizopoda) as a method for bioindication. *Acta Zoologica Fennica* 196, 267–270.

Wanner, M. (1999) A review on the variability of testate amoebae: methodological approaches, environmental influences and taxonomical implications. *Acta Protozoologica*, 38, 15–19.

Wanner, M., Naehring, J.M. and Fischer, R. (1997) Molecular identification of clones of testate amoebae using single nuclei for PCR amplification. *European Journal of Protistology* 33, 192–199.

Wardle, D.A. (1998) Controls of temporal variability of the soil microbial biomass: a global-scale synthesis. *Soil Biology and Biochemistry* 30, 1627–1637.

Wardle, D.A. (2002) *Communities and Ecosystems: Linking the Aboveground and Belowground Components*. Princeton University Press, Princeton, New Jersey.

Wardle, D. and Yeates, G. (1993) The dual importance of competition and predation as regulatory forces in terrestrial ecosystems – evidence from decomposer food-webs. *Oecologia* 93, 303–306.

Watanabe, T. (1994) *Pictorial Atlas of Soil and Seed Fungi.* Lewis Publishers, London.

Wells, C.E. and Eissenstat, D.M. (2001) Marked differences in survivorship among apple roots of different diameters. *Ecology* 82, 882–892.

Werner, R.A., Bruch, B.A. and Brand, W.A. (1999) ConFlo III – an interface for high precision delta ^{13}C and delta ^{15}N analysis with an extended dynamic range. *Rapid Communications in Mass Spectrometry* 13, 1237–1241.

Westover, K.M., Kennedy, A.C. and Kelley, S.E. (1997) Patterns of rhizosphere microbial community structure associated with co-occurring plant species. *Journal of Ecology* 85, 863–873.

Wheeler, W.C. and Hayashi, C.Y. (1998) The phylogeny of the extant Chelicerata orders. *Cladistics* 14, 173–192.

White, D. (1995) *The Physiology and Biochemistry of Prokaryotes.* Oxford University Press, New York.

White, D.C. (2000) Rapid detection/identification of microbes, bacterial spores, microbial communities, and metabolic activities in environmental matrices. In: Glasser, J.A. and Sasek, V. (eds) *The Utilization of Bioremediation to Reduce Soil Contamination: Problems and Solutions.* NATO Advanced Research Workshop, Liblice Castle, Prague, Czech Republic.

Whitehead, D.C. (2000) *Nutrient Elements in Grassland: Soil–Plant–Animal Relationships.* CAB International, Wallingford, UK.

Whitford, W.G., Freckman, D.W., Elkins, N.Z., Parker, L.W., Parmelee, R., Phillips, J. and Tucker, S. (1981) Diurnal migration and responses to simulated rainfall in desert soil microarthropods and nematodes. *Soil Biology and Biochemistry* 13, 417–425.

Wicklow, D.T. and Carroll, G.C. (1992) *The Fungal Community: its Organisation and Role in the Ecosystem*, 2nd edn, Marcel Dekker, New York.

Williams, F.M. (1980) On understanding predator–prey interactions. In: Ellwood, D.C., Edger, Latham, Lynch and Slater (eds) *Contemporary Microbial Ecology.* Academic Press, New York, pp. 349–375.

Williams, S.T. and Laning, S. (1984) Studies of the ecology of streptomycete phage in soil. In: Ortiz-Ortiz, L., Bojalil, L.F. and Yakoleff, V. (eds) Biological, Biochemical and Biomedical Aspects of Actinomycetes. Academic Press, New York, pp 473–483.

Whittaker, R.H. (1975) *Communities and Ecosystems*, 2nd edn, MacMillan, New York.

Wolff, M. (1909) Der Einfluss der der Bewasserung auf die Fauna der Ackerkrume mit besonderer Berucksichtingung der Bodenprotozoen. *Mitteilungen der Kaiser Wilhelm Institut Landw. Bromberg.* 1, 382–401.

Wong, P.T.W. and Griffin, D.M. (1976a) Bacterial movement at high matric potential. I. In artificial and natural soils. *Soil Biology and Biochemistry* 8, 215–218.

Wong, P.T.W. and Griffin, D.M. (1976b) Bacterial movement at high matric potential. II. In fungal colonies. *Soil Biology and Biochemistry* 8, 219–223.

Wood, J.M. (1999) Osmosensing by bacteria: signals and membrane based sensors. *Microbial Molecular Biology Reviews* 63, 230–262.

Woodring, J.P. and Cook, E.F. (1962a) The internal anatomy, reproductive physiology and molting process of *Ceratozetes cisalpinus* (Acarina: Oribatei). *Annals of the Entomological Society of America* 55, 164–181.

Woodring, J.P. and Cook, E.F. (1962b) The biology of *Ceratozetes cisalpinus* Berlese, *Schleroribates laevigatus* Koch, and *Oppia neerlandica* Oudemans (*Acarina: Oribatei*) with a description of all stages. *Acarologia* 4, 101–137.

Worland, M.R. and Lukesova, A. (2000) The effect of feeding on specific soil algae on the cold-hardiness of two Antarctic micro-arthropods (*Alaskozetes antarticus* and *Cryptopygus antarticus*). *Polar Biology* 23, 766–774.

Yeates, G.W. (1981) Soil nematode populations depressed in the presence of earthworms. *Pedobiologia* 22, 191–195.

Yeates, G.W. (1998) Feeding in soil. In: Perry, R.N. and Wright, D.J. (eds) *Physiology and Biochemistry of Free-living and Parasitic Nematodes*. CAB International, Wallingford, UK.

Yeats, G.W. and Coleman, D.C. (1982) Role of nematodes in decomposition. In: Freckman, D.W. (ed.) *Nematodes in Soil Ecosystems*. University of Texas Press, Austin, Texas, pp. 55–80.

Yeates, G.W. and Foissner, W. (1995) Testate amoebae as predators of nematodes. *Biology and Fertility of Soils* 20, 1–7.

Yeates, G.W., Bongers, T., de Goede, R.C.M., Freckman, D.W. and Georgieva, S.S. (1993) Feeding habits in soil nematode families and genera: an outline for soil ecologists. *Journal of Nematology* 25, 315–331.

Zelles, L. (1999) Fatty acid patterns of phospholipids and lipopolysaccharides in the characterisation of microbial communities in soil: a review. *Biology and Fertility of Soils* 29, 111–129.

Zheng, D.W., Bengtsson, J. and Agren, G. (1997) Soil food webs and ecosystem processes: decomposition in donor-control and Lotka–Volterra systems. *American Naturalist* 149, 125–148.

Zimmermann, G. and Bode, E. (1983) Untersuchungen zur Verbreitung des insektenpathogenen Pilzes *Metarhizium anisopliae* (Fungi imperfecti, Moniliales) durch Bodenarthropoden. *Pedobiologia* 25, 65–71.

Zischka, H., Oehne, F., Pintsch, T., Ott, A., Keller, H., Kellerman, J. and Schuster, S.C. (1999) Rearrangement of cortex proteins constitutes an osmoprotective mechanism in *Dictyostelium*. *EMBO Journal* 18, 4241–4249.

Zou, X. and Gonzalez, G. (1997) Changes in earthworm density and community structure during secondary succession in abandoned pastures. *Soil Biology and Biochemistry* 29, 627–629.

Index